Entrepreneurs in High Technology

ENTREPRENEURS IN HIGH TECHNOLOGY

Lessons from MIT and Beyond

EDWARD B. ROBERTS

New York Oxford
OXFORD UNIVERSITY PRESS
1991

Oxford University Press

Oxford   New York   Toronto
Delhi   Bombay   Calcutta   Madras   Karachi
Petaling Jaya   Singapore   Hong Kong   Tokyo
Nairobi   Dar es Salaam   Cape Town
Melbourne   Auckland

and associated companies in
Berlin   Ibadan

Copyright © 1991 by Oxford University Press, Inc.

Published by Oxford University Press, Inc.,
198 Madison Avenue, New York, New York 10016-4314

Oxford is a registered trademark of Oxford University Press

All rights reserved. No part of this publication may be reproduced,
stored in a retrieval system, or transmitted, in any form or by any means,
electronic, mechanical, photocopying, recording, or otherwise,
without the prior permission of Oxford University Press.

Library of Congress Cataloging-in-Publication Data
Roberts, Edward Baer.
Entrepreneurs in high technology: lessons from MIT and beyond/Edward B. Roberts.
p.   cm.   Includes bibliographical references and index.
ISBN 0-19-506704-5
1. High-technology industries—Massachusetts—Boston Metropolitan Area.
2. Entrepreneurship—Massachusetts—Boston Metropolitan Area.
3. New business enterprises—Massachusetts—Boston Metropolitan Area.
4. Massachusetts Institute of Technology.   I. Title.
HC108.B65R62   1991   338.4′76200097446—dc
2090-26256

5 7 9 8 6 4

Printed in the United States of America
on acid-free paper

TO MY PARTNERS

In Entrepreneurship
Alexander Pugh and Henry Weil
Neil Pappalardo, Morton Ruderman,
Jerome Grossman, and Curtis Marble
Paul Kelley, Gordon Baty, Joseph Lombard,
and Jerome Goldstein

In Life
Nancy Roberts

PREFACE

When I was a child growing up in a suburb of Boston, my parents often took me to the outdoor concerts of the Boston Symphony Orchestra at the Esplanade along the Boston shores of the Charles River. Looking across the Charles toward Cambridge in the evenings I was repeatedly awed by the looming majesty of M.I.T.'s bulk and, alongside it, the bright blue flashing roof-sign and logo of Electronics Corporation of America, an early high-technology firm located a few doors down Memorial Drive from what is now the MIT Sloan School of Management. My first strongly formed images of MIT were thus intimately interwoven with a fascination for technological entrepreneurship. Little did I realize then that my life's work would be at that interface of MIT and entrepreneurship.

This, then, is a book about entrepreneurs. But it is mostly about a very special group of entrepreneurs who were nurtured at or nearby MIT in the post-World War II explosion of science and technology and its applications to industrial and societal advance. Trained in high-technology in MIT's labs and academic departments or in the local industrial marvel that became known as the "Route 128 phenomenon", these entrepreneurs took their technical and innate skills with them to found their own new companies. The book explains the origins of these people and of the companies they founded and grew. It focuses on people, technology, money, and markets and their interplay in the formation, development and success or failure of hundreds of high-technology companies in the Greater Boston area.

The formal studies that led to this book began in 1964 and continue to the present. But three years earlier, out of a gnawing curiosity while I was still an economics doctoral student at MIT, I had cross-registered at the Harvard Business School to enroll in their New Enterprises subject, the only related subject then available in the Greater Boston area. And in 1963, just one year before this research began, I recruited my close MIT System Dynamics colleague, Jack Pugh, to join me in forming Pugh-Roberts Associates, my first act of business entrepreneurship. Over a quarter-century has passed since these beginnings and they have been exciting and fulfilling years, made whole especially by the combination of new enterprises research and action that have paralleled and become integral with my family life.

This book fuses my work with many close working colleagues, including research associates, graduate research assistants, and many thesis students. But it also draws from the unique environment of MIT and Greater Boston, and the generous willingness of the entrepreneurial community to share their experiences, their pains and their successes. The research could not have been carried out in a less thriving, less self-assured, less open

community. My hope is that the insights provided from the findings presented in this book somehow contribute to fulfilling other entrepreneurs'
dreams and other communities' hopes.

Cambridge, Massachusetts   E. B. R.
April 1991

ACKNOWLEDGMENTS

Many people enabled the underlying research embodied in this book to be accomplished and the book itself to be written. Most significant at the outset was my close research colleague of many years Herbert Wainer. Herb was my first graduate research assistant on the project and then stayed with me as a research associate for several critical years while much data gathering and analysis was undertaken and many of the needed additional phases of research conceived. Nearly forty other graduate students contributed importantly to the multi-phase research program, all identified in the book's Appendix in regard to their individual areas of effort. During the early years I benefited from the encouragement and research insights of the late Donald Marquis, founder of the MIT Program on the Management of Research and Development, who inspired a focus on ambitious technology-related studies and careful measurement of results. And Jay Forrester, who attracted me from Electrical Engineering to a career in the MIT Sloan School of Management, constantly inspired entrepreneurial thoughts and acts by his own pioneering examples and insights.

As the research work emerged into written form I gained enormously from thoughtful comments from many colleagues. Of special note, F. Michael Scherer, Ian MacMillan and Andrew Van de Ven produced detailed assessments of all aspects of the book which I deeply appreciate. Ralph Katz, Marc Meyer and Steven Ruma also provided significant help. Karl Vesper, Jeffry Timmons, Rosabeth Moss Kantor and David Morgenthaler gave important encouragement and commentary on the work. Of course, my editor Herbert J. Addison supplied insightful guidance and the persuasiveness for me to make substantial changes, as well as unfailing support throughout the publication process.

My wife Nancy tried for years to get me to turn those piles of notes and drafts into a book, perhaps just to clean up the mess. But then she read and marked up all parts of the manuscripts and helped assure me that the effort was credible. Nancy was also involved in at least the background of all of my personal entrepreneurial ventures so she became a second source of memory for many details and perspectives. She shares more than anyone else in the overall credits.

CONTENTS

Entrepreneurs in High Technology

CHAPTER 1

High-Technology Entrepreneurs

An upper floor in an old factory building or a converted warehouse some-where in Cambridge, Massachusetts, housing a new technical company founded by several people associated with the Massachusetts Institute of Technology (MIT) and driven by the spirit of entrepreneurship—this de-scribes the beginnings of numerous high-technology enterprises in the Boston area. Most of this book focuses on lessons learned broadly from investigating several hundreds of these firms. But aggregate statistics on the formation of spin-off companies from a great research university or even extensive details on the personal backgrounds of their founders do not provide a sufficient picture of the formation and growth of a new technological enterprise. A technical idea and a set of circumstances are not enough. The formation and the survival of an organization depend on unique people who take the risks of leaving their established organizations to start and build a new firm. This chapter presents a brief backdrop of early entrepreneurship at MIT, followed by four in-depth histories of an entrepreneurial founder and his technical company. In each case I have combined objective research with personal involvements as co-founder, director, or consultant. Each company is substantially different from the others. Together, they reflect the diversity that is high-technology entre-preneurship. The main themes of the book follow these cases along with a preview of the chapters.

Founders and Firms

In the Beginning

The first modern technology-based companies in the Boston area seem inevitably linked to MIT. A number of unique faculty, who sensed needs or opportunities, or both, to transfer their technological skills and know-how to the marketplace, became the early technological entrepreneurs of Greater Boston. EG&G, Inc., for example, is a case of a "pure" and early MIT spin-off, with all three founding partners associated with the Institute as staff or faculty both before and after the start-up of their company. Faced with a dearth of job opportunities when he graduated from MIT, in 1931, in the midst of the depression, Kenneth J. Germeshausen accepted the offer of his professor, Harold E. "Doc" Edgerton, to form a consulting partnership. In his doctoral dissertation Edgerton had pioneered the development of stroboscopic photography. He and Germeshausen began their firm with the application of "strobe" analysis to industrial problems and progressed to the development of related inventions that they then licensed to other companies for their commercial exploitation. Two years later, Herbert Grier, a 1933 MIT graduate in electrical engineering, joined the partnership, continuing work on high-speed motion picture techniques and the related flash lamps and cameras.

The partnership's work was carried out in space furnished by MIT, with the company supporting the lab, buying the supplies, paying technicians and all other out-of-pocket costs. In return the company was expected to be on call to help any other part of MIT that had problems in its field. Today, this arrangement would certainly be regarded as irregular, but this is evidence of the long tradition at MIT of encouraging entrepreneurial activities, even nurturing them physically under its own roof. Partnership activities were interrupted by the advent of World War II when Germeshausen was asked to join the MIT Radiation Laboratory, the center of U.S. radar development efforts, and Grier and Edgerton joined other MIT laboratories engaged in wartime research. In 1945, Germes, as his friends called him, began work on a secret contract with a large government agency through MIT and found himself spending almost full time on it. Since MIT did not want to be so heavily involved in classified work, the partnership was revived to take on this contract work. This government project, which became the detonation device and broader instrumentation support for the entire U.S. nuclear bomb program, led the original three partners to decide to start Edgerton, Germeshausen and Grier, Inc. (later changing its imposing name to EG&G, Inc.), formally incorporated in 1947 with the three partners investing $5,000 each with equal ownership.

Nearly 35 years have past. Typical of most of the faculty-initiated enterprises, Edgerton remained at MIT and gradually phased out of active involvement in EG&G. Under Germeshausen's long leadership the company grew dramatically, heavily from its early and enduring work for the U.S. Atomic Energy Commission (which became part of ERDA and is now in the Department of Energy). Gradually, Bernard O'Keefe, who joined the company shortly after the war, assumed increasing responsibilities and eventually took over as president and chief executive. Barney built EG&G much further, largely on the basis of effective technology-based acquisitions. By 1990, with none of its founders still active, the company's sales have grown to over $1.5 billion, including many acquisitions that now account for half the annual growth.

The EG&G story is paralleled by many other academic spin-offs from MIT. Pre-World War II activities of Vannevar Bush, professor of electrical engineering and later vice president of MIT, include co-founding what became an initiating part of the Raytheon Corporation. Right after the war the entrepreneurial trend accelerated. In 1946, John G. Trump founded High Voltage Engineering Corporation to build and apply atomic particle accelerators and electrostatic generators developed by his MIT colleague and co-founder, Robert J. Van de Graaff. Denis M. Robinson, recruited to head the start-up, recalls: "I was very doubtful whether this was something too 'chancy' for somebody nearly 40, who really had only one more throw of the dice. He (Trump) wanted to make accelerators for generating x rays for cancer treatment. I went to some people I knew in that field and they were so intransigently opposed and narrow-minded about it that I decided to throw in my lot with him just on that" (*International Science and Technology*, 1965). High Voltage Engineering eventually grew to over $100 million in sales.

In 1948, Richard H. Bolt, an MIT professor in the Physics Department and director of the MIT Acoustics Laboratory, and Leo L. Beranek, an electrical engineering faculty member and technical director of the Acoustics Lab, formed the partnership of Bolt and Beranek to offer consulting services in acoustics, responding to a call from the architects of the United Nations Headquarters buildings. When they recruited Robert B. Newman, a graduate student in architecture, to join them as a full partner in 1950, the name of the firm was changed to Bolt Beranek and Newman (now BBN Inc.). With emphasis gradually shifting over three decades from acoustics and noise control toward signal processing and computing, BBN, in 1989, had sales of $292 million.

Each of these faculty-based enterprises exemplifies the importance of direct transfer of advanced technology to the commercial marketplace. But although these professors and their early new enterprises became visible very quickly, most firms in the Greater Boston area that have been formed and grown in areas of high-rate-of-change technology are not founded by academics. Rather, they are created by engineers who had worked for a major MIT or industrial lab. The first in-depth example that follows, Digital Equipment Corporation (DEC), was founded by two former employees of MIT's Lincoln Laboratory, a major government-sponsored research and development organization established by MIT in the late 1940 to focus on the problems of air defense of the United States. Like the faculty-based companies DEC reiterates the importance of transfer of advanced technology to the initial product lines of an entrepreneurial firm. But DEC also demonstrates the critical contribution toward corporate "take-off" of a continuing flow of highly skilled professionals from a closely connected university into a rapidly growing and exciting organization.

Kenneth Olsen and Digital Equipment Corporation

Born in Bridgeport, Connecticut, in 1926, Kenneth Harry Olsen was the son of a machinery designer/sales engineer and was brought up in an evangelical Scandinavian Protestant family. His father's religious values and commitments influenced Ken deeply, his activities even today being strongly church-related. With a tool shop in the basement, Ken and his younger brother Stan became inventive gadgeteers, working with both mechanical and electrical devices. Joining the Navy directly from Stratford High School near the end of World War II, Ken received further training as an electronics technician before entering MIT, where he majored in electrical engineering.

Upon receiving his bachelor's degree Olsen joined Jay W. Forrester's MIT Digital Computer Laboratory group in July 1950 as a research associate. Forrester's team was just beginning to tackle the problems of upgrading the pioneering MIT real-time Whirlwind Computer into the basis for the SAGE (Semi-Automatic Ground Environment) system, the nation's first continental air defense system. MIT had recently established Lincoln Laboratory as prime contractor to the Air Force for this major program. Olsen's responsibilities quickly grew in that organization, taking on the project engineer's role for the first memory test computer (MTC) for the magnetic core memories created by Forrester and his associates, while also completing MIT requirements for his master's degree. Soon after IBM received the contract to supply the SAGE system's AN/FSQ-7 computer, Olsen and his wife Aulikki and young child moved to Poughkeepsie, New York, to become MIT's on-site liaison to the IBM development and manufacturing organization. The year-plus of day-to-day work with IBM taught Olsen much about the ways of a large bureaucratic organization, most of it not especially to the tastes of this young man who had grown up in MIT's far more free-wheeling environs. But he also learned to appreciate the disciplines of a well-run company. Returning to Lincoln Lab, Olsen took on leadership of the TX-2 computer development project, a small fast experimental machine designed around new transistorized circuits.

In late 1956, Olsen was approached by some other engineers who wanted him to join them in forming a new company. Ken remembers them as "vague in their thinking—they didn't have any specific products in mind; they just wanted to go into business". When their proposal failed, the germ was planted in Ken's mind to do something on his own. Despite having moved up to a section chief's job, the challenge of Lincoln Lab was wearing out and he felt he had to move. During the spring of 1957, discussions with Harlan Anderson, a Lincoln Lab engineer since 1952, who had worked in Olsen's early MTC project group, evolved the concept of a new computer company. They intended to design and build machines that reflected the Whirlwind/TX-2 real-time interactive approach, in contrast to the large number-crunching data processing computers that IBM and Univac already had in the marketplace.

With few financing alternatives apparent in the summer of 1957 Olsen and Anderson contacted American Research and Development Corporation (ARD) to fund their proposed Digital Computer Corporation (see Chapter 5 for further discussion of AR&D's pioneering role in the venture capital industry). In his commencement address at MIT in 1987, Olsen reminisced on this experience. "AR&D told us that this was not the right time for starting a company... [The AR&D staff] gave us three pieces of advice: Don't use the word computer, because *Fortune* magazine said no one is making money in computers... The promise of five per cent profit on sales in our initial presentation is not high enough; raise it... [so] we promised ten per cent.... Promise fast results, because most of the [AR&D] board members were over 80. So we promised to make a profit in one year" (Olsen, 1987, p. 8). With the business plan thus revised to meet the investors' stated prejudices, the AR&D board approved the investment in Digital Equipment (note, *not Computer*) Corporation of $70,000 for which they took 70 percent of the authorized stock. Intrigued by the young technologists and their objectives, but skeptical about their lack of management training and experience, AR&D insisted that stock be reserved to hire an experienced manager for this start-up. As that third key person was never brought on-board, the AR&D holdings turned out to be 78 percent of the total equity issued in DEC, the single investment that "made" the ARD portfolio so successful over the years, perhaps the best investment in the history of venture capital. But Olsen has never expressed regret about the limited initial funds or the disproportionate stock holdings by ARD. Ken has often remarked, "The nice thing about $70,000 is that you can watch every dollar", and he has frequently praised his investors for their long view and their lack of interference in DEC's efforts.

Incorporating DEC on August 28, 1957, were two young married couples, Kenneth and Eeva-Liisa Aulikki Olsen and Harlan and Lois Jean Anderson, somewhat prototypical of American entrepreneurial beginnings; but certainly not typical in the outcomes they achieved over the next three decades. Olsen was 31 and Anderson just 28 when the company was started, leaving Lincoln Lab to occupy inelegant but more importantly inexpensive space (25 cents per square foot per year, including watchman service and heat!) in an old textile mill in Maynard, Massachusetts, not far out into the countryside from their previous MIT lab location. On their first day at the mill they were joined by Stan Olsen, Ken's younger brother who had graduated from Northeastern University in Boston and had worked at Lincoln Lab as a technician, the only Olsen family member ever to work at Digital. For several years that initial threesome provided much of the functional leadership at Digital, with Ken handling engineering, "Andy" doing finance, and Stan running production.

The entrepreneurs began by developing a line of high speed transistorized circuit modules, similar to what had been created at Lincoln Lab for the TX-0 and TX-2 computers, but redesigned to accommodate the newest transistors available. Olsen says he saw the need for these packaged modules while at the lab and believed that DEC could develop a better line of products than was available at the time. Dick Best, Lincoln Lab's top-notch circuit designer, joined DEC as employee #5 to move this project forward. These products, soon called DECblocks, were well received by digital systems developers in both industry and academia and company sales began to take off. Ted Johnson, a Lincoln Lab engineer previously but now with a Harvard MBA to boot, was hired to become the first module salesman, eventually heading up all marketing and sales for Digital for the next two decades.

In addition to the early flow of technology and key people from Lincoln Lab and the MIT campus, Olsen began to develop other close ties to MIT. Jay Forrester, who had headed up the computer and systems efforts on the SAGE system, had moved in 1956 to a professorship in the MIT School of Industrial Management (now the MIT Alfred P. Sloan School of Management). Ken felt he had learned important organizational skills from experiencing Forrester's way of running the Digital Computer Lab and the SAGE program. One "gem" Olsen observed is that Forrester practiced "pulsed management", periodically focusing deeply on one part of the organization, turning it topsy-turvy, and ending up with substantial improvement. Later Forrester would "pulse" an entirely different aspect of work. At Olsen's invitation Forrester became a member of the Board of Directors of DEC and began to interest Olsen in the new field of industrial system dynamics modeling that Forrester was creating. In the summer of 1959, with his first full year of product sales behind him, Ken enrolled in an intensive two-week summer course at MIT to learn about system dynamics. Company sales for the year ending July 1 had reached $756,000 and profits after taxes were $112,400, clearly bettering the 10 percent promise Olsen and Anderson had made to ARD. As one of Forrester's research assistants I remember Olsen well from this first encounter in 1959. He was curious about our approach to modeling corporate structure and performance, but rather shy about asking questions. Ken was obviously readily able to handle the technical simulation modeling work, and came across much more like the bright young engineer he was than as a company president, despite the very impressive early signs of success of his new firm. One other student among the sixteen in that class remains memorable to this day, T. Vincent Learson, then vice president for Military Products of IBM, soon thereafter to become the IBM CEO who made and implemented the decision to develop the IBM 360, a dramatic and risky decision that vaulted IBM high over its then competitors. Upon Learson's retirement from IBM in 1966, *Fortune* magazine presented a retrospective on the IBM 360 decision: "Along with a group of computer users, [Learson] had recently attended a special course on industrial dynamics that was being given at the Massachusetts Institute of Technology. Much of the discussion had been over his head, he later recalled; but from what his classmates were saying he came away with the clear conviction that computer applications would soon be expanding rapidly..." (Wise, 1966). In contrast Olsen clearly understood those new directions in computers and was already committed to leading those changes, albeit from a far less lofty post than Learson's.

As DEC grew its connections to MIT also increased. The second product line was memory testers, reflecting Olsen's own experiences in running the MTC project. DEC's third product line, finally, was computers, what Olsen and Anderson had really wanted to do from the outset. Ben Gurley, another Lincoln Lab alumnus, designed and was responsible for much of the success of the PDP-1 (programmed data processor), first delivered in November 1960 to Bolt Beranek and Newman, Inc., another MIT spin-off firm mentioned earlier in the chapter. The machine's intellectual origins are clearly Lincoln Lab, conceptually flowing from the TX-0 and TX-2 computers. But MIT did not pursue Digital Equipment in regard to ownership of the "intellectual property" that went into the PDP-1, no doubt reflecting both MIT's general historic attitude of encouraging spin-off of its technology to commerce and government as well as the fact that no specific devices or designs were transferred from MIT to DEC. Shortly after the PDP launch DEC hired another MIT electrical engineering graduate who was to influence dramatically the technological growth of the company and its spectacular business success. Attracted to the excitement of this three-year-old firm, Gordon Bell quit his Ph.D. program in 1960 to join Digital, affecting from his arrival the design and development of DEC's computers.

Despite Digital's technological heritage being military-sponsored research and development (R&D) efforts at MIT, Olsen was clear from the outset that DEC would not undertake government contract work, preferring the discipline of financing his own development efforts with company money. Yet product sales to government markets, essentially on the same terms as to other markets, were visible from the outset, accounting for about 30 percent of DEC's business during its first decade. With the company finally in the computer business, sales accelerated, reaching $9.9 million for the fifth full year of existence, with profits of $1.2 million, still keeping ahead of its original targets.

Olsen also looked to MIT for management help, hiring Professor Edgar Schein of the Sloan School's Organization Studies Group as a consultant in 1966 to work with his key executive team. That relationship became and for more than two decades now continues to remain intimate, leading among other things to Digital being featured as the rather obvious Action Company case in Schein's book, *Organizational Culture and Leadership* (1985). Schein was a critical link in helping DEC to develop and articulate its own unique philosophy and style over the years.

The ties between DEC and the MIT Sloan School brought many of the faculty into close contact with DEC's people and problems, and many MIT management graduates joined MIT engineers in a steady flow into that attractive and growing company. Win Hindle, one of those early graduates referred to Digital by Jay Forrester, grew quickly in his responsibilities, emerging as a key member of the executive committee for many years, and surviving many waves of change in the organization. At an early stage he headed Personnel and instituted a series of middle management development seminars. My first seminar occurred three weeks after Data General Corporation had been formed by three departing key DEC engineers. DEC top management was uniformly bitter about this overt "lack of loyalty", with strong feelings expressed about deceitful behavior. But surprisingly the middle managers at the seminar focused most of their questions on the subjects of new enterprise initiation, financing, and success, almost ignoring my other lecture topics on engineering project organization and management. Despite the apparent fascination with entrepreneurial spin-off, in fact relatively few Digital employees left to become new company founders, and only Data General ended up becoming an important competitor. One new enterprise that included a DEC employee in its founding team, Medical Information Technology, Inc., is discussed later in this chapter.

Digital under Olsen's continuing leadership has always been a managerial dilemma, simultaneously seeming to be both in the backwash as well as in the vanguard of managerial ideas. In the late-1960s I organized and ran for five years a major technical management development program for the top managers at Digital, with hundreds of DEC managers eventually going through many series of nine-day intensive sessions with me and my Sloan School faculty colleagues. These contacts provided ongoing opportunities to watch the excitement and growth of Digital and its key players. Olsen opened the first session with a combination of encouragement but wariness. "Just remember", he warned, "we're just a bunch of engineers. Let's not get too sophisticated." But Gordon Bell, the brilliant VP of Engineering, who with many others later created the VAX architecture that became the product base for DEC's continuing rise, delighted in grabbing new ideas and turning them into vehicles for intellectual engagement with "his troops". Soon DEC was awash with Sloan School ideas about "technological gatekeepers", "critical innovation functions", "real versus paper matrix organization structures", and "internal entrepreneurship".

Olsen's outspoken resistance to business school ideas and methods was even stronger at the first several days seminar on corporate planning that MIT faculty conducted in the early 1970s for DEC's top eighty managers. Ken's management style leans away from formal planning approaches and more toward decision making through "technical shootouts" and consensus building through internal conflict. These attitudes are deeply rooted. Olsen says that even DEC's original business plan stated that it would never have a central planning group, not the sort of thing usually even mentioned in a business plan. Schein explains his Action Company's behavior: "The founding group comes from an engineering background, is intensely practical and pragmatic in its orientation, has built a strong and loyal 'family' spirit that makes it possible to confront and have conflict without risk of loss of membership, and clearly believes that 'truth' lies not in revealed wisdom or authority but in 'what works', both technologically and in the marketplace" (Schein, 1985, p. 10). For whatever reason, even today, Ken takes advantage of many speaking opportunities to criticize business schools, as if he resents what they do or perhaps what they "stand for" in his mind. Some observers have wondered whether this is an expression of hidden resentment of General Doriot, his ARD mentor who was a professor at the Harvard Business School, and whom Olsen has always praised lavishly. Or might it reflect resentment of Forrester, Olsen's self-assured role model at Lincoln Lab and early DEC board member, who was a professor of management at MIT? Others, perhaps more wisely, suggest that Olsen has learned well from both "his professors", themselves outspoken critics of their own and other business schools, and that Olsen's teasing of business schools is intended to lift their standards.

In fact, much of Olsen's values and managerial style seems to resemble Jay Forrester's. Each is primarily a brilliant and hard-driving engineer, committed to self-initiated and self-determined outcomes; each focuses on reality, but creates his own boldly stated theories to explain behavior; each espouses deep convictions about both personal and product integrity. "To an overwhelming degree, people prefer to be ethical and to work for an ethical company ... people are honest with suppliers and customers when they know that the company is not interested in unethical profiteering", Olsen exclaimed (Olsen, p. 9). Ken once told me his greatest source of anguish at DEC was in turning out an occasional poor quality product.

Both Forrester and Olsen have argued repeatedly for the importance of entrepreneurship within organizations. But each is awesome in personal intellectual and argumentative strength and stomp out less assertive and less capable subordinates. Looking backward over a long span of company history, Olsen testified to dealing with this issue at a critical point in DEC's evolution: " ... we had to face the question of how to introduce the spirit of entrepreneurship throughout the company. We were doing well ... But there was only one entrepreneur at the top" (*Ibid.*). Forrester had already proposed what to do. In his "A New Corporate Design" (1965), Professor Forrester argued: " ... moving away from authoritarian control ... can greatly increase motivation, innovation, and human growth and satisfaction.... The nonauthoritarian structure implies internal competition for resource allocation.... One wishes to combine the stability and strength of the large ... business organization with the challenge and opportunity that the small company offers to its founder-managers.... The profit center concept of the proposed organization brings into the corporation the same free-enterprise profit motive that we believe is essential to the capitalist economy." Olsen had apparently listened to his mentor's treatise on internal entrepreneurship, among many other arguments bubbling up throughout Digital to cope with Ken's concerns. "To solve the problem we broke the company up into a number of entrepreneurial product lines. Each one had a manager with complete responsibility for that segment of the business, and everyone else served that manager. This went over like a lead balloon. Everybody thought they were being demoted. Many people quit; even some of the board resigned. But the results were magnificent. Within a year we had doubled our profit without hiring anybody. For many years afterwards we grew at the rate of 20, 30, 40 percent a year and made very good profit" (Olsen, p. 9). But a true entrepreneurial environment permits failures as well as successes, and many key departures occurred, including Harlan Anderson, Olsen's co-founder who apparently felt squeezed out of the company at this time (1966). Most of the "old guard" hung on, however, continuing to build Digital, until the early 1980s when many senior officers left, including Ken's brother Stan.

The Olsen record of performance at Digital Equipment Corporation is unparalleled in American history. For over thirty years he has driven and led the company he founded to over $13 billion in revenues in 1990, clear leader in the minicomputer field. Although still a distant second to IBM in the overall computer industry, DEC is now even playing on IBM's court with the VAX 9000 entry into the mainframe business. Named "America's Most Successful Entrepreneur" by *Fortune* in 1986 when Digital's sales were slightly more than half today's numbers, *Fortune* pointed out: "DEC today is bigger, even adjusting for inflation, than Ford Motor Co. when death claimed Henry Ford, than U.S. Steel when Andrew Carnegie sold out, than Standard Oil when John D. Rockefeller stepped aside" (Petre, 1986). Olsen's stockholdings in Digital are worth over $200 million, although he has given one-third of it to Stratford Foundation, a charitable foundation he has established. Ken has commented on the dilemma of his success: "It's almost an impossible combination to survive ... to become flattered by the world and to become wealthy." Yet his life-style is still modest, he and his wife still living in the home they moved to shortly after Digital was established. Controversial as a manager, as evidenced throughout his unauthorized biography, *The Ultimate Entrepreneur* (Rifkin and Harrar, 1988), the success produced by Olsen's leadership is undeniable.

Ken Olsen has claimed that his greatest source of satisfaction at DEC has been in watching people develop. He echoes this thought in his final words at the MIT Commencement: "My ambition is to leave ... when I can be remembered as someone who challenged them, who influenced them to be creative and enjoy work and have fun for a long time" (Olsen, p. 10).

Not only is Digital Equipment Corporation the most successful MIT spin-off company, in many ways it also reflects the most direct transfers from MIT. Both founders and all key early employees were from MIT. Their principal, in most cases sole, prior work experience was at MIT. The general skills and knowledge of those engineers as well as the technologies that were the basis for the first three product lines all came from MIT Lincoln Lab. But management know-how was also transferred from MIT into DEC, both early and often. Jay Forrester's ideas seem to have had critical early influence. Ed Schein's involvement in top management decision making still persists over the decades. No other company that I have observed in over twenty-five years of entrepreneurial research so strongly reflects an MIT heritage.

Arthur Rosenberg and Tyco Laboratories, Inc.

Although Tyco Laboratories also has a tie to Lincoln Labs, its history shows that technology-based companies are also financial creatures. Tyco demonstrates the creative use of capital both to launch and then to build a large diversified technological firm. Its founder, Arthur Rosenberg, was born in Boston, the second son of middle-class parents. His father came to this country from England as a boy and was trained as a civil engineer at MIT. Arthur's older brother was very bright with a flair for physics, entering college at age 16. As Arthur grew up, he decided to go into the sciences for no particular conscious reason. He reports now always feeling technically in the shadow of his older brother, although he admired him very much and never really tried to compete with him. An average high school student, Rosenberg at one time aspired to medicine and enrolled in a college that had a good graduate medical school. He left college for the Army after one term, but returned one year later and reenrolled in college, concentrating in biology. He graduated at age 21, at which time he abandoned any ideas of entering medical school.

While Arthur was a teenager, his father, a professional engineer who had always apparently entertained ideas of self-employment, went into the construction business, but his venture did not prove successful. He subsequently suffered a nervous breakdown and died shortly thereafter. Although Arthur and his father were not particularly close, in retrospect Rosenberg sees his father and in particular his father's entrepreneurial experience as having had a profound effect on him. He remembers his mother as the chief disciplinarian and homemaker and as having a dominant influence on the family.

After graduation from college Rosenberg enrolled in graduate school at Harvard in physics. He feels that he was unprepared for his experience at Harvard, but that it was the greatest personal challenge he ever faced. Receiving the Ph.D. degree from Harvard still represents a great accomplishment to him, although he feels that he did not have the ability to embark on an academic career. Arthur is sure that the Harvard experience prepared him for the eventual presidency of Tyco Laboratories, Inc., although he feels he would have been some other type of entrepreneur without it.

With doctorate in hand Rosenberg joined a large chemical company in New Jersey as a research physicist, but started to make waves within the organization and soon became, in his own words, "notorious". Looking back he feels that with this job came his first realization of the frustration that can accompany work at a large organization. With his new perspective he began the search for another job, but felt he was spotted by people as a troublemaker from the beginning. He wanted to return to Boston, he thinks probably to be near Harvard, and was also attracted to the growing semiconductor field. After missing one attractive job opportunity due to a misunderstanding, Rosenberg finally settled on employment at MIT's Lincoln Laboratory. He was passed over for promotion early during his time at the lab because he was judged not to be good managerial material. He believes this to have been an honest appraisal based on his lack of sensitivity to people. Arthur's desire and attempts to manage people were not successful nor was he taken seriously. During his last couple of years at the Lab, Rosenberg began to take part in the outside scientific community primarily because he realized that he could not remain with the Lab forever. By the time he left his MIT job he had established a good reputation in the local scientific community, a step that provided him with excellent prospects for a new job.

At this point, Rosenberg decided to join a fledgling company on Route 128, the circumferential highway around Boston that has become famous for its concentration of high-technology companies. He was recruited for a key technical–managerial role and upon joining was given options for 5 percent equity in this young firm. The initial aim of the company was to create subsidiary companies based on advanced technology, something that had been done earlier in the Boston area by National Research Corporation, founded by Richard Morse, an MIT alumnus who later became a faculty member teaching entrepreneurship in the MIT Sloan School of Management. But several early developments and manufacturing ventures that were set up in this young firm proved unsuccessful. Rosenberg was director of the company's contract research center, which had secured several government contracts in the materials field. Anxiety on the part of the firm's founders over a contract that did not come through, however, resulted in the beginnings of some ill will between Rosenberg and the founders.

In view of the trouble besetting the parent company, compared with the success of the contract research center he headed, and motivated in part by his desire to be independent, Rosenberg decided that he wanted to split off his group into a separate organization. At this time he felt certain that the parent organization was going to go under, but that his research department could survive independently.

Aided by a young and aggressive New York underwriting firm, L. M. Rosenthal and Co., Rosenberg devised a plan for financing his spin-off research center. The new company, Tyco Laboratories, Inc. was incorporated with Rosenberg putting up $10,000, giving up part of his equity to several others in the company. A small but successful public offering by Rosenthal raised a few million dollars for Tyco Laboratories. Most of the original company's rights in Tyco were purchased with part of the proceeds of this offering but some problems between Rosenberg and some of the original management continued over ownership issues and managerial responsibility for the new organization. Thus, Tyco Labs became a very early example of what today is known as a management buy out (MBO).

Although imbued with a desire to manage people, Rosenberg was most confident of his ability to do research in the materials field and of his intuitive ability. Initially he had joined what became Tyco Laboratories not as an entrepreneur, but rather to try his hand as a research director. The notion of entrepreneurship and the feeling of being an entrepreneur came much later in conjunction with his assuming the leadership of the company and in part because of the resultant conflict with some of the founders of the original organization. Rosenberg reports that although he had enough ideas to be a research director he did not realize, at this point, that the job of running a contract research organization was primarily that of being a salesman. He now closely associates entrepreneurship and salesmanship.

During the first few years of Tyco Laboratories, Rosenberg did not foresee having an actual manufacturing capability, thinking of the firm only as a research organization that would continue to grow as a research organization. The firm had been beset by certain internal stresses as well as by the job of obtaining and maintaining external contracts and customers. Rosenberg was able to hire many good people in building up his research company, but states that he also had the strength to kick them out when necessary, a characteristic he believes important for a successful entrepreneur.

In the early stage of its existence, Tyco was a little company fighting for survival in the game of contract research. While the company became known quickly after securing several good basic materials research contracts, its growth potential was limited since it engaged in no hardware development or product sales. It was at this point that the notion of acquiring manufacturing companies took shape, and a year or so later Rosenberg approached the first company. He undertook this step somewhat reluctantly, not knowing exactly how to go about it and having seen his former associates attempt acquisitions and fail. However, the responsibility for bringing earnings to the public stockholders made requisite some change in the completely basic research nature of the business. I became a Board member of Tyco Labs on the night of approval of its first acquisition, Mule Battery, a small low-technology producer of industrial batteries, and thus observed as a participant the early boom period of the firm. The stock market rewarded this apparent "promise" of a great commercial future with a substantial jump in Tyco's stock price. That fueled Rosenberg's appetite and made further acquisitions easier and less expensive to accomplish. Subsequently, Rosenberg made several successful acquisitions that brought Tyco Laboratories into the sphere of manufacturing and added several million dollars to its sales base. In slightly over two years of this approach, twelve acquisitions were made and the company's stock became a glamorous speculation. Rosenberg said he began to see the ability to average risks as one advantage of a sizable as opposed to a small corporation. He says he thought of Tyco Laboratories as "an experiment in management", not as a holding company.

But the acquisitions brought with them a dramatic shift in the ownership of Tyco Laboratories. Entrepreneurial founders of acquired companies held large blocks of stock and demanded active participation in Tyco management. Rosenberg continued to follow his strategy of bigger and bolder acquisitions, aiming to become a billion-dollar firm. Each next step required more risk, including as it turned out to Rosenberg himself.

Eventually, with the company doing over $100 million in mostly relatively recently acquired sales, but with operational problems becoming bigger day-by-day, a "palace coup" occurred. The Board of Directors, entirely appointed by Rosenberg, but with several acquired entrepreneurs in its membership, and with me no longer a member, voted to dismiss Rosenberg as president and chief executive officer. Interestingly, after a brief pause for consolidation, the successor CEO who was recruited from the outside continued to follow Rosenberg's strategy, bringing Tyco Laboratories to its present size of more than $2 billion.

Rosenberg had been personally concerned with the growth mechanisms of the company much more than with internal operations, which he entered largely as an arbitrator. He feels that judgment is very important and has confidence in his own. His greatest rule is: "Never be doctrinaire." He is now relaxed about his ouster as president and philosophizes that if an entrepreneur loses internal visibility within his firm, he can lose control of his enterprise. "It's clear Tyco Laboratories grew too rapidly." He sees as important "the building of a system with positive interactions between the various parts." He was looking for new techniques and feels that the problem of human relations must not be neglected, especially in a diverse firm doing over $100 million of sales, with multiple sources of power within the company.

On the satisfactions of being an entrepreneur, Rosenberg states that he had achieved at Tyco a "sense of complete fulfillment of my own talents", in an environment that he created himself. He does not have the desire to be "very, very rich", although he stated that despite being "comfortably well off", he would like to have his income more separate from his professional activities. Nor does he profess to be much interested in public recognition. Today, Rosenberg spends his time investing in and building up companies founded by others, like many other successful ex-entrepreneurs. He regards himself now more as a "dabbler" in young technology-based firms than as a "real" entrepreneur.

One of his regrets is not having spent more time with his family, but Rosenberg admits that this is largely due to the fact that his work has always been extremely important to him and that he enjoys working the long hours that are demanded. He sees this problem as one that is not peculiar to entrepreneurs, but one that poses difficulties for many types of professionals and their families.

Samuel Morris and Transducer Devices, Inc.*

Many entrepreneurs try more than once, often with the second try being

---

* At his request the founder's name, the company name, and some details have been disguised to assure anonymity. The basic history is essentially unchanged, however.

far more successful than the first new company. Raymond Stata and Matthew Lorber co-founded Solid State Instruments, which performed modestly and was sold out after a few years to Kollmorgen Corporation. They later formed Analog Devices, which has become a great success, now doing nearly $500 million in revenues under Stata's continuing leadership. After leaving Analog in 1970, Lorber started Printer Technology, which failed rather quickly, and then in 1983, formed Copley Controls, whose outcome is not yet clear. Other founders have their big hit with their first attempt, like Philippe Villers whose Computervision did extremely well, going public, and eventually selling out to Prime Computer in 1988 for $435 million. But Villers' second firm, Automatix, is still struggling to survive, no longer with Villers at the helm, and his third company, Cognition, has already closed down, with Automatix picking up its pieces. Still other entrepreneurs seem to try and try again, without ever achieving success, as is illustrated in this next case history.

While Transducer Devices further indicates the diversity of entrepreneurial backgrounds, its most significant characteristic in my judgment is its continued attempts at "technology push" rather than "market pull", a lack of market orientation leading to its eventual demise. Let's begin with the early life of Samuel Morris, co-founder and president of Transducer Devices, Inc. (and of several other businesses before that).

The third child of parents who immigrated to this country from Russia, Sam Morris describes himself as a "bright youngster" and "not too well restrained." He grew up in the Bronx and Manhattan during financially difficult years for his family, when they were forced to move many times due to their inability to pay the rent. Sam was brought up mainly in Jewish neighborhoods where scholastic standards were high and demanding. He proved to have an inventive bent at an early age, building a ferris wheel from scratch when he was only ten years old.

He describes his parents as hardworking, honest and good natured and recalls that they expected great things of him. His father had been trained in Europe as a leather scraper and, after he came to this country, was involved in several businesses for which he had to put in long hours of hard work. Sam was frequently left to his own devices while his mother helped his father with the business. Sam's older brother Charlie went into business for himself at an early age. Sam sees Charlie as having had a significant influence on his own life, and as having provided him with the initial, although perhaps unconscious, inspiration to become a businessman. He was also personally motivated in terms of scientific things.

After attending a high school for gifted children for a few years, Sam enrolled in City College where he began to take engineering courses, driving a taxi on weekends to support himself. A job at the New York Board of Transportation in an non-air-conditioned office provided him with the impetus to construct a small air conditioner for which he soon got orders from friends. This work was initially carried out in his parents' apartment while they were away for the summer. But the orders for air conditioners picked up so quickly that he rented a loft and hired one other person to assemble them. Believing he could make more money selling air conditioners, Morris quit his job and started to contact department stores, actually receiving prototype orders from a few of them. However, safety measures and the laboratory testing required by some stores posed problems. Morris says that he never had enough capital (he had started with about $500 of his own), and was finally forced out of business due to continuing problems. Although he reports that his profits were not really great enough to cover his time, Morris describes this episode, which lasted about two seasons, as an interesting experience.

As a consequence of this entrepreneurial experience Morris took some management courses and went to work on a consulting basis for his brother Charlie who was in the dry cleaning business. There, he tried to introduce piece work, but had some difficulty with the unions to whom this practice was unacceptable. At the same time Sam invented a special clothes hanger for which he obtained a patent. He then began a business to build machines to produce the hangers. However, other hanger companies threatened to drop their prices so Morris was forced out of business again, this time without actually making any hangers.

Next, he went to work as a consulting engineer for a "body shop" firm and was leased out to several companies, including an MIT laboratory. By this time Morris, who already had two children, decided that his education was obsolete so he got a full-time job at the MIT Instrumentation Laboratory, taking electrical engineering courses at the same time. He also took some courses in investment analysis at Boston University and became involved in the stock market.

His interest in the stock market stemmed in part from the fact that he had received a cash settlement for the development of a machine, a side project that he had managed while working as a consulting engineer. For this development, which he describes as his first profitable venture, Morris received $30,000. Successful investment of this money in the stock market resulted in quite a sizable sum and precipitated his decision to go again into business for himself. Sam sees the stock market as an enjoyable game at which he has been very successful.

Morris' original entrepreneurial philosophy had been to take a going company and build on it. His brother's profitable dry cleaning business and his own successful investment provided the two brothers with some capital with which to buy a business. However, his development of an analog-to-digital conversion device for advanced military applications, and the interest of two MIT laboratory colleagues, led to the decision to use this device as the basis for a new company. Their idea was to develop one or more products to the point where they could generate sufficient income to allow the company to go into still other products, and thus build up a combination development and manufacturing company. This "philosophy" has prevailed throughout the history of Transducer

Devices, though its implementation has often been lacking.

At the time the company was founded Morris says he naively figured he could build another IBM. He was 35 years old and he believed that he would make a million dollars in but a few years. The first problems encountered by the founders were figuring out how to fund the business and at what level to operate. Although several anticipated original investors backed out of the deal, a small office near MIT was rented, and within a reasonable time a working model of the first device was produced. Starting out on a part-time basis with only two employees and three founders the company got a $130,000 order almost immediately from a large electronics firm. The founders decided to sell a limited private offering of stock to several individuals, raising $200,000 for a very small part of the company plus some additional options. Soon afterward, however, a government cutback on some advanced development work forced the termination of the project. Although the founders decided to continue development of the machine, this contract termination served to prolong the part-time phase of the business, while Morris and his co-founders continued to work "full-time" at the Instrumentation Laboratory.

As in the case of EG&G, ownership of the company was originally split evenly between the three MIT scientists/engineers and Morris' older brother who had put up most of the capital. In return for his efforts on their behalf, their attorney also received some stock. When all three founders eventually quit their jobs at MIT and began to work full time, a clash over their respective positions in the company caused some emotional stress. The three men were very different in personality and work style and the resulting arrangement did not prove to be a completely compatible one. Eventually one of the three founders left the firm, Sam solidified his position as president, and the other co-founder remained as chief engineer.

The company experienced some problems in regard to marketing since it did not have a catalogue of items to sell. Transducer Devices was in reality selling techniques. The engineers attempted to adapt to customer requirements more than vice-versa, but at the same time were semicommitted to maintaining a standard price. Sales did begin to go up as the firm came out with support products. However, delays were later caused by a combination of increased government source inspection requirements and some latent defects in systems in the field. They also encountered some problems with their suppliers.

The primary and continual problem besetting Transducer Devices was loss of money and the need for additional capital. It was kept in operation with the help of supplier credit, some bank loans, additional capital from Charlie Morris, and contacts made by some of the initial individual investors. The firm was at one point dropped by one bank, but picked up by another for credit on a limited basis. The lack of adequate financing interfered in many areas, even though the continued faith of Sam's brother led him to supply the firm with additional capital when it was necessary.

As president of the company Morris took care of marketing, internal policies, administrative functions, and customer contacts. At the time he became president he was putting in 60–70 hours a week. He says that although his wife thought it was admirable at first, she eventually tired of his working that many hours.

Transducer's sales and profits history is shown in Table 1–1. The losses might have been significantly larger if Morris and his co-founders had paid themselves higher salaries. Seven years after the company was started Sam and his co-founder were receiving little more than their final salaries at the MIT lab they left.

With losses growing and Charlie's ability and willingness to finance them waning, Transducer Devices was sold during the next year for an amount less than its total debt. Sam found a job as chief engineer with an invention development company, closer to his "true love" activity than his role as company president.

Morris feels that the company really had too much talent that was not channeled and contained effectively. He also feels that they could have built up a significant sales volume, if several of the product lines had been exploited more fully. He admits that his firm made some mistakes, the most far-reaching of which were, in his opinion: not enough money, not sticking closely enough to schedule, and the founding of a company by a noncompatible triumvirate. To quote Morris, "In equality there is weakness." He feels now that the personal difficulties that beset the firm would have been averted had he carried out his original plan and gone into business with his brother.

Sam Morris seems to have been determined to become a technical entrepreneur from an early age. His initial unsuccessful attempts to start businesses merely led to further attempts. His hope for a more successful future never really seemed to flag, nor did his older brother's faith in him and financial support for his venture. Morris expressed a desire to build a relatively large company and was also interested in obtaining significant

Table 1–1
Transducer Devices' Sales and Profits

Fiscal Year	Sales ($000)	Profits ($000)
1	0	− 66.2
2	118	− 77.4
3	260	−170.0
4	260	−194.0
5	456	+26.0
6	1,118	+37.0
7	1,110	−220.0

financial returns. Perhaps in conflict with his entrepreneurial instincts was his continued inventive bent and interest in being part of the technical side of the business. He cites as a distinct source of pleasure the delivery of unique technical problem solving to customers. As he has observed, further exploitation of a few of the company's product lines, rather than the expansion of the product line to meet all customers' specific requirements, might have reaped more benefits for the firm.

Morris reports that his wife was very much in favor of his entrepreneurial venture and so encouraged it. She was not, however, warm to his alternative idea of going into the stock market on a full-time basis, which he says would have made him a much richer man today. Both she and he are happy with Sam's new job, which demands fewer hours but still has the opportunity for technical creativity and some new opportunities for striking it rich.

Incidentally, Transducer Devices continues to make slow but steady progress under its new management. Sales in 1990 were above $5 million and the company is profitable.

Neil Pappalardo and Medical Information Technology, Inc.

A portrait of modern high-technology entrepreneurship demands inclusion of at least one computer software entrepreneur. The list of software spin-offs from MIT labs and departments alone is almost without bounds: Charles Adams and Associates, General Computer, Gold Hill Computer, Index Technology, Interactive Data Corporation, Intermetrics, Logo Computer Systems, Lotus Development, SofTech, to mention just a few, with probably a hundred more in my various research studies. I focus here on one software entrepreneur and his company with which I personally have been closely involved. Meditech's success to date certainly builds heavily on its technological leadership. But, in contrast with Transducer Devices, Meditech's growth has come from market-oriented savvy, focus, and dedication to customer service, qualities that are instrumental to growing great companies.

As a child in Rochester, New York, Neil Pappalardo got lots of exposure to business activities from his father's legal practice and the many small businesses of his uncles and cousins in their Sicilian "extended family" community. Neil's mother was in complete control of the home and also managed the family finances. She invested conservatively in local blue chips like Eastman Kodak and IBM, but also put some money into a smaller Rochester company named Haloid Corporation, better known today as Xerox. Perhaps as a result of this background Neil feels that he never had concerns about money, either having it or making it.

When it came time to apply to college Neil knew he wanted to be an engineer. So Neil applied to MIT, only, indicating at an early age the self-confidence he would exude later as an entrepreneur. Though technology intrigued him, classes at MIT· did not, and Papp reports spending far more time playing bridge with his fraternity brothers than studying or even attending classes. But he got through, with a passable C+ average.

"When I was an EE undergraduate at MIT, I realized that I wanted to create a company. But I wasn't in any panic," Neil recalls. As a result of thesis work at the Peter Bent Brigham Hospital on a computerized system for monitoring various cardiac conditions, Neil accepted a job upon graduation, in 1964, that led him first into programming and then into running a programming group in the Massachusetts General Hospital's (MGH) newly formed Laboratory of Computer Science. His frustration with the large costs and the slow progress encountered in the experimental information systems project being conducted at the hospital stimulated Pappalardo to "bootleg" his own approach. Working primarily with another MIT graduate, Curtis Marble, he conceived and developed on the DEC PDP-7 computer an integrated data-base and time-shared operating system which in true programmer "nerd" fashion they named MUMPS (MGH Utility Multi-Programming System), a system that eventually revolutionized health care data processing. The task completed and MUMPS diffusion well underway at MGH, Neil concluded, in 1968, that it was time to leave MGH and set up his own company. He informed his boss that he planned to depart as soon as he finished whatever was needed at the hospital in terms of loose ends, like documentation and hiring his own successor, despite the fact that his company ideas were still vague.

While Neil was in this transition phase I began discussions with him. I had also made a decision to establish some type of medical computing firm, growing out of the market needs I sensed from my teaching, research, and consulting activities with medical schools and hospitals. But I was looking for full-time partners, as I had no desire or intention to leave MIT. When I met with Neil he was nearly ready to depart MGH, with his colleague Curt Marble expected to join him as soon as something was underway. Dr. Jerome Grossman, another MIT graduate who had received his M.D. degree from the University of Pennsylvania, was working with Neil and Curt on medical applications of computers, and also planned to participate on a part-time basis. We four hit it off well together and began weekly evening meetings to plan the company in detail and to try to raise funding. Our conclusion that we needed a strong marketing person in the group led us to recruit Morton Ruderman, who had been DEC's product line manager for its Laboratory Instrument Computer (LINC, incidentally, as the name suggests, a development licensed by Digital from MIT's Lincoln Laboratory), and later DEC's bio-medical marketing manager, to join us as a co-founder and initial president. With four of the five co-founders from MIT we decided to name the company Medical Information Technology, Inc. (the initials were the key), calling it Meditech for short.

During the months of planning and slow-going fund-raising Neil, who was the "principal" entrepreneur of this company and had already left

MGH, "got itchy" and ordered the company's first PDP-9 computer from DEC and signed a five-year lease for 5,000 square feet of space behind MIT in East Cambridge, pledging his own personal stock investments as collateral. With five co-founders on the team, side issues kept getting raised during the planning stage. Neil wanted to bring into our founding group a cousin who had been very supportive of Neil's ideas. His cousin had some family money, which we desperately needed, and an undergraduate degree in business administration, and claimed to be more practical about business than the rest of us. Eventually we talked Neil out of this proposal. Two other MIT alumni, whom all five of us knew, were in the process of starting a different medically oriented software firm, called Cyber, to do contract programming and to develop a system for multiphasic health screening. After serious discussions about possible merger of our efforts, before actual incorporation of either group, we decided to go our own ways. Incidentally, Cyber lasted only a few years before closing down.

The most demanding issue was posed by Mort Ruderman. Out of loyalty to Digital Equipment he insisted that Meditech make no financing commitment until we offered DEC a chance to become our "corporate sponsor". Although we felt we needed to raise a half million dollars from some venture capital company, all of us finally consented that we would be willing to accept a DEC computer and a much smaller amount of funds in exchange for 20 to 40 percent of Meditech, if Ken Olsen would agree to our proposal. At a sensitive point in our search for financing, when we mistakenly thought we were about to reach closure with a group of investors for our needed capital, Mort went to DEC and presented our offer. It was our 1969 version of what today is labeled a "strategic alliance". The Digital executives told Mort that they could not encourage this kind of defection, but Olsen wished Mort well and invited him to return to Digital if Meditech did not work out. According to *The Ultimate Entrepreneur* (Rifkin and Harrar, 1988), Olsen's offer to Ruderman was only rarely made to anyone leaving DEC.

Having passed over these preformation hurdles, finally, at a public seminar I taught during the summer, I met two corporate staff members from EG&G, Inc., another much larger MIT spin-off company described earlier, who were searching for investment opportunities for their company. Amazingly, within ten days Ken Germeshausen, EG&G's chairman, and Barney O'Keefe, the president, decided that EG&G should invest $500,000 in Meditech for 30 percent of the company, beginning a set of close corporate and personal relationships that continue productively to this day. The five co-founders invested a total of $14,000. Meditech was formally incorporated on August 4, 1969, the same day Jane Pappalardo gave birth to their fourth child. Neil was 27 years old; I was the oldest of the group at 33. Incidentally, Pappalardo had drawn no salary during the nine months from leaving MGH until Meditech was incorporated.

Meditech began by doing what it still does today, developing and selling applications software products for hospitals, as well as the operating system and programming language for them. Neil and Curt immediately started to upgrade the MUMPS operating system and renamed it MIIS (Meditech Interpretive Information System), and later went on to MIIS Standard and currently MAGIC. From an applications perspective most hospital data processing at the time was concerned with financial matters like payroll and billings. Meditech turned its attention instead to the medical care delivery side of the hospital, focusing initially on clinical laboratory modules, setting up interfaces between the laboratory instrumentation and Meditech's remote-accessed time-shared computers. The ability to just pay a monthly rental fee to Meditech and essentially "try before you buy" was a key selling point for small- and medium-sized hospitals to use the latest technology, and created a viable marketing approach for the emerging company. But high communications costs for remote time sharing required Meditech to establish regional sales and/or computer service centers around the country, in New York, Washington, D.C., St. Louis, Phoenix, and San Francisco. Eventually, the dramatic changes in computer hardware technology and the rapid cost declines for both computers and long-distance communications led to recentralizing all offices and machines back to Massachusetts and shifting almost entirely to stand-alone computers in the hospitals instead of time sharing.

By the time Neil took over as CEO, in 1973, Meditech was well established but still quite small, and profit margins were low. The company's technological directions were solid but its product and market approach needed further tuning. Neil decided to decrease significantly the software customization being done for individual hospital clients by making each application product far more powerful and standardized. He and the company officers and board soon decided to begin the investments needed to broaden the applications areas to include financial aspects of hospital operations as well, enabling the gradual development of fully integrated but modular hospital information systems. These decisions provided the keys to Meditech's continuing growth and profitability over the past decade and a half, generating fairly smooth compound annual sales growth in excess of 20 percent and even higher growth of net income. In 1990, Meditech's revenues were nearly $70 million and profits after taxes close to $19 million.

Larry Polimeno was the first hire into Meditech, coming from BBN's computer time-sharing operation where he had had great working relationships with the co-founders from MGH. Larry joined as operations manager, but moved up to become general manager and then executive vice president and chief operating officer. Roland Driscoll entered soon after founding as manager of finance, rising to senior vice president of finance. Once Neil became president, he, Larry, and Roland became a triumvirate, an informal "office of the president" that conferred closely on all matters of importance to Meditech, but with Pappalardo clearly in charge. The value system at Meditech has from the beginning been comparable in many ways to that of DEC, but even more so. Total integrity in dealing with customers is demanded. "Deliver what you promise" became such a fetish that for years new products were not even announced until the finished product was already working satisfactorily in one or two beta sites.

Meditech's employment practices are what I would call "Mafioso management", but this term obviously needs careful explanation. Neil's interpretation serves well: "We view ourselves as a family. We always hire entry level people, whether in software or sales. We don't hire new people into managerial slots. All of our managers have been promoted from within. We want our people to join us when they're still young and get their training here in our culture. They don't have much experience when they start but at least they're malleable at that age. And they learn quickly."

"The essence of the Meditech organization", Pappalardo continues, "is that we're concerned with our own and we take care of our own. We teach people, hold on to them, try to develop them. Sometimes we may hold on for too long, but this policy works well in the long haul. We want to reward people by achieving continuing company growth, paying them well, and building up their financial assets. But in turn we expect clear loyalty, respect for our corporate values, good performance. And so far we've gotten all this and more from our employees. Not everyone who comes to work here loves it. But if someone lasts through the first three years, they'll stay forever." Meditech employee turnover is one of the lowest in the rather turbulent software industry, with about 12 percent of the employees quitting each year.

Financial management is extremely conservative at Meditech, in line with the triumvirate's personal conservatism. Orders are not booked until the contract has been signed and a 10 percent deposit received. Initial company office buildings were purchased for cash. More recent real estate has been bought using five-year commercial loans, not the conventional thirty-year mortgages, fully secured by preferred stock held in Meditech's investment portfolio. Financial leverage is treated as inappropriate and current shareholders are expected to carry full costs of operations, not building a debt burden that future shareholders might have to repay. Going public or selling the company is seen as highly unlikely, especially by Neil, in that either act might place Meditech's family culture into jeopardy.

But in terms of technology Meditech has been far from conservative, always trying to be the pacesetter, and this has been almost entirely of Neil's doing. Meditech started with interactive time sharing on minicomputers, when the hospital information systems (HIS) industry was dominated by batch processing on mainframes. Its software was based on high level interpretive languages when most everyone else used compiled code generated by old standards like Fortran or Cobol. Meditech has continuously upgraded and then obsoleted its own software, with Pappalardo personally driving the operating system changes and managing the system software group. The company has repeatedly borne the high cost of rewriting all of its applications programs to serve its customers optimally. It provides interfaces to the widest variety of customer hardware, including both computer systems and various types of laboratory equipment and hospital data generating devices. Meditech provided real networking capabilities, developed and introduced proprietary full color terminals, and demonstrated compatibility with new reduced instruction set computer (RISC) architecture either first among HIS vendors or within months of first introduction. Fortunately, these changes have not been "bet your company" investments for Meditech, but their absence might have resulted in a "betting" of the company's future. As to the future Neil says, "What we want is more of the same, in every way", the essence of a "stick to your knitting" philosophy (Peters and Waterman, 1982).

Preview of the Book

What?

As these four sets of founders and firms reveal, a series of in-depth histories does not by itself provide a basis for generalizations about origins or makeup of high-technology entrepreneurs or their companies. More important, several cases can illustrate but cannot alone determine the causes of success or failure of these firms. To attempt generalization and explanation, data are needed on far more technology-based companies. The data must include measures that describe many critical aspects of the firms and their founders. And those measures must be carefully analyzed to test for possible relationships and to illumine the sources and successes of technological entrepreneurs.

For those who might be interested in the background history, general approach and contents of the research, I strongly urge reading of the Appendix before going to Chapter 2. The Appendix presents the empirical bases for the book, a program of research studies that I began in 1964 and that continues through this writing. With my graduate research assistants and master's and doctoral students I have now conducted over forty research studies on many aspects of technological entrepreneurship, described in the Appendix as falling into five "research tracks" or lines of study. Aggregated together the analyses of these data cover formation, transition and growth, and success or failure of several hundred high-technology firms. Many of these companies derive directly from MIT academic departments or research laboratories, and almost all of them were initiated in the Greater Boston area. The synopsis of studies in Table 1–2 indicates the general subject matter of each of the five research tracks and the resulting data analyses.

The research findings presented throughout the book highlight the origins of founders and firms, and the determinants of their success and failure. Some of the conclusions are:

- Entrepreneurs are very likely to have had self-employed fathers. But first-born sons are not more likely than their siblings to become high-technology entrepreneurs.
- Technical entrepreneurs come far more frequently from development work than from research, where they excelled as high performers.
- Entrepreneurs are not all alike; they display wide ranges of personalities, motivations, and goals for starting new enterprises.
- The key initial technologies of the new firms were transferred primarily from development projects carried out by the entrepreneurs at their previous employers.
- Initial capitalization is typically very small and provided from the entrepreneurs' personal savings. Multiple co-founders raise larger amounts of initial capital.

Table 1–2
Synopsis of Research Program

Research Track	Focus of Research	Data Analyses
1	Comprehensive studies of spin-off companies founded by MIT staff and faculty, from four major MIT labs and five academic engineering departments	156 companies
2	Comprehensive studies of spin-off companies from four non-MIT "source organizations," including MITRE, Air Force Cambridge Research Lab, and two major industrial corporations	82 companies
3	More focused studies of eleven independent samples of new enterprises from different high-technology industries, plus one sample of low-technology consumer-oriented manufacturers	196 companies
4	Six focused studies on personal characteristics of technological entrepreneurs and prospects, control studies of engineers and scientists at two major MIT labs, and two studies of MIT technical faculty	129 entrepreneurs 299 engineers and scientists 73 faculty
5	Fifteen focused studies on various aspects of high-technology financings, including investigations of business plans, searches for capital, processes of going public, venture capital investment decisions, bank lending decisions, underwriting decisions, institutional investor attitudes, and venture capital fund performance	99 companies 106 financial institutions

- Specific plans and an initial product help generate greater initial capital.
- Widespread deficiencies in business plans and in team composition hurt the new enterprise's ability to raise "outside" capital.
- Companies that go public later in their corporate lives undergo far easier and less expensive financing.
- Family background has no impact on entrepreneurial success. Successful entrepreneurs are made, not born!
- Prior supervisory, managerial, and especially sales experience by founders contributes to successful enterprises.
- Entrepreneurs with a high need for achievement are more likely to succeed.
- Multifounder teams generally perform far better than single founders, and the greater their number the more likely is their success.
- Firms that start with products significantly outperform those that begin as consultants or R&D contractors.
- The more technology transferred initially from the entrepreneurs' "source" organization, the greater the eventual company success.
- Firms that begin with a marketing orientation, and/or evolve one early in their development, are more likely to succeed.
- Companies that focus on core technologies and markets do much better than those that diversify into multiple technologies and markets.
- "Founder's diseases" are widespread, but not universal, with two-thirds of the founders of successful technological enterprises being displaced before their companies achieve "super-success".
- The future for high-technology entrepreneurship in the United States, and indeed increasingly throughout the world, is very promising.

Where?

The book chapters comprise three connected sections—treating birth, transition and growth, and success or failure—all developed from extensive analyses of my research data, combined with perspectives gained from my own quarter century of involvement as a multicompany founder, director, and venture capitalist. The research data provide the information on which the chapters rest, but my personal experiences often influence the interpretations.

Chapters 2 through 5 cover the birth of high-technology companies, treating in turn the founding people, their base technology, and their initial capital. Chapter 2 begins with a broad discussion of the influences of the Greater Boston environment, and especially MIT, on the formation of new firms. Chapter 3 then compares in detail the founders of a large number of high-technology companies with representative samples of engineers and scientists who remained as employees of the laboratories that had spawned the entrepreneurs. This type of "controlled" comparison provides higher reliability for drawing conclusions as to the unique "makings" of a technological entrepreneur. Chapter 4 focuses on the extent to which the technology used to start new firms is drawn from previous key employers of the entrepreneurs, and assesses the influences upon this technology transfer. Chapter 5 discusses the typical financial requirements of high-technology companies over their full life cycle, examines the general preferences of various types of investors and financial institutions, and then displays the initial financing experiences of several hundred technological companies, analyzing the causes of that initial financing.

Chapters 6, 7, and 8 examine the early transition and growth of technological firms. Chapter 6 documents the beginning market and product orientation of the founders and their companies and the early changes that occur in business objectives, management time allocation, and operations. Chapter 7 investigates the need for additional financing of high-technology firms, and the time-consuming efforts by their entrepreneurs to raise the funds. Chapter 8 assesses the decision to "go public" by a fraction of these technological firms, and their search for underwriters. It also presents analyses of the underwriters' decision making, followed by data on the process of going public and the consequences to founders and firms of this action.

The next section of the book, Chapters 9, 10 and 11, relates in depth evaluations from three different points of view of the keys to success and failure of high-technology companies. Chapter 9 presents comprehensive statistical analyses of the broad data base of the companies studied in search of significant linkages with company success. That chapter is divided into sections that focus on possible ties to background and characteristics of the entrepreneurs themselves, the technology base of the start-up firms, their financing, and general management approaches. Many ties to success are uncovered and some keys to poor performance. Chapter 10 takes a hard look at the product strategy of high-technology companies and its connection to success, using a separate group of studies to trace the full multiproduct sequences developed and marketed by a number of computer-related companies. Chapter 11 reports a final attempt to determine corporate success, examining another special cluster of companies, those firms that were among the most successful high-technology companies in Massachusetts. Within that elite group, the factors leading to "super-success" are sought, emphasizing a number of dimensions of strategic management.

The final chapter summarizes the findings developed throughout the book in regard to our knowledge of birth, growth, and success of high-technology firms. What we know is highlighted, but additional desirable insights are also described. I end with some evidences of recent trends and my opinions on what is likely to happen in the future to technological entrepreneurship in this country and abroad. Rest assured—while the future is never certain and storm clouds loom for some aspects of technological enterprise, high technology entrepreneurship remains a continuing and ever more important part of the American dream and reality, now to be shared with the rest of the world.

References

International Science and Technology. "The Business of Science", August 1965, 52–53.

J. W. Forrester. "A New Corporate Design", *Industrial Management Review* (now published as *Sloan Management Review*), Fall 1965.

K. H. Olsen. "The Spirit of Entrepreneurship", *Technology Review*, August/September 1987, MIT 8–10.

T. J. Peters and R. H. Waterman. *In Search of Excellence* (New York: Harper and Row, 1982).

P. Petre. "America's Most Successful Entrepreneur", *Fortune*, October 27, 1986, 24–32.

G. Rifkin and G. Harrar. *The Ultimate Entrepreneur: The Story of Ken Olsen and Digital Equipment Corporation* (Chicago: Contemporary Books, 1988).

E. H. Schein. *Organizational Culture and Leadership* (San Francisco: Jossey-Bass Publishers, 1985).

J. A. Timmons. *New Venture Creation: A Guide to Entrepreneurship*, second edition (Homewood, IL: Richard D. Irwin Inc., 1985).

T. A. Wise. "I.B.M.'s $5,000,000,000 Gamble", *Fortune*, September 1966.

CHAPTER 2

An Environment for Entrepreneurs

Global pursuit of technology-based industrial development has mush-roomed in the past decade. Greater Boston's Route 128 and California's Silicon Valley have become prototypes for other regions' and other nations' visions of their own futures. Research and writing about the Technopolis (Dorfman, 1983; Miller, 1985; Rogers and Larsen, 1984; Segal Quince Wickstead, 1985; Smilor, Gibson and Kozmetsky, 1989; Tatsuno, 1986) have accompanied actions by cities and states through-out the United States and Europe to launch entrepreneurial centers, often based on newly established university incubators and venture capital firms. In Asia, Japan has committed major funding to create a network of "science cities" (going beyond its own Tsukuba), the Republic of China has coupled tax incentives and subsidies to help grow its technology park, and Singapore has linked sophisticated local industrial development planning with government-funded venture capital invest-ments in overseas start-ups to attract high-technology opportunities. Even the Soviet Union has established a joint venture with U.S. and Japanese corporations to generate a center for new technology-based industry.

None of these kinds of governmental programs contributed to the growth of high-technology industry in the Greater Boston area. But what did cause this original American Technopolis to develop? What forces continue today to encourage young local scientists and engineers to follow entrepreneurial paths? In his review of entrepreneurial decision making, Cooper (1986) argues for six different potential environmental influences: economic conditions, access to venture capital, examples of entrepreneurial action, opportunities for interim consulting, availability of support personnel and services, and access to customers. This chapter traces the evolution of Boston's high-technology community, providing support for all of Cooper's variables. But I also identify even more critical aspects of culture and attitude that have built a local environ-ment that fosters entrepreneurship.

31

Early Influences: The Heritage of
World War II Science and Technology

The atomic bomb, inertially guided missiles and submarines, computer-based defense of North America, the race to the moon, and the complex of high-technology companies lining Route 128 outside of Boston are phenomena that became prominent in the postwar years. This was a time marked by a plethora of scientific and technological advances. World War II had defined technology as the critical element upon which the survival of the nation rested. That war brought scientists from the shelter of their labs into the confidence of those in the highest levels of government. And in the postwar years their power and their products and by-products began to shape society, the economy, and the industrial landscape.

How had this started? The sudden need for war research in the early 1940s transformed universities like MIT into elite research and development centers where the best scientific and technological talent was mobilized for the development of specific practical devices to win the war. Virtually whole universities redirected their efforts from pure scientific inquiry to the solving of critical problems. While many scientists had to neglect their previous research in favor of war-related innovations, the scientists themselves were not neglected. Science and its offspring technology had become the property of the whole nation with an immediate relevance for all the people.

In addition to the urgent expansion and redirection of university research, the war made necessary the reorganization of research groups, the formation of new working coalitions among scientists and engineers, between these technologists and government officials, and between the universities and industry. These changes were especially noteworthy at MIT, which during the war had become the home of major technological efforts. For example, the Radiation Laboratory, source of many of the major developments in wartime radar, evolved into the postwar Research Laboratory for Electronics. The Servomechanisms Lab, which contributed many advances in automatic control systems, started the research and development project that led to the Whirlwind Computer near the end of the war, created numerically controlled milling machines, and provided the intellectual base for undertaking, in 1951, the MIT Lincoln Laboratory. After the war the Servo Lab first changed its name to the Electronic Systems Lab and continues today as the Laboratory for Information and Decision Systems. Lincoln Lab focused initially on creating a computer-based air defense system (SAGE) to cope with the perceived Soviet threat. To avoid continuing involvement in production and operations once the SAGE system was ready for implementation, MIT spun off a major group from Lincoln Lab to form the nonprofit MITRE Corporation, chartered to aid in the later stages of SAGE and undertake systems analysis for the government. Lincoln then reaffirmed its R&D thrust on computers, communications, radar and related technologies primarily for the U.S. Department of Defense. The

Instrumentation Lab, growing out of the wartime gunsight work of Dr. Charles Stark Draper, its founder and director throughout his career at MIT, continued its efforts on the R&D needed to create inertial guidance systems for aircraft, submarines, and missiles. It followed up with significant achievements in the race to the moon with developments of the guidance and stellar navigation systems for the Apollo program. The former Instrumentation Lab now bears Draper's name in its spun-off-from-MIT nonprofit status. Draper testified as to the scope of these endeavors: "Personal satisfaction ... was greatest when projects included all essential phases ranging from imaginative conception, through theoretical analysis and engineering to documentation for manufacture, supervision of small-lot production, and finally monitoring of applications to operational situations" (Draper, 1970, p. 9). All these MIT labs, included as major potential "source organizations" in my entrepreneurship research studies (as outlined in the Appendix), were spawned during a period in which little debate existed about a university's appropriate response to national urgency. They have been successful in fulfilling their defined missions, while also providing a base of advanced technology programs and people for other possible societal roles.

Building on a Tradition

The war efforts and the immediate postwar involvements of MIT with major national problems built upon a much older tradition at MIT, enunciated by its founder William Barton Rogers, in 1861, when he created an institution to "respect the dignity of useful work". Its slogan is *Mens et Manus*, the Latin for mind and hand, and its logo shows the scholar and the craftsman in parallel positions. MIT "for a long time ... stood virtually alone as a university that embraced rather than shunned industry" (*The Economist*, 1987, p. 7). From its start MIT developed close ties with technology-based industrialists, like Edison and Alexander Graham Bell, then later with its illustrious alumnus Alfred P. Sloan during his pioneering years at General Motors, and also with close ties to the growing petroleum industry. In the 1930s, MIT generated The Technology Plan, to link industry with MIT in what became the first and is still the largest university–industry collaborative, the MIT Industrial Liaison Program.

These precedents were accelerated by the wartime leadership of its distinguished president, Karl Taylor Compton, who brought MIT into intimacy with the war effort just as he himself headed up all national R&D coordination in Washington. In the immediate postwar years Compton pioneered efforts toward commercial use of military developments, among other things helping to create the first institutionalized venture capital fund, American Research and Development (ARD). "ARD was, in part, the brain-child of Compton, then head of MIT. In discussions with Merrill Griswold, Chairman of Massachusetts Investors Trust, and Senator Ralph

Flanders of Vermont, then President of the Federal Reserve Bank of Boston, Compton pointed out that some of the A-bomb technology which had been bottled up for four years had important industrial applications. At the same time, it was apparent to Griswold and Flanders that much of New England's wealth was in the hands of insurance companies and trusts with no outlet to creative enterprises. Griswold and Flanders organized ARD in June 1946 to supply new enterprise capital to New England entrepreneurs. [Compton became a board member, MIT became an initial investor, and a scientific advisory board was established that included three MIT department heads. General Georges] Doriot, who was Professor of Industrial Management at Harvard, was later asked to become president" (Ziegler, 1982, p. 152). ARD's first several investments were in MIT developments, and some of the emerging companies were housed initially in MIT facilities, an arrangement that even today would be seen as a source of controversy and potential conflict at most universities. Compton's successor as president of MIT, James R. Killian, furthered the encouragement of entrepreneurial efforts by MIT faculty and staff, as well as close ties with both industry and government. At various times Killian served on the boards of both General Motors and IBM, and as President Eisenhower's Science Advisor.

The traditions at MIT of involvement with industry had long since legitimatized active consulting by faculty of about one day per week, and more impressive for its time had approved faculty part-time efforts in forming and building their own companies, a practice still questioned at many universities. Faculty entrepreneurship, carried out over the years with continuing and occasionally heightened reservations about potential conflict of interest, was generally extended to the research staff as well, who were thereby enabled to "moonlight" while being "full-time" employees of MIT labs and departments. The result is that approximately half of all MIT spin-off enterprises, including essentially all faculty-initiated companies and many staff-founded firms, are started on a part-time basis, smoothing the way for many entrepreneurs to "test the waters" of high-technology entrepreneurship before making a full plunge. These companies are obvious candidates for most direct movement of laboratory technology into the broader markets not otherwise served by MIT. Incidentally, few of the faculty founders ever resign their MIT positions, preferring to remain at MIT like Amar Bose, founder of Bose Corporation, or Harold Edgerton, co-founder of EG&G, while turning over the full-time reins to their former graduate students and lab colleagues. George Hatsopoulos, founder of Thermo Electron Corporation, Jay Barger, co-founder with another faculty colleague of Dynatech, and Alan Michaels, founder of Amicon, are among the few faculty who left to pursue their entrepreneurial endeavors on a full-time basis, with great success achieved in all three cases.

Although today regional and national governments on a worldwide basis seek to emulate the Boston-area pattern of technological entrepreneurship, in the early years the MIT traditions spread to other institutions very slowly. The principal early disciple was Frederick Terman, who took his Cambridge experiences as an MIT Ph.D. student back to Stanford University, forsaking a faculty offer by MIT to eventually lead Stanford into technological excellence. Terman had gained first-hand exposure to the close ties between MIT and industry, made more important to him by his being mentored by Professor Vannevar Bush, later dean of engineering and then vice president at MIT, who participated in founding the predecessor of the Raytheon Corporation. The attitudes he developed at MIT led Terman to encourage and guide his former students, such as William Hewlett and David Packard and the Varian brothers, to start their high-technology firms and eventually to locate them next to the university in Stanford Research Park (Rogers and Larsen, 1984, p. 31). While these efforts obviously founded what has become known as "Silicon Valley", the resulting proliferation of firms there came from multiple spin-offs of other companies, and did not follow the dominant Greater Boston pattern of direct fostering of new firms from MIT labs and departments. Only eight out of 243 new technical firms studied in the Palo Alto area have their origins in Stanford University (Cooper, 1971), probably due in part to Stanford's lack of major government-sponsored laboratories. Indeed, despite the distance from their alma mater, MIT alumni are surprisingly the founders of over 175 companies in northern California, accounting for 21 percent of the manufacturing employment in Silicon Valley (Chase Manhattan, 1990). Similarly, our MIT study of major technology-based regions in North America and Europe (Sirbu et al., 1976) determines that Research Triangle Park in North Carolina has little evidence of local entrepreneurial activity and few ties between entrepreneurship and the three major universities in that area. And in 1989, only 23 firms in total are documented as "spin-outs" of UT-Austin, including faculty, staff, students, and technology transferred out to other entrepreneurs (Smilor, Dietrich and Gibson, 1989). Feeser and Willard (1989) find far fewer university spin-offs, just one, in their national sample of 108 computer-related founders. Cambridge University, England, is seen as heavily responsible for the development of the several hundred high-technology firms in its region, and yet only "17 percent of new company formation has been by individuals coming straight from the University (or still remaining in it)" (Segal Quince Wickstead, 1985, p. 32). Thus, the MIT–Route 128 model still today remains unusual in its degree of regional entrepreneurial dependence upon one major academic institution. Perhaps other regions need other "models" if they are to achieve technology-based industrial growth (Cooper, 1985).

The Neighboring Infrastructure

Yet MIT has not been alone over the past several decades in nurturing the technology-based community of Boston, now sprawling outward beyond

Route 128 to the newer Route 495. Northeastern University, a large urban institution with heavy engineering enrollment and an active cooperative education program, has educated many aspiring engineers who provide both support staff and entrepreneurs to the growing area. Wentworth Institute educates many of the technicians needed to support the development efforts at both the university labs as well as the spin-off companies. Boston University and Tufts University, both with strong science and engineering faculties, also play important roles. Even small liberal arts Brandeis University has participated, with Professor Orrie Friedman, in 1961, starting Collaborative Research, Inc., forerunner of the much later biotechnology boom in the Greater Boston area.

Possibly surprising to readers from outside of the Boston area, Harvard University has not had a substantial role in entrepreneurial endeavors until the recent biotechnology revolution. In many ways Harvard, over the years, has looked down its "classics" nose with disdain at the "crass commercialism" of its technological neighbor a few miles down the Charles River. An Wang, who had worked at the Harvard Computation Laboratory, is the most prominent exception to this rule. Change in regard to encouraging entrepreneurship is in the wind, even at Harvard. The outpouring of excellent research and discovery from Harvard's Chemistry and Biology Departments, as well as from the Harvard Medical School across the river in Boston, has caused Harvard faculty and staff recently to become much more active and successful participants in entrepreneurial start-ups, although not without voiced reluctance and controversy at the university. In fact, in a dramatic revolution of its policies Harvard asked Professor of Biochemistry Mark Ptashne to start Genetics Institute in 1979, a company in which Harvard would hold 15 to 20 percent equity. But protest by critics as to possible influence of such ownership caused Harvard to pull out. Ptashne went ahead and formed the company, while still remaining on the Harvard University faculty (*Boston Business Journal*, March 23, 1987). In 1989, the Harvard Medical School took the far-reaching step of organizing a venture capital fund to invest in new companies whose founders relate in some manner to Harvard Medical, in some ways mimicking MIT's much earlier activities in regard to AR&D, but nevertheless a pioneering step among academic institutions. Indeed, a recent survey of life sciences faculty (Louis et al., 1989) places Harvard tenth in the nation, with 26 percent, in percentage of "faculty members holding equity in a company whose products or services are based on their own research". MIT life sciences faculty place first in that same survey with 44 percent, such as Professors Alex Rich and Paul Schimmel who co-founded Repligen Corporation. Some of these biotech ventures involve faculty from both Harvard and MIT, such as Biogen, co-founded by Harvard's Walter Gilbert and MIT's Phillip Sharp.

Encouraged no doubt by the exemplary venture capitalist role of Professor Doriot, and separated by a river from main campus influence, many Harvard Business School graduates, joined after its 1951 founding by MIT

Sloan School of Management alumni, found welcome homes even in the early company developments. These business school graduates got involved in start-up teams initially as administrators and sales people, and in more recent years participating frequently as primary founders. Thus, Aaron Kleiner, from the MIT School of Management, shares the founding of three high-technology companies with his MIT computer science undergraduate roommate Raymond Kurzweil. And Robert Metcalfe combined MIT educational programs in both engineering and management prior to his launch of 3Com. The Greater Boston environment has become so tuned to entrepreneurship that even student projects with local companies, a part of routine course work in every local management school, have ended up helping to create numerous entrepreneurial launches. Several firms are claimed to have been generated from feasibility studies done as part of Doriot's famed *Manufacturing* course at the Harvard Business School. And *INC.* magazine founder Bernard Goldhirsch credits a Sloan School marketing course with confirming for him the huge market potential for a magazine targeted toward entrepreneurs and small business managers (*INC.*, 1990, pp. 39–40).

Boston entrepreneurs also have benefited from understanding bankers and private investors, each group setting examples to be emulated later in other parts of the country. The First National Bank of Boston (now Bank of Boston) had begun in the 1950s to lend money to early stage firms based on receivables from government R&D contracts, a move seen as extremely risky at the time. Arthur Snyder, then vice president of commercial lending of the New England Merchants Bank (now Bank of New England), regularly took out full page ads in the *Boston Globe* showing himself with an aircraft or missile model in his hands, calling upon high-technology enterprises to see him about their financial needs. Snyder even set up a venture capital unit at the bank to make small equity investments in high-technology companies to which he loaned money. Several scions of old Boston Brahmin families became personally involved in venture investments even in the earliest time period. For example, in 1946, William Coolidge helped arrange the financing for Tracerlab, MIT's first nuclear-oriented spin-off company, eventually introducing William Barbour of Tracerlab to ARD, which carried out the needed investment (Ziegler, 1982, p. 151). Coolidge also invested in National Research Corporation (NRC), a company founded by MIT alumnus Richard Morse to exploit advances in low-temperature physics. NRC later created several companies from its labs, retaining partial ownership in each as they spun-off, the most important being Minute Maid orange juice. NRC's former headquarters building, constructed adjacent to MIT on Memorial Drive, in Cambridge, now houses the classrooms of the MIT School of Management. Incidentally, long before the construction of Route 128, Memorial Drive used to be called "Multi-Million Dollar Research Row" because of the several early high-technology firms next to MIT, including NRC, Arthur D. Little Inc., and Electronics Corporation of America. The comfortable and growing ties between Boston's worlds of academia and finance helped create bridges to the large Eastern family fortunes—the Rockefellers, Whitneys, and Mellons, among others— who also invested in early Boston start-ups.

And by the end of the 1940s, when space constraints in the inner cities of Boston and Cambridge might have begun to be burdensome for continuing growth of an emerging high-technology industrial base, the state highway department launched the building of Route 128, a circumferential highway (Europeans would call it a "ring road") around Boston through pig farms and small communities. Route 128 made suburban living more readily accessible and land available in large quantities and at low prices. MIT Lincoln Lab's establishment in 1951 in the town of Concord, previously known only as the site of the initial 1776 Lexington–Concord Revolutionary War battle with the British, "the shot heard round the world", or to some as the home of Thoreau's Walden Pond, helped bring advanced technology to the suburbs. Today, Route 128, proudly labeled by Massachusetts first as "America's Technology Highway" and now as "America's Technology Region", reflects the cumulative evidence of forty years of industrial growth of electronics and computer companies. Development planners in some foreign countries have occasionally been confused by consultants and/or state officials into believing that the once convenient, now traffic-clogged, Route 128 highway system actually caused the technological growth of the Greater Boston area. At best Route 128 itself has been a moderate facilitator of the development of this high-technology region. More likely the so-called "Route 128 phenomenon" is a result and a beneficiary of the growth caused by the other influences identified earlier.

Accelerating Upward From the Base: Positive Feedback

A critical influence on entrepreneurship in Greater Boston is the effect of "positive feedback" arising from the early role models and successes. Entrepreneurship, especially when successful, begets more entrepreneurship. Schumpeter observed: "The greater the number of people who have already successfully founded new businesses, the less difficult it becomes to act as an entrepreneur. It is a matter of experience that successes in this sphere, as in all others, draw an ever-increasing number of people in their wake" (1936, p. 198). This certainly has to be true at MIT. The earliest faculty founders, Edgerton and his colleagues, Bolt, Beranek, and Newman of the MIT Acoustics Lab and then of the company bearing their names (now BBN, Inc.), and John Trump of High Voltage, were senior faculty of high academic repute at the times they started their firms. Their initiatives as entrepreneurs were evidences for others at MIT and nearby that technical entrepreneurship was a legitimate activity to be undertaken by strong technologists and leaders. Karl Compton's unique role in founding ARD furthered this image, as did the MIT faculty's efforts in bringing early-stage developments to ARD's attention. Obviously, "if they can do it, then so can I" might well have been a rallying cry for junior faculty and staff, as well as for engineers in local large firms. Our comparative study of Swedish and Massachusetts technological entrepreneurs finds that on average the U.S. entrepreneurs could name about ten other new companies, three or four of which were in the same general area of high-technology business. Few of the Swedish entrepreneurs could name even one or two others like them (Utterback et al., 1988). A prospective entrepreneur gains comfort from having visibility of others like himself, this evidence more likely if local entrepreneurship has a critical mass, making the individual's break from conventional employment less threatening.

The growing early developments also encouraged their brave investors, and brought other wealthy individuals forward to participate. As example of the spiraling growth of new firms, even in the early days, Ziegler (1982) shows the proliferation of thirteen nuclear-related companies "fissioning" within fifteen years from Tracerlab's 1946 founding, including Industrial Nucleonics (now Accuray), Tech Ops, and New England Nuclear (now a division of DuPont). With forty years of activity, a positive feedback loop of new company formation can generate significant outcomes, even if the initial rate of growth is slow. In the mid-1960s, through dramatic proliferation of spin-off companies, Fairchild Semiconductor (founded by MIT alumnus Robert Noyce) gave birth to similar and rapid positive feedback launching of the semiconductor industry in Silicon Valley (Rogers and Larsen, 1984). And Tracor, Inc. seems to be providing a comparable impetus to new company formation in Austin, Texas, leading to 16 new firms already (Smilor, Gibson and Kozmetsky, 1989). Exponential growth starting in the early–middle 1970s has generated the several hundred firm Cambridge, England high-technology community (Segal Quince Wickstead, 1985, p. 24).

A side benefit of this growth, also feeding back to help it along, is the development of supporting infrastructure in the region—technical, legal, accounting, banking, real estate, all better understanding how to serve the needs of young technological firms. In Nancy Dorfman's (1983) assessment of the economic impact of the Boston-area developments she observes "a network of job shoppers that supply made-to-order circuit boards, precision machinery, metal parts and sub-assemblies, as well as electronic components, all particularly critical to new start-ups that are developing prototypes and to manufacturers of customized equipment for small markets. In addition, dozens if not hundreds of consulting firms, specializing in hardware and software populate the region to serve new firms and old." Of course, this massive network is itself made up of many of the entrepreneurial firms I have been investigating. Within this infrastructure in the Boston area are new "networking" organizations, like the MIT Enterprise Forum (to be discussed later) and the 128 Venture Group, which serve to bring together on a monthly basis entrepreneurs, investors, and other participants in the entrepreneurial community, contributing further positive loop gain (Nohria, 1990).

This positive feedback effect certainly occurred in the Greater Boston region as a whole and, as illustrated by the Tracerlab example, also occurred at the single organizational level. As one individual or group departs a given lab or company to form a new enterprise, the phenomenon may mushroom and tend to perpetuate itself among others who learn about the spin-off and also get the idea of leaving. Sometimes one group of potential entrepreneurs feels it is better suited than its predecessors to exploit a particular idea or technology, stimulating the second group to follow quickly. Perhaps as a result, four companies were formed from Instrumentation Laboratory employees to produce "welded module" circuits, a technique developed as part of the Instrumentation Lab's Polaris guidance system project. Remember also, from Chapter 1, that Ken Olsen being approached by others to start a company was his first recollection of thinking about entrepreneurship as a career. The "outside environment" can help this process by becoming more conducive to additional new enterprise formation. In particular, venture capitalists, learning more about a source organization from its earlier spin-offs, may actively seek to encourage further spin-offs from the same source. This certainly played an important role in the 1980s proliferation of biotechnology spin-offs from MIT and Harvard academic departments.

Other "Pulls" on Potential Entrepreneurs

In addition to the general environmental encouragements on Greater Boston technological entrepreneurship, specific "pulls" are at work on some of the people, making entrepreneurship an attractive goal to attain. Such influences may inhere in the general atmosphere of a particular organization, making it more conducive to the new enterprise spin-off process. For example, until his recent death Stark Draper, visionary leader of the MIT Instrumentation Laboratory (now renamed the Draper Lab), was a key source of encouragement to anyone who came in contact with him. No wonder that the National Academy of Engineering established the Draper Prize to be the equivalent in engineering of the Nobel Prizes in science. With the good fortune to fly coast-to-coast with him one night on a "red eye" from Los Angeles, I learned much about Draper's unique attitudes toward developing young technologists. "I try to assign project managers who are just a bit shy of being ready for the job. That keeps them really hopping when the work gets underway, although the government officials usually want to wring my neck." "I break up successful teams, once they've received their honors. That way every one remembers them for their success, rather than for some later failure. Also, this causes every young person in the Lab to be sitting within one hundred feet of someone who's had his hand shaken by the President of the United States." "The Lab is a place for young people to learn. Then they can go someplace else to succeed." "When I give speeches I single out those who have already left the Lab—to become professors elsewhere, VPs of Engineering in industry, or founders of their own companies. Staying behind in the lab is just for a few old beezers like me who have no place else to go!" His environment was one of high achievement, but with negative incentives for remaining too long. Salaries flattened out quickly, causing the income gap between staying and leaving to grow rapidly as an engineer gained experience. Engineers completing a project had a sharp breakpoint, a good time for someone confident from the success of his or her project to spin-off. In retrospect Stark Draper seemed consciously trying to encourage spin-off of all sorts from his laboratory, perhaps the highest attainment achievable by an academic scientist.

"They were looking for excitement. They weren't just looking for a more logical way to make software: they wanted to be part of another major breakthrough. After all, Margaret Hamilton had helped send a man to the moon by the time she was 32. 'Apollo changed my life,' she said. 'It had a profound effect on us. Some people never got over it. And there have been other spin-offs from Draper because of it.' The follow-up for Hamilton, who was in charge of more than 100 software engineers at Draper, was going to have to be something big. She seems to have found it by starting her own business. To Hamilton, 'A growing high-technology company is like a mission.' With theory in hand, Hamilton and [Saydean] Zeldin founded HOS [Higher Order Software, the only company in my research sample founded by two women] in 1976" (*Boston Business Journal*, August 20–26, 1984, p. 7).

No questions were asked if Instrumentation Lab employees wanted to borrow equipment to take home over the weekend, and many of them began their new companies "moonlighting" with this kind of undisguised blessing. Draper wanted reasonably high levels of turnover, and constant introduction to the Lab of bright eager young people. Over a fifteen-year period during which I traced Lab performance, the average age of Instrumentation Laboratory employees remained at 33 years, plus or minus six months. This young-age stability, maintaining the lab's vitality and fighting off technological obsolescence, was not true at most of the other MIT labs studied.

Draper apparently produced similar effects in his teaching activities at MIT. Tom Gerrity, founder of Index Systems, which in turn later created Index Technology and Applied Expert Systems as sponsored spin-outs, reports that Draper's undergraduate elective subject showed him the importance of being able to put together lots of different skills and disciplines to produce a result. Gerrity adopted this systems point of view in founding Index several years later, after three MIT degrees and a stint as a faculty member in the MIT Sloan School of Management.

Some other MIT laboratory directors followed similar patterns of entrepreneurial "sponsorship" in smaller less well-known labs. For example, the head of the Aeroelastic and Structures Laboratory of the MIT Department of Aeronautics and Astronautics had the attitude that the lab provided an internship type of position and that staff members were more or less expected to move on after a reasonable period. In other labs the environment just seemed to breed entrepreneurism. Douglas Ross, who left the Electronic Systems Lab with George Rodrigues to found SofTech, Inc., comments: "The entrepreneurial culture is absolutely central to MIT. The same mix of interests, drives and activities that make a [Route] 128-type environment is the very life blood of MIT itself. No other place has the same flavor" (Simon, 1985, p. 20). Ross epitomizes this "life blood" quality. When SofTech was established MIT took the exceptional step for that time of making a direct equity investment in his ground zero company, joining a large number of us who shared great confidence in Doug Ross's vision.

Indeed, the challenging projects underway at most of the labs create a psychological "let-down" for their participants when the projects end. Many of the entrepreneurs indicate that they became so involved with their work on a given project that when these projects were completed they felt that their work too was completed. Several of the entrepreneurs attest that their sense of identification with the source lab began to wane as the project neared completion. As Margaret Hamilton indicates earlier, only through the challenge of starting their own enterprises did they think they could recapture the feelings that they were doing something important.

Beyond the labs other activities at MIT have over the years encouraged entrepreneurship. The MIT Alumni Association, not the central MIT administration, undertook special efforts to encourage entrepreneurship among its members. Beginning in the late 1960s, the Alumni Association initiated a series of Alumni Entrepreneurship Seminars. Intended to serve an expected small group of 40 to 50 Boston-area young alumni, the effort escalated when over 300 alumni signed up for the first weekend. Over a two-year period the Alumni Association then launched a pattern of weekend seminars targeted for MIT alumni all around the country. Over 2,000 attended the initial national series, and called for more follow-ups. The alumni committee got ambitious and wrote a book on how to start a new enterprise, the only book ever jointly published by the MIT Alumni Association and the MIT Press, and distributed it widely to interested alums (Putt, 1974). Directories were assembled and widely distributed of alumni interested in the possibility of starting a firm, who might be willing to meet with similarly interested alums, thus beginning a rudimentary matching service. Ongoing monthly programs were started in several cities across the country, including The MIT Venture Club of New York City and then the MIT Enterprise Forum in Cambridge. The latter still continues to stimulate and help new enterprises, and to provide the networking needed to build start-up teams and linkages with prospective investors and advisors. And now the MIT Enterprise Forum has expanded to chapters in fourteen major cities across the U.S. and even in other countries where MIT alumni are concentrated.

All of these efforts spread the word, and legitimatize the activities of entrepreneurship. And they have produced results. Over the years many entrepreneurs have introduced themselves to me, saying they remember hearing me talk at the MIT Alumni Entrepreneurship Seminars. My first meeting with Neil Pappalardo, with whom I much later participated in founding Meditech (see Chapter 1), occurred at the first MIT Alumni Entrepreneurship Seminar. Bob Metcalfe, the principal inventor of Ethernet and later the founder of 3Com, a great success in the computer networking market, reports that after attending an MIT alumni luncheon on starting your own business, he resigned from Xerox's Palo Alto Research Center, returned to Boston and established his company with two other engineers (Richman, 1989, p. 37). Similarly, the founders of Applicon, now the CAD division of Schlumberger, decided to create their firm after listening to a seminar at Lincoln Lab that reported on the characteristics of the previous Lincoln spin-off entrepreneurs.

And most recently new policies instituted by John Preston, head of MIT's Technology Licensing Office, further encourage entrepreneurship, especially by faculty and research staff. In addition to conventional technology licensing to mainly large corporations for fees, still dominating the MIT technology transfer portfolio, Preston now is willing to license MIT-originated technology in exchange for founder stock in a new enterprise based on that technology. In the first year of this new practice, 1988, six new companies were born based on licensed MIT technology, with sixteen firms started in the second year of policy implementation. Matritech is one example, based on technology developed by Professor Sheldon Penman and researcher Edward Fey to employ antibodies to find proteins within cells, a new approach for detecting certain cancers. Entrepreneur Steve Chubb, Matritech's president, received a license from MIT and raised $3.5 million in early outside venture capital in exchange for giving MIT an equity participation in the new venture (Gupta, 1989).

"Pushes" on Entrepreneurship

Some environmental forces affecting the "would-be" entrepreneur are the "negatives" about his present employer, rather than the "positives" of going into business for himself. The uncertainties due to the ups and downs of major projects have often been cited as a source of grief, and sometimes even led to expulsion of individuals into a reluctant entrepreneurial path. The evidence suggests that a stable work environment would probably produce far fewer entrepreneurial spin-offs than one marked by some instability. For example, the entrepreneurs who emerged from one large diversified technological firm rank most frequently "changes in work assignment" as the circumstance that precipitated formation of their companies, followed by "frustration in job". One fourth of the companies from that firm were founded during the three years that the firm suffered some contract overruns and laid off some technical people, although none of those actually layed off from this firm became entrepreneurs. The "worry about layoff" and seeing the parent firm in a terrible state are cited by many of that period's spin-offs. Even at the Draper Lab staff was cut by about 15 percent through layoff and attrition after the completion of the Apollo program, stimulating a number of new firms. Ninety-two percent of the spin-offs from the MIT Electronic Systems Lab (ESL) occurred during an eight-year period, when only 28 percent would have been expected if spin-offs occurred randomly over time as a function only of total employment. The large number of ESL projects completed during that period is one explanation for the "lumpiness" of new company creation.

Frustration with the noncommercial environment at the MIT labs and academic departments bothered some of the potential entrepreneurs. Margaret Hamilton, already mentioned in regard to her formation of HOS, exclaims: "The Draper non-profit charter was frustrating, especially if you wanted to get into something exciting. There was always the sense of living in a no-man's land" (*Boston Business Journal*, August 20–26, 1984, p. 6). Many of the entrepreneurs had specific devices or techniques that they wanted to market. Others had no definite products in mind but saw clear prospects for further applications of the technology or skills they had learned at their source organizations. The prospective entrepreneurs usually felt they could not exploit these possibilities at MIT labs, because the labs concentrated on developing new technology rather than finding applications for existing technology. Unfortunately for their industrial employers, many of the spin-offs from industrial companies report the same frustration, despite the not unreasonable presumption that their large firm employers should welcome at least some of these new ideas. In another geographic area Cooper (1986) finds that 56 percent of the new company founders had been frustrated in their previous jobs. Yet frustration should manifest itself more reasonably with just job changing, not company creating, behavior. Clearly the overall environment promoting entrepreneurship in Greater Boston makes the new company option an active choice, if other conditions are right. To gain insight into those "other conditions", in the next chapter I examine the high-technology entrepreneurs themselves, in contrast with their nonentrepreneurial technical colleagues.

Summary

Although quantitative evidence is lacking to support this assertion, an overwhelming amount of anecdotal data argues that the general environment of the Greater Boston area, beginning during the postwar period, and in particular the atmosphere at MIT, have played a strong role in affecting "would-be" local entrepreneurs. The legitimacy of "useful work" from MIT's founding days was amplified and directed toward entrepreneurial expression by prominent early actions taken by administrative and academic leaders like Compton and Edgerton. Policies and examples that encouraged faculty and staff involvement with industry and, more important, their "moonlighting" participation in spinning off their ideas and developments into new companies, were critical early foundation stones. MIT's tacit approval of entrepreneurism, to some extent even making it the norm, was in my judgment a dramatic contribution to the Greater Boston culture. Key individual and institutional stimulants like Stark Draper and the MIT Enterprise Forum reinforced the potential entrepreneurial spin-off that derived from a wide variety of advanced technology development projects in MIT labs and in the region's industrial firms. These actions fed into a gradually developing positive loop of productive interactions with the investment community that in time created Route 128 and beyond.

References

Boston Business Journal. August 20–26, 1984.

Boston Business Journal. "Corporate Album: Genetics Institute", March 23, 1987.

Chase Manhattan Corporation. *MIT Entrepreneurship in Silicon Valley* (Privately published, April 1990).

A. C. Cooper. "Spin-offs and Technical Entrepreneurship", *IEEE Transactions on Engineering Management,* EM-18, 1 (1971), 2–6.

A. C. Cooper. "The Role of Incubator Organizations in the Founding of Growth-Oriented Firms", *Journal of Business Venturing,* 1 (1985), 75–86.

A. C. Cooper. "Entrepreneurship and High Technology", in D. L. Sexton & R. W. Smilor (editors), *The Art and Science of Entrepreneurship* (Cambridge, MA: Ballinger Publishing, 1986), 153–167.

Nancy S. Dorfman. "Route 128: The Development of a Regional High Technology Economy", *Research Policy,* 12 (1983), 299–316.

Charles Stark Draper. "Remarks on the Instrumentation Laboratory of the Massachusetts Institute of Technology". Unpublished paper (January 12, 1970).

The Economist. "A Survey of New England: A Concentration of Talent", August 8, 1987.

Henry R. Feeser & Gary E. Willard. "Incubators and Performance: A Comparison of High- and Low-Growth High-Tech Firms", *Journal of Business Venturing,* 4, 6 (1989), 429–442.

Udayan Gupta. " How an Ivory Tower Turns Research Into Start-Ups", *The Wall Street Journal,* September 19, 1989.

INC. "After the Sale", August 1990, 39–50.

Karen Seashore Louis, David Blumenthal, Michael E. Gluck, & Michael A. Stato. "Entrepreneurs in Academe: An Exploration of Behaviors Among Life Scientists", *Administrative Science Quarterly*, 34 (1989), 110–131.

Roger Miller. "Growing the Next Silicon Valley", *Harvard Business Review*, July-August 1985.

Nitin Nohria. "A Quasi-Market in Technology Based Enterprise: The Case of the 128 Venture Group". Unpublished paper (Boston: Harvard Business School, February 1990).

William D. Putt, Editor. *How to Start Your Own Business* (Cambridge, MA: The MIT Press and the MIT Alumni Association, 1974).

Tom Richman. "Who's in Charge Here?", *INC.*, 11 (June 1989), 36–46.

Everett M. Rogers & Judith K. Larsen. *Silicon Valley Fever* (New York: Basic Books, 1984).

Joseph A. Schumpeter. *The Theory of Economic Development* (Cambridge, MA: Harvard University Press, 1936).

Segal Quince Wickstead. *The Cambridge Phenomenon*, second printing (Cambridge, England: Segal Quince Wickstead, November 1985).

Jane Simon. "Route 128", *New England Business*, July 1, 1985, 15–20.

M. A. Sirbu, R. Treitel, W. Yorsz, & E. B. Roberts. *The Formation of a Technology Oriented Complex: Lessons from North American and European Experience* (Cambridge, MA: MIT Center for Policy Alternatives, CPA 76–8, December 30, 1976).

R. W. Smilor, D. V. Gibson, & G. B. Dietrich. "University Spin-out Companies: Technology Start-ups from UT-Austin", in *Proceedings of Vancouver Conference* (Vancouver, British Columbia: College on Innovation Management and Entrepreneurship, The Institute of Management Science, May 1989).

R. W. Smilor, D. V. Gibson, & G. Kozmetsky. "Creating the Technopolis: High-Technology Development in Austin, Texas", *Journal of Business Venturing*, 4, 1 (1989), 49–67.

Sheridan Tatsuno. *The Technopolis Strategy* (New York: Prentice Hall, 1986).

J. M. Utterback, M. Meyer, E. Roberts, & G. Reitberger. "Technology and Industrial Innovation in Sweden: A Study of Technology-Based Firms Formed Between 1965 and 1980", *Research Policy*, 17, 1 (February 1988), 15–26.

Charles A. Ziegler. *Looking Glass Houses: A Study of Fissioning in an Innovative Science-Based Firm*. Unpublished Ph.D. dissertation (Waltham, MA: Brandeis University, 1982).

CHAPTER 3

The Makings of an Entrepreneur

Creation of a high-technology entrepreneur occurs over a long time, critically affected by many elements of society: The entrepreneur's family molds her or him, schools and work organizations help the entrepreneur to mature and gain knowledge and skills, the surrounding community provides influencing role models and resources. Beyond those largely intangible influences provided by the neighboring environs, as discussed in Chapter 2, my quarter century of research provides concrete data on hundreds of individuals (almost all men) who have become entrepreneurs, their personal background, education, work experience. This chapter reviews the literature on both general and technical entrepreneurs and then evidences some primary influences on becoming a high-technology entrepreneur.

The Entrepreneurial Mystique

Until recently the creators of new enterprises have been treated in the literature only in the folkloric tradition of Horatio Alger. Extensive accounts of the lives of men like J.P. Morgan, Andrew Carnegie, and the Rothchilds produce a feeling for the spirit and mystique of these capitalist giants. Entrepreneurship had not been subjected to much more careful scrutiny. The few empirical scholars, however, do provide some important concepts to consider for extension to the new high-technology entrepreneur.

Classical Perspectives

Earlier scholars of entrepreneurship have backgrounds that range from economics to psychology. Joseph Schumpeter (1966), a great economist, glorifies the entrepreneur as the motivating force behind technological change and economic development. David McClelland (1961), primarily a social psychologist, also ties the entrepreneur to the elements of economic change and growth, but his writings are strongly oriented to those psychological characteristics of entrepreneurs that make them likely to become business innovators. With a strong empirical and psychological orientation Collins and Moore (1964) discuss the origins and experience of entrepreneurs.

Schumpeter looks to the entrepreneur as the key to innovation and thus economic change:

> ... the function of entrepreneurs is to reform or revolutionize the pattern of production by exploiting an invention or, more generally, an untried technological possibility for producing a new commodity or producing an old one in a new way, by opening up a new source of supply of materials or a new outlet for products, by reorganizing an industry and so on (Schumpeter, 1966, p. 132).

Entrepreneurs stimulate the economy by introducing innovations with a view to making money. They assume risk, manage the activities and efforts of people besides themselves, receive profits or cover losses, discover new ways of doing things, and find new products and markets. But in explicating his theory, Schumpeter has little to say about the entrepreneur's characteristics.

McClelland sees the entrepreneur as the one who translates his individual need for achievement (n-Ach) into economic development. The entrepreneur in McClelland's scheme is "the man who organizes the firm (the business unit) and/or increases its productive capacity" (McClelland, 1961, p. 105). McClelland's underlying assumption is that entrepreneurs have a high n-Ach that leads them to behave in certain ways in business situations. A crucial element in most business situations is decision making under uncertainty, a circumstance in which some degree of risk is necessarily present. The entrepreneur, driven by his or her need for achievement, undertakes actions that have moderate risk characteristics, rather than those at either end of the risk continuum. The rationale for such behavior lies in the satisfaction expected from the different risk-associated decisions. In a situation with complete certainty as to outcome, the decision maker derives little satisfaction of his need for achievement because of the predetermined nature of the result. McClelland claims that entrepreneurs thrive on situations in which they can get a sense of personal achievement through taking responsibility for success and failure. In the pure chance situation, like winning a lottery, the individual making the decision again derives little satisfaction of his need for achievement because of the lack of influence of skill on the outcome. By definition failure is more probable in extremely high risk settings. Therefore, the entrepreneur would also be unlikely to attain his goals in such cases, again frustrating a high n-Ach. Outcomes of decisions with moderate risk depend on a mixture of skill and chance and thus are the situations most apt to satisfy the high n-Ach entrepreneur, according to McClelland.

Entrepreneurs, according to McClelland, work hard and do things in an innovative rather than a traditional manner. They work harder when challenged and when the work to be done requires ingenuity rather than standard procedures. But they require concrete feedback in the form, for example, of production volume or profit as measures of how well or how poorly they are doing. Entrepreneurs are future oriented in that they think ahead more in their decision making. Lastly, entrepreneurs coordinate the efforts and activities of other people. In most cases they must be effective in leading the work of others if they are to be effective at all in their entrepreneurial endeavors.

In a study of eight-year-old boys Winterbottom determines that mothers of high n-Ach boys have different attitudes toward child raising (McClelland, p. 46). Mothers of high n-Ach boys expect their sons to master earlier such activities as knowing their way around the city, being active and energetic, trying hard for things for themselves, making their own friends, and doing well in competition. Mothers of low n-Ach boys, on the other hand, impose more restrictions such as not wanting their sons to play with children not approved by the parents and not wanting them to make important decisions by themselves. Many such studies indicate that careless or indulgent parents, who do not expect great things from their children, clearly contribute to a child developing a low n-Ach, a somewhat self-fulfilling prophecy. Furthermore, a child growing up in a strictly controlled or rigidly authoritarian environment develops a lower n-Ach than the child who is reared in a less structured manner (McClelland, pp. 351, 352).

McClelland argues that parental values, indicated strongly by their religious orientation, are among the strongest factors directly associated with the development of n-Ach. His data and those of others demonstrate differences in n-Ach among the three primary religions in the United States. In *The Achieving Society* McClelland concludes (pp. 361, 365):

(1) More traditional Catholics appear to have some of the values and attitudes that are associated with lower need for achievement.

(2) Other groups of Catholics exist, at least in the United States and Germany, which have moved away from some of these traditional values toward the 'achievement ethic'.

(3) There is little doubt that the average need for achievement among Jews is higher than for the general population in the United States at the present time.

In addition to associating religious differences with levels of n-Ach, McClelland considers several other factors as possible but less direct influences of n-Ach. For instance, evidence suggests "that first-born children tend to have a higher n-Ach presumably because their achievement-oriented parents can set higher standards, be more affectionate, with one child than with several" (McClelland, p. 374). Socioeconomic status of the parents appears to be an important determinant of n-Ach in offspring. More specifically, middle-class offsprings tend to have a higher n-Ach than either upper or lower class offsprings (p. 378). One of McClelland's most interesting findings (p. 380) is the existence of what I call an "entrepreneurial heritage". He reports a slight tendency in entrepreneurial families to emphasize child-rearing practices that stimulate development of a high n-Ach. While this tendency is statistically insignificant in his U. S. studies, Fraser finds strong support of the tenet that entrepreneurial families in India produce an entrepreneurial orientation in their offspring.

One of the more modern empirical investigations of U.S. entrepreneurs is Collins and Moore's *The Enterprising Man*. In-depth interviews and Thematic Apperception Tests (codings based on verbal interpretations of fuzzy sketches) are used to determine the psychological motivation behind the entrepreneurs' behavior. Unfortunately from my interest, their 150 business initiators are seldom involved in technology-based companies. Collins and Moore find that entrepreneurs tend to subscribe to critical aspects of the Protestant Ethic (Weber, 1956, p. 67), a value system that stresses hard work and striving to produce an earthly, that is, pre-heavenly, reward. Entrepreneurs do not aspire to moving up the social hierarchy and do not seem to need to achieve positions of authority. They have an obsessive drive to push themselves even harder, what we often call "workaholics". They tend to overextend themselves in their activities within the business, but rarely have feelings of love for it. Accomplishment of goals within the business are rarely satisfying for long and the need to undertake new problems or endeavors is obsessive. Collins and Moore find entrepreneurs to be patronizing with their subordinates, usually seeing subordinates as either eager and industrious or slothful and rebellious. Authority is a difficult area for the entrepreneur. He or she is unwilling to submit to it, unable to work with it, and has a strong need to escape it. This is thought by Collins and Moore to arise from the entrepreneur's childhood perception of the male authority figure as cold and unsupporting but possessing awesome power. Collins and Moore conclude that the entrepreneur's relationship to adult figures, more than any other single factor, makes him different from an organization man. The entrepreneur cannot easily accept another's leadership and cannot exist in a situation where behavior is controlled and dictated by others.

Recent Research on Technological Entrepreneurs

These classical researchers provide a profile of the entrepreneur that makes an interesting base for rereading Chapter 1. Note how much of their theories are evidenced in the several case histories of high-technology entrepreneurs. But much is also missing. Among other issues, none of these early works examines technological entrepreneurs. Fortunately, increasing numbers of studies are focusing upon the personal backgrounds of technological entrepreneurs. A thorough review of the literature, including the annual conference volumes of *Frontiers of Entrepreneurship Research* , all issues of the *Journal of Business Venturing*, a review book (Sexton and Smilor, 1986) and a review article (Hisrich, 1990), has identified many works on the characteristics of entrepreneurs, with a smaller number of empirical studies on the backgrounds of technological entrepreneurs. The pre-1980 work provides a foundation for theorizing what influences lead to the establishment of new technology-based companies (Cooper, 1971; Cooper and Bruno, 1977; Roberts, 1968; Roberts and Wainer, 1971). The prevalent view (see Cooper, 1986) is that multiple factors affect the "making" of a technological entrepreneur, including background aspects of the individual (e.g., family, religion, education, work experience), motivational and other psychological aspects of the individual, and environmental influences (including aspects of the "incubating organization" where the entrepreneur had worked previously, external economic conditions, and technological opportunities). The recent empirical works on technical entrepreneurs build on and refine these perspectives, and are cited specifically at each point in this chapter where their findings apply.

Unfortunately many studies of technical entrepreneurs leave unresolved four methodological issues that limit the extent to which previous conclusions can be deemed reliable. (1) Sample size in some studies is quite small, raising questions as to generalizability of findings. (2) Response rate, especially in mail questionnaire studies, is frequently very low, causing lingering doubts about the representativeness of those who did agree to participate. Follow-up studies of non-respondents are usually absent. (3) The samples are often restricted to surviving or "successful" firms, providing little basis for knowing whether the non-survivors are similar or different in character. For example, use of Dun and Bradstreet listings of small companies in compiling entrepreneurial samples necessarily omits the large number (from my own research experience) of startup firms that never reach the stage of such listing. Indeed, similar bias occurs when telephone "Yellow Pages" are used to determine company samples. (4) The most important methodological issue is that only two prior studies use any form of control group for comparison with the technical entrepreneurs. In general, the prior research reveals certain profile characteristics of technical entrepreneurs, but not whether these are the same or different from the characteristics of the technical people with whom the entrepreneurs had worked before starting their own firms. What is unique about those who leave a lab to set up a new firm? McQueen and Wallmark (1984) compare the undergraduate grades of 47 technical company founders who graduated from Chalmers University of Technology with the grades of 49 randomly selected nonfounder graduates, and incidentally find no difference between the groups. Smith, Bracker, and Miner (1987) contrast test data from 118 company founders with 41 manager-scientists among applicants for National Science Foundation Small Business Innovation Research program grants, and find significant differences on several psychological dimensions. None of the other studies examined, including my own early work (Roberts, 1968; Roberts and Wainer, 1971), compares high-technology entrepreneurs with matched groups of nonentrepreneurial scientists and engineers.

Gartner, Mitchell, and Vesper (1989, p. 170) argue persuasively that "No 'average' or 'typical' entrepreneur can represent all entrepreneurs."

This is no doubt true even when limiting ourselves to high-technology entrepreneurs. Nevertheless, this chapter now seeks to provide insights on the backgrounds of advanced technology entrepreneurs, using the data collected in my research program described in the Appendix of this book.

Figure 3–1 is a four-factor model of the development of a technical entrepreneur, reflecting the research literature. The first influence on an individual is his or her family background, which no doubt affects all other aspects of personal development, including certainly goal orientation, personality, and motivation, as well as "growing up" elements such as educational attainment and age before becoming an entrepreneur. Growing up experiences may also affect goals, personality, and motivation. The nature and extent of work experience are no doubt influenced by all three of these clusters, but work may also feed back to alter goals and motives, and

Figure 3–1
A Model of Entrepreneur Development

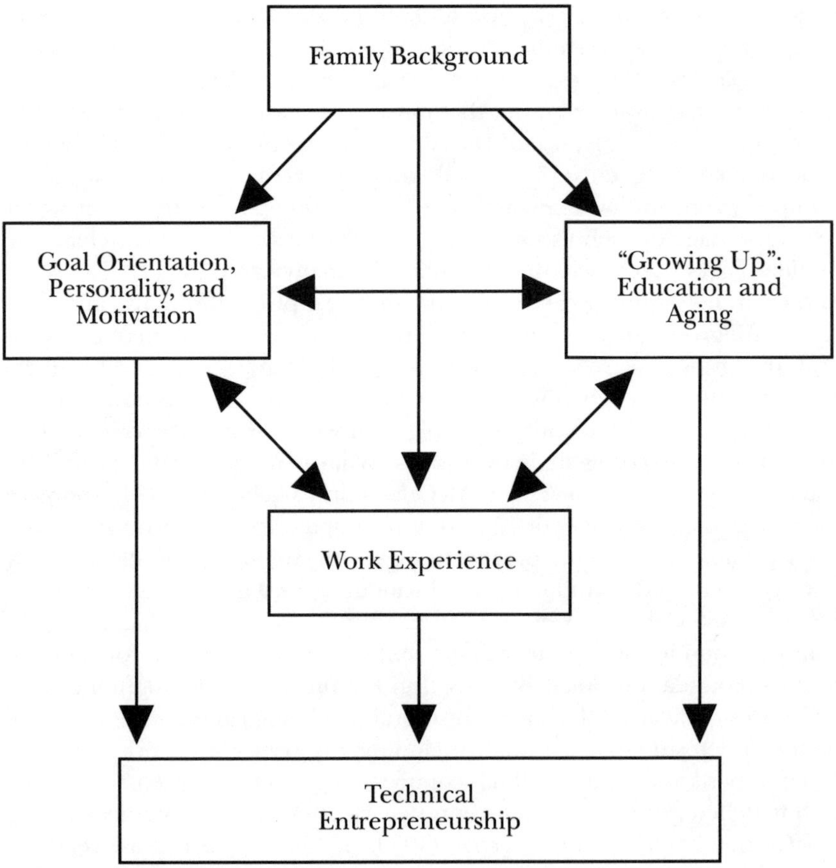

can affect education too. Elements of each of these four dimensions are shown in the figure to affect the "career choice" of starting a new company, with probably different effects on each individual. This chapter presents research results for each dimension of possible influence on becoming an entrepreneur. Whether or not each dimension affects the entrepreneur's success or failure is left for discussion in Chapter 9.

Breeding of the New Entrepreneurs: Family Background

Understanding the new technology-based entrepreneurs begins most logically with an examination of their home environments or family backgrounds. These provide the first influences that help mold the personal development, attitudes, and orientation of the future entrepreneurs. Based on the research literature, three areas of family background are expected to influence an individual's later decision to become an entrepreneur: (1) parental role models, arising especially from the father's occupation, in particular whether or not he is self-employed; (2) effects on individuality from small family size, and especially from whether or not the entrepreneur was a first-born son (unfortunately, very few women are contained in my research data, thereby preventing tests of any hypotheses regarding women); and (3) learned values and aspirations, expected to be embodied in the family's religious background.

My studies of technical entrepreneurs collected data on few family characteristics: father's occupational status, whether or not the entrepreneur's father was self-employed, and the entrepreneur's religious background. Sometimes my assistants and I also found the number of brothers and sisters in the entrepreneur's family and the birth order among them. Wherever possible, throughout this chapter, to place the distributions of the entrepreneurs in more meaningful perspectives, I compare data on the entrepreneurs with information collected from a control sample of scientists and engineers still employed at the MIT Lincoln Laboratory and Instrumentation Laboratory, two of the largest entrepreneurial "source organizations" that generated the primary research samples. These two sources account for 72 of the 129 MIT "spin-offs" included in the primary data sample analyzed here (see Appendix for further details).

"The Entrepreneurial Heritage": Father's Occupational Status
The largest percentage (59 percent) of technical entrepreneurs came from families where the father was either a professional or a manager. Comparison of the entrepreneurs with the control group of MIT lab employees indicates little difference between the two groups on this dimension. When farmer parents are added to both groups, the totals are nearly identical. A

closer look at Table 3–1, however, shows four times as many entrepreneurs with professional fathers than one might expect based on the control sample of employed scientists and engineers. These results raise the possibility that children of professionals are more likely to become entrepreneurs than the offspring of managers. To the extent that parents influence their children through the example of their own behavior, the findings are not surprising in light of the nature of the work of the professional as opposed to that of the manager. A professional, such as a lawyer or a physician, is seldom a member of a large hierarchical organization. And even when in a large organization (corporation, law firm, or hospital), the professional typically possesses a degree of independence not held by a manager who is almost always part of a very structured organization. After witnessing his professional father's independence, the child is more likely to find it appealing to obtain some type of occupational independence himself. (Use of male pronouns here is an empirical observation rather a male chauvinist statement. Only three of the 113 technical entrepreneurs in this particular group are women.) The entrepreneurs studied had parents of the previous generation, during which time the primary occupational roles in the United States were being served by the father. As women of the present generation move more into both professional and managerial roles, they will also serve increasingly as career role models for their children.

Whether or not the entrepreneur's father was self-employed is a second and critical aspect related to the father's occupation. As shown in Table 3–2, the difference between technical entrepreneurs and the control

Table 3–1
Father's Occupational Status

Father's Occupational Status	Technical Entrepreneurs $(n = 113)^*$ (%)	Employed Scientists and Engineers (S&E) (Control Group) $(n = 296)$ (%)
Professional	32	8
Managerial	27	44
Farmer	4	8
Clerical and sales	9	7
Skilled labor	18	21
Unskilled labor	10	11
Totals	100	99†

---

* n = 113 means that data on 113 entrepreneurs are used for the comparison shown here. Total sample sizes vary from possible maximum due to missing data. This occurs throughout this chapter.
† Round-off error.

group of employed scientists and engineers is quite striking. Entrepreneurs, as differentiated from their non-founder technical colleagues, tend strongly to come from families in which the father was self-employed (p=.0001). These results on entrepreneurs who spun-off from MIT labs and academic departments are strengthened by my studies of technical entrepreneurs who originated from other source organizations: 48 percent of the entrepreneurs who spun-off from a large electronic systems firm, 57 percent of those from a large diversified technological corporation, 61 percent of the biomedical entrepreneurs investigated, and 65 percent of a sample of computer-related entrepreneurs had fathers who were in their own businesses. As further comparison U.S. census data indicate that only about 25 percent of sons of self-employed fathers should be expected by chance alone. Presented statistically, the probability that a particular engineer or scientist will form his own company is significantly greater in cases where his father had his own business (.01; the computed probability appears in parentheses throughout this book, as a shorthand notation for saying the chance, in this case, is only 1 out of 100 that the distribution of questionnaire answers leading to this finding results from random factors alone; see Appendix for further details on statistical testing). This is still further supported by my control study of MIT faculty which finds that even professorial sons of self-employed fathers more frequently claim a serious interest in being in business for themselves (.05). All these evidences demonstrate that entrepreneurial fathers produce entrepreneurial sons disproportionately, supporting the effect of an "entrepreneurial heritage".

This finding is also supported by prior research in other geographic settings, although "controls" are absent to validate the uniqueness of the data. Hisrich (1990, p. 212) cites five studies, none relating to technolo-

Table 3–2
Whether or Not Father was Self-Employed

Father Self-Employed	Technical Entrepreneurs (n = 119) (%)	Employed Scientists and Engineers (n = 296) (%)
Yes	51	30
No	49	70
Totals	100	100

X^2 = 15.06, p = 0.0001. (These statistics measure the degree of similarity of two sets of data, here "technical entrepreneurs" and "employed scientists and engineers"; X^2 indicates that the chi-square test is used; the p value indicates that the strong differences in the data that are shown here might arise from purely random causes in only one case out of 10,000 chances. Further information on statistical testing is provided in the Appendix.)

gists, that strongly support the prevalence of self-employed fathers for male entrepreneurs and five more showing similar effects on female entrepreneurs. Cooper (1986) lists five different studies of technical entrepreneurs that find from 20 to 50 percent parental business ownership. Twenty-four of 53 Swedish technical entrepreneurs came from homes in which parents were self-employed (Utterback et al., 1988). Most recently 20 Japanese high-technology entrepreneurs were discovered to have self-employed fathers of 38 company founders whose fathers could be classified (Ray and Turpin, 1987, p. 565).

Yet seldom is the father's business at all related to his son's specific entrepreneurial activities, nor are many of the father's businesses (or occupations generally) even technical in nature. Most frequently parental businesses (as illustrated in Chapter 1) are small retail stores, farms, and small nontechnical manufacturing firms. And many of the sons are in businesses that could not have existed in their father's time, such as computer software, electronic systems, or biotechnology. The general image and example of a father as a self-employed professional or as an independent business owner provides the role model for the child, not any specific technical or managerial knowledge. Indeed it may be that simply familiarity with a business environment, growing from "table talk" at home, is the key to increasing the probability that an offspring will later become an entrepreneur. The likelihood of "helping out" in the family business no doubt also provides increased understanding of what is involved in running a business. I recall vividly the discussions at home during my childhood and the experiences from age five in helping my father in his small business. They provide my earliest and lasting instruction relating to pricing, competition, rationing during World War II, labor shortages, profits and taxes. No doubt self-employed parents serve their children well in providing "close encounters" with entrepreneurship.

Family Size and Birth Order

Some of the research literature assumes that children in smaller families may get more attention and become more inclined to independent careers, including starting their own companies. But my data show that family size has no direct bearing on the incidence of technical entrepreneurship: 78 percent of the entrepreneurs and 75 percent of the employee control group came from families with three or less children. However, splitting the populations in regard to parental employment does shows some effect in regard to family size. For those with self-employed fathers exactly 71 percent of the control group of employed scientists and engineers and 71 percent of the entrepreneur group came from families with three or less children. For those whose fathers were not self-employed, 88 percent of the entrepreneurs came from families with three or less children compared with only 76 percent of the control group, suggesting a slight tendency for entrepreneurs to come from smaller sized families.

But what about the oft-claimed effect of being the first-born son? "Being first born or an only child is postulated to result in the child's receiving special attention and, thereby, developing more self-confidence" (Hisrich, 1990, p. 211). Is this phenomenon evident among technological entrepreneurs? Indeed, 55 percent of the entrepreneurs are first-born sons, seemingly supporting the folkloric prediction (as well as Winterbottom's and others' research findings). But when compared with the control group of employed scientists and engineers, 54 percent of that nonentrepreneur group also are first-born sons. Careful reanalysis of the birth order data, searching for possible effects related to parental employment patterns, produces no new insights. To do this analysis correctly, expected frequencies of birth order distributions need to be calculated, based on the family size data. Much of the literature is misleading in this regard, not accounting for family size in the populations studied. My results show no "first-born" effect, at least for high-technology entrepreneurs. Much of the general clamor about the important role of first-born sons probably arises from a lack of careful statistical comparison with the family groups from which these sons originate.

The entrepreneurs were born and brought up all over the world, but with a heavy bias toward New England origins, reflecting the location of the incubating sources of new enterprises in the research samples. Of note is that 10 to 25 percent of each subsample of new high-technology firms are formed by someone born outside of the United States, such as Thermo Electron Corporation, founded by Greek immigrant George Hatsopoulos. Utterback et al. (1988) determine similarly that 12 percent of Swedish technical entrepreneurs have migrated from other countries. Technical entrepreneurship continues the "melting pot" nature of opportunities for personal growth and development in the Boston area, attracting and retaining productive talent for the region.

Family Background and Religion

As cited earlier, the cultural values and aspirations taught in the home environment are seen as primary molding influences upon entrepreneurial choices. Especially McClelland and his followers argue that these goals are reflected from the family's religious background. Despite these expectations no readily discernible religious differences exist between the overall group of Greater Boston technical entrepreneurs and the control group of employed scientists and engineers. A little more than half of both the entrepreneurs and the control group are Protestant, about 25 percent of both groups are Catholic, and slightly more than 20 percent in each group are Jewish. At first pass it appears that simply being of a certain religious background does not directly increase or decrease the likelihood that an individual engineer or scientist will become a technical entrepreneur.

Relative to an individual's family background and the incidence of technical entrepreneurship, the only significant and differentiating findings thus far relate to the father's occupational status. This is true when the entire sample is considered, that is, the entrepreneurs and control group members with and without self-employed fathers. As indicated earlier, the hypothesis of an "entrepreneurial heritage" can explain why a disproportionate number of entrepreneurs are the sons of entrepreneurs. But what explains the entrepreneurial activity of those individuals, comprising almost half of the samples, whose fathers were not self-employed and who, therefore, cannot be said to have an entrepreneurial heritage? The somewhat obvious step was therefore taken, of splitting the sample by parental employment background and reexamining the other family background data.

Table 3–3 indicates approximately equal percentages of entrepreneurs and control group subjects in each religious category for the subset of the entire population whose fathers were self-employed. However, for those whose fathers were not self-employed, Table 3–4 shows that the Catholic group has one third fewer and the Jewish group five times more entrepreneurs than expected based on the control group distribution (.01). In the absence of the entrepreneurial heritage syndrome, religious differences do affect the incidence of technical entrepreneurship. More specifically, confirming McClelland's predictions (p. 365), more Jews and fewer Catholics can be expected to go into business for themselves from a mixed religious population of U.S. technologists that have no self-employed fathers.

Does the greater preponderance of Jewish technical entrepreneurs just represent a general "upward-striving minority" effect? Bearse (1984) examines this question for other potentially upward striving minorities and demonstrates for the entire United States "that the incidence of nonagricultural entrepreneurial types in the labor force is significantly *less* for: a) Blacks than Hispanics, b) Hispanics than Asians, c) minorities than nonminorities, and d) females than males." Thus, McClelland's explanation for these findings

Table 3–3
Religion for Those with Self-Employed Fathers

Religious Background	Technical Entrepreneurs ($n = 51$) (%)	Employed Scientists and Engineers ($n = 83$) (%)
Protestant	41	40
Catholic	27	25
Jewish	31	35
Totals	99*	100

* Round-off error.

Table 3–4
Religion for Those with Fathers who were Not Self-Employed

Religious Background	Technical Entrepreneurs ($n = 50$) (%)	Employed Scientists and Engineers ($n = 168$) (%)
Protestant	70	68
Catholic	20	30
Jewish	10	2
Totals	100	100

$X^2 = 6.33$, $p = .01$.

seems plausible, resting on dominant cultural values rather than on minority status alone. The rapidly growing number of Asian Americans becoming technological entrepreneurs might similarly reflect cultural predispositions, not just minority status.

Thus, the major measurable influence of a technical entrepreneur's family background upon his decision to become an entrepreneur, rather than remain as an employed technologist, is his father's career. In the absence of a father whose career provides a role model with a high degree of independence or autonomy, that is, especially from self-employment or perhaps professional status, aspirations inculcated primarily through the family's religious background, may then affect the breeding of entrepreneurs.

The Entrepreneur's "Growing Up": Education and Aging

The nature of technological enterprise somewhat mandates that the entrepreneur be reasonably well-educated, to the extent of a bachelor's or master's degree in some technical discipline. Doctoral education seems superfluous for entrepreneurship, except in certain fields, such as biotechnology, where the doctorate may provide the "entry card" for competitive technical know-how. In the same manner the technical entrepreneur is presumed to bring to the marketplace new technology, rather than accumulated wisdom and experience. Consequently, technical entrepreneurs are assumed to be relatively young in age. These hypotheses are tested in the following sections of this chapter.

The Education of Technical Entrepreneurs

Many personal characteristics of high-technology entrepreneurs are probably true of all entrepreneurs. However, their education provides one of the most prominent differences. The Collins and Moore study of Michigan entrepreneurs reports that only 40 percent of the entrepreneurs had any education beyond the high school level (1964). My somewhat comparable study of Massachusetts consumer-manufacturing entrepreneurs produces the same 40 percent post high school educational attainment. In contrast, however, the technical entrepreneurs in the MIT spin-off group have a median educational level of a master's degree, generally in engineering. Only 1 percent of these high-technology entrepreneurs lack some college education and only 9 percent do not have at least a bachelor's degree. But the comparison posted in Table 3–5 indicates that at their median (M.S. degree) the technological entrepreneurs are not educated differently from their former "control" lab co-workers. At least among the MIT spin-offs listed here, however, the entrepreneurs perhaps surprisingly include a higher percentage of Ph.D.s than their employed counterparts. I shall return to this issue soon.

The relative lack of differentiation between technical entrepreneurs and employed scientists and engineers (similar medians) suggests that the most likely important reason for these entrepreneurs' educational level is the nature of the source laboratories at which they worked prior to their enterprise formation and the training needed for employment there. The technical entrepreneurs in two samples of industrial spin-off companies and in the computer-related new enterprises (see the Appendix for more information on these studies) average a bachelor's degree plus some course work, reflecting the slightly lower educational base of the industrial labs relative to the MIT laboratories.

Table 3–5
Educational Level

Educational Level	Technical Entrepreneurs ($n = 124$)		Employed Scientists and Engineers ($n = 299$)	
	(%)	Cumulative (%)	(%)	Cumulative (%)
High school	1	1	1	1
College without degree	8	9	5	6
B.S.	7	16	8	14
B.S. plus courses	20	36	31	45
M.S.	18	54	12	57
M.S. plus courses	11	65	21	78
Professional engineering degree	3	68	2	80
Ph.D.	31	99*	20	100

* Round-off error

A comparison of the educational levels of the technical entrepreneurs with the general population, business leaders, and fathers of technical entrepreneurs appears in Table 3–6. Both the entrepreneurs and business leaders are much better educated than the general population, and also better educated than the fathers of the entrepreneurs, with the technical entrepreneurs heavily skewed toward the highest levels of education. Supporting evidence of a similar median education level of a Master's degree plus is found in my study of 29 biomedical entrepreneurs and by Van de Ven et al. (1984, p. 93) of 14 educational software start-ups. Teach et al. (1985) find 47.3 percent of their microcomputer software entrepreneurs with advanced degrees and Smilor, Gibson, and Dietrich (1989) encounter 12 of 23 University of Texas "spin-out" entrepreneurs with Ph.D.s.

Family Background and Education

A detailed analysis of a smaller subset of the MIT spin-off entrepreneurs sought to explain further these educational levels. Parental income probably affected both the timing and extent of their children's education, but unfortunately no data were collected on parental income. However, one might expect that occupational status levels generally reflect income differences. The occupational status groupings of the fathers of the entrepreneurs do correlate significantly with the educational levels of the MIT entrepreneurs (Kendall tau=0.19, p=.06, n=58). This finding is confirmed by the spin-offs of the industrial electronics firm (.02). Entrepreneurs from lower occupational status families probably did not have enough money to go to college as early or for as long on a full-time basis as did those from higher status groups. They would have needed to work, at least part-time, to produce money for their schooling. Indeed, an inverse statistical relationship was found between paternal occupational status and the age of the entrepreneur when he finished the B.S. (tau=minus .23) and M.S.

Table 3–6
Educational Distribution of Technical Entrepreneurs
Compared to other Groups (%)

Educational Level	General Population	Fathers of Technical Entrepreneurs	Business[*] Leaders	Technical Entrepreneurs
Less than high school	58	23	4	0
Some high school	15	—	10	0
High school graduate	15	53	11	1
Some college	5	—	20	9
College graduate	5	25	55	90

[*] Warner and Abegglen, 1955, 35.

degrees (tau=minus .30). Entrepreneurs from families of lower occupational status and, therefore, probably lower income received their B.S. and M.S. degrees at older ages than did entrepreneurs from higher occupational status families (.01).

The same data indicate a nonlinear effect of whether or not an entrepreneur's father was in his own business on the education level of the entrepreneur. Table 3–7 shows for entrepreneurs at each level of educational attainment the number and percent of their fathers who were in their own businesses. A statistical test (using paired group percentages) confirms that entrepreneurial sons of self-employed fathers tend to be educated to at least the B.S. degree but not more than the M.S. degree and course work (p=.07). One possible explanation for this clustering is that technical entrepreneurs whose fathers were in their own businesses had been planning to go into business for themselves from an earlier age. Their education was therefore targeted (consciously or unconsciously) to a level appropriate to establishing a technically based enterprise. Going beyond the B.S. or M.S. degree was inappropriate because these sons of entrepreneurs long had in mind the specific goal of starting a company, not of doing research, teaching, or any other activity that might demand the still higher education of a Ph.D. Incidentally, no correlation exists between the educational level of the fathers and the education of their entrepreneurial sons.

Degree Disciplines and Sources

Search for relevant educational data in regard to degree discipline reveals

Table 3–7
Entrepreneurs within Each Educational Level Whose
Fathers Were Self-Employed (n = 96)

Entrepreneur's Educational Level	Entrepreneurs in Each Educational Group (n)	Fathers Were in Own Business (n)	(%)
No school beyond high school	1	0	0
College without any degree	8	3	38
B.S. degree	2	2	100
B.S. degree and course work	16	8	50
M.S. degree	13	8	62
M.S. degree and course work	10	6	60
Professional engineering degree	3	1	33
Ph.D. or greater	11	4	36
Totals	64	32	50

relatively little, the entrepreneurs looking more-or-less like their technically employed counterparts. About two-thirds of the technical entrepreneurs have degrees in engineering, 30 percent in science, and 3 percent in other fields. (As would be expected, my separate study of biomedical entrepreneurs determined that more of the founders, who include several M.D.s, came from initial education in the natural sciences.)

Of greater interest is that of the 217 degrees earned by the 106 technical entrepreneurs in one carefully checked sample, only three are in management (at the Master's degree level), a similar percentage of management degrees as found in the group of employed scientists and engineers. In fact, relatively few of the technical entrepreneurs had even taken business courses before company formation. Of course, some of the primarily technical entrepreneurs in the research had co-founding partners with management education. More recent research samples of entrepreneurs, both my own and others', include an increasing percentage of engineers with graduate management education, reflecting the explosive growth in popularity of the M.B.A. degree, frequently acquired in the evenings while employed as engineers.

Influenced no doubt by the high concentration on MIT departments and labs as research sources, the largest number by far of the technical entrepreneurs studied earned their degrees at MIT. This dominance is also true, however, in the research data on spin-offs from non-MIT source organizations, as well as in the special industry-related new enterprise groups, in which less bias toward MIT backgrounds is expected. The Greater Boston area concentration of the samples no doubt also explains a less-well-known phenomenon of high prevalence of entrepreneurs trained at Northeastern University, a large urban school with the largest "private" engineering enrollment in the country. In most research sub-samples Northeastern-educated entrepreneurs account for far more companies than Harvard or other local or nationally known educational institutions, although a wide diversity of college backgrounds is represented.

Education of Faculty Entrepreneurs

To this point no distinction has been made between entrepreneurs who were MIT faculty members and those who had worked as research and engineering staff members at MIT or elsewhere. In fact no significant differences exist between these groups in regard to family background, father's occupation, and religion. The first major difference arises in regard to education. Nearly all the faculty entrepreneurs have Ph.D.s, reflecting MIT's faculty recruitment and selection criteria. For example, all nine faculty entrepreneurs from the Mechanical Engineering Department have their doctorates (eight of nine from MIT). Four of seven Aeronautical Engineering faculty founders have doctorates, the other three cases being unique situations of age and/or circumstance. In contrast, for example, the non-faculty Aeronautical entrepreneurs (departmental staff members) include two doctorates and nine less-well-educated company creators. Indeed, the faculty account for the larger fraction of Ph.D.s among the technical entrepreneurs listed in Table 3–5: a far larger percentage of engineering faculty become entrepreneurs than do regular nonfaculty lab employees.

Influence of Multiple Founders on Education Levels

The research studies focus for the most part on the entrepreneur who departed from selected source organizations or, in many of the studies, on the entrepreneur/chief executive. But most companies were in fact founded by more than one person. For example, Figure 3–2 shows that 75 of 118 firms (64 percent) in one cluster of companies had two or more co-founders. A second sample contains 17 of 21 (81 percent) multifounder firms, with an average of 3.2 founders per company. Twenty-four of the 39 spinoffs (or 49 percent) from the electronic systems firm were co-founded by two or more entrepreneurs. Similarly, ten of 18 companies (56 percent) started by faculty and staff of the MIT Aeronautics and Astronautics Department were multifounded, with companies formed by faculty involving significantly more founders than those initiated by research staff members (.01). This pattern of a majority of multifounder teams seems to be universal in other studies of technological entrepreneurship (Cooper, 1986; Doutriaux, 1984; Feeser and Willard, 1989; Olofsson et al., 1987; Teach et al., 1985; Tyebjee and Bruno, 1982).

One small study of eighteen firms gathered detailed data on all the co-

Figure 3–2
Number of Founders ($n = 118$)

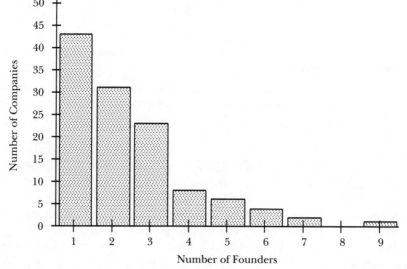

founders, reflecting a rather highly educated sample. Only two companies are limited to co-founders with no more than a high school education. Another company has two co-founders who dropped out of college without receiving their degrees. An additional company was formed by two men with high school education and one college dropout. (Let me caution that the college dropout status is neither a pejorative nor a predictor of entrepreneurial failure. While not included in this particular sample, Edwin Land, the famed founder of Polaroid Corporation, had dropped out of Harvard College before eventually receiving his undergraduate degree. And David Boucher, founder of Interleaf, is an MIT dropout.) All other firms in this specific small sample have at least one founder with a college degree. But the addition of his co-founders to the principal entrepreneur (who averages a master's degree education, as shown previously) slightly reduces the average of founders' education level to bachelor's degree plus.

How Old Are Technical Entrepreneurs?

Across all my research, with few exceptions, I find rather remarkable similarity in the age patterns of technical entrepreneurs at the time of company founding. Table 3–8 shows data from twelve of my studies, indicating the range and median of entrepreneurial ages. The range of ages in 270 companies is 23 to 69, with an overall median of 37 years. On an "eyeball" judgment basis, the MIT laboratory spin-offs are somewhat younger, averaging 34 against a representative 38 years for their industrial and mixed

Table 3–8
Age Distribution of Technical Entrepreneurs

	Sample Size	Age Range	Median
MIT Laboratory Spin–off Studies			
Electronic Systems Laboratory	11	27–43	35
Instrumentation Laboratory	27	24–55	33
Lincoln Laboratory	47	25–65	34
Research Laboratory for Electronics	13	29–64	36
Other New Enterprises Research Studies			
Spin-offs from diversified technological company	23	25–54	39
Computer–related firms (2 studies)	42	24–51	37
Recently formed high–technology firms	18	26–52	39
Biomedical companies	28	—	36
Analyses of business plans	20	23–48	37
Search processes for raising venture capital	21	23–43	35
In–depth analyses of venture capital investment decisions	20	23–69	39

source counterparts. As suggested in Chapter 2, this may be due to more positive encouragement at MIT for spin-off company formation as well as more access to advanced technological bases for new firms.

Going beyond MIT and the Greater Boston area to other regions, entrepreneurs who started 14 high-technology firms in Ottawa, Canada average 35.3 years old at their time of company formation (Doutriaux, 1984). The 60 Swedish entrepreneurs in the sample by Utterback and colleagues (1988) have a median age of 34, and a second study of Swedish technical ventures identifies the median age of the "central" founder as 38 years (Olofsson et al., 1987). Cooper (1986) reports a median for seven studies of 37 years, whereas California technical entrepreneurs are a bit older, indicated by the 42-year-old median in one study (Tyebjee and Bruno, 1982) and the 39.7 average for recent Silicon Valley "lead" founders (Bruno and Tyebjee, 1984). But University of Texas-Austin technology spin-off entrepreneurs match their MIT counterparts at 34 years (Smilor et al., 1989).

To provide more detail on ages, Figure 3–3 shows the age distribution of 119 MIT spin-off entrepreneurs, including some from academic departments as well as the laboratories (range = 23–65, median = 34). Two thirds of the entrepreneurs started their companies when they were between the ages of 28 and 39. Only 7 percent of the entrepreneurs were younger than

Figure 3–3
Age Distribution of MIT Spin-Off Entrepreneurs at the
Time they Started their New Enterprises ($n = 119$)*

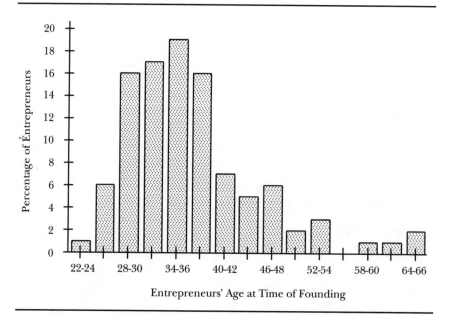

Entrepreneurs' Age at Time of Founding

* Round-off error; adds to 103%

28 and less than 10 percent were older than 48. Those who were older at founding had a higher educational level (.04). Indeed two of this MIT spin-off group were 65 years old, faculty placed into retirement at that age who felt they still had worthwhile ideas and the entrepreneurial energy to pursue them. The overall youth attributable to the entrepreneurs must be fundamentally more related to attitude than chronology alone, although this is hard to prove.

But mentioning the faculty does identify one of the exceptions to the generally observed "rule" of mid-1930s entrepreneurial ages. Examined separately, MIT faculty form companies at significantly older ages than their staff colleagues ($p=.05$). For example, in the Department of Aeronautics and Astronautics the ages of faculty founders range from 26 to 65 with a median of 44 at the time of company formation. Their nonfaculty departmental colleagues who became entrepreneurs range from 26 to 40 with a median of 32. The difference no doubt reflects the longer educational periods (disproportionately more Ph.D.s, $p=.10$) required by faculty members for entry into their primary career roles, combined with the long time then needed to secure their faculty position. Academic tenure tends to be awarded by age 36 and few faculty take much time away from teaching and research before the tenure decision. Presence of faculty entrepreneurs in each of the MIT lab spin-off groups listed in Table 3–8 slightly increases the median ages shown for those samples. Without faculty entrepreneurs included the MIT lab spin-offs would be even a bit younger in age pattern.

A second exceptional group are the consumer-oriented manufacturers whom I studied for intended contrast with the high-technology entrepreneurs. That group ranges from 28 to 55 at the time of forming their companies, with a median age of 45. In that regard, but no other, they look like MIT faculty entrepreneurs. (The Collins et al. general entrepreneurs have an average age of 52.)

Most of the technical enterprises were founded by teams of several people (as indicated earlier), but the median ages of team members match the individual figures shown in Table 3–8. In one group of 20 technical companies studied, the median age of the youngest person on each team is 30, the median of the oldest is 41, with a median overall age of 37. Another research sample of 20 companies produces a median youngest team member age of 34, a median oldest of 43, and a median overall founder age of 39.

Why So Young?

One probable cause of the young age of the technical entrepreneurs is the youthful age structure of the technical organizations at which they previously worked. A comparison of the entrepreneurial age distribution with the employed scientists and engineers at the MIT laboratories used as control studies shows roughly comparable age patterns, with the entrepreneurs about two to three years younger on average. Technical organizations such as the MIT departments and labs, Air Force Cambridge Research

Laboratory, or the large technological industrial organizations in the Greater Boston/Route 128 area, such as Raytheon and DEC, are the breeding grounds for new enterprises. Their age base sets the bounds within which entrepreneurship can take place. Yet on average the younger people relative to these organizations as a whole are the ones who leave and eventually start new enterprises.

A second but related influence on the young age of the technical entrepreneurs is the newness of the technology they are using. An older person first has to learn a new and emerging field that is the entrepreneur's training grounds in terms both of formal education and work experience. This is a time-consuming and arduous task, even when possible. The older person is accustomed to the existing uses for the existing technology and is less likely to see new uses. In contrast, the older technologist might have a better sense for market needs and be able to relate technological change to the marketplace. A lower bound on the technical entrepreneurs' ages is set by their almost universal college education, probably a prerequisite for acquiring the knowledge needed to form a technical company (despite exceptional college dropouts like Edwin Land, cited earlier, who created Polaroid). One violation of that lower age bound is illustrated by Douglas Macrae and his two co-founders who developed the Ms. PacMan video-game in their MIT undergraduate dormitory room, establishing General Computer Corporation (now GCC Technologies) to commercialize it (see Chapter 6 for more on GCC).

Depending on the particular research sample, anywhere from 75 to 100 percent of the entrepreneurs are married at the time of company establishment, most of them with children (two on average). This suggests a third possible influence on the youthful entrepreneurial ages found. Financial requirements of the family with very young children are not usually so great in the United States, at least not in comparison with needs as the children grow up and begin approaching their college years. Free elementary and pre-college education in contrast with very expensive college education in the United States might well make a difference. The burden of loss of income from giving up a well-paying engineering job and taking less or no pay as a start-up entrepreneur can be more readily absorbed by the young family. A prospective entrepreneur in his 40s, with mid-teen-aged children, is likely to be more risk averse than his younger counterpart. Also those growing teenagers are going to be more time demanding, competing with the heavy time requirements involved in getting a new company off the ground.

Supporting these perspectives are the common findings of research on job turnover. The older a person, the less likely he or she is to change jobs; the more longevity in an individual's employment with one organization, the less likely is job turnover; and the higher a person's position, the less likely is job changing. Age, employer longevity, and position are all highly correlated, but each acts independently in the same direction of discouraging voluntary departure from an organization. These general observations are bolstered by the control studies at two major MIT laboratories: Staff members who are older have been at the labs longer and have higher positions there (p=.001). An unpublished MIT study (Marquis and Rubin, 1966) draws the same conclusions for a broad industrial sample of engineering project managers, themselves likely prospects for entrepreneurial roles. Still older people feel more locked-in by pension considerations, too little work time left to find a job and build a new career should the enterprise fail, and reluctance to try something new. Risk taking eventually ages out!

Note that many entrepreneurs form more than one company during their lifetime. For example, William Poduska was a co-founder of Prime Computer, then founded Apollo Computer, and later formed Stellar Computer, now merged with Alliant to form Stardent Computer. And Henry Kloss became a legend with audiophiles through his entrepreneurial pioneering with new speaker systems from four different Boston-based firms he co-founded: Acoustics Research in the 1950s, KLH in the 1960s, Advent Corporation in the 1970s, and Cambridge SoundWorks in the late 1980s. Raymond Kurzweil and Aaron Kleiner have jointly founded three companies to further develop and commercialize Ray's evolving technology in the area of pattern recognition/ artificial intelligence: Kurzweil Computer Products, in 1974, when Ray was just 25, Kurzweil Music Systems in 1982, and Kurzweil Applied Intelligence a few years later. As noted in Chapter 1, Phil Villers formed three CAD/CAM-oriented firms, beginning with Computervision and going on to Automatix and then Cognition. Indeed, a recent MIT survey of 99 alumni entrepreneurs of very successful Massachusetts companies (not limited to technical fields) finds that on average these superstars have created 2.9 firms each, with the record held by one faculty member who participated in starting twelve firms. The age effects described relate to the first-time entrepreneur. The multi-time entrepreneur's "career" has become "forming companies", to be continued until retirement, or up to the semi-retirement practiced by a number of ex-entrepreneurs of investment management and/or college teaching.

Thus, as suspected, the major growing up characteristics identifiable in the technical entrepreneur are his advanced technical education, median of an engineering master's degree, not different from his nonentrepreneurial work colleagues, and his relatively young mid-30s age, somewhat younger than his employed colleagues.

Work Experience

The research literature generally documents the importance of the technical entrepreneur's work experience on the new company he forms. He is seen as having performed many years of engineering and some supervisory work before setting up his own company. My earlier studies of smaller numbers of technical entrepreneurs emphasize that his prior work is skewed toward development, not research, and I expect this pattern to be sustained in larger samples. But this section explores the entrepreneur's work experience in much more depth than earlier studies.

Phases of Experience

The median education level of the technical entrepreneurs (about master's degree, usually obtained at age of 23 or 24), coupled with their age characteristics discussed previously, leads to the logical deduction that the typical entrepreneur had about 13 years of work experience before starting his own company. Actually the mean number of years of work experience for one group of 111 carefully studied entrepreneurs is 12.7 years. (The variance from this mean is almost eight years, reflecting the inclusion in the sample of several MIT faculty members who started their companies when they were over 60 years of age, thus adding more than 40 years of work experience in several cases.) This mirrors the mean of 13.2 years of experience in my biomedical sample and the average 12 years experience of 60 Swedish technological entrepreneurs (Utterback et al., 1988). California entrepreneurs have significantly more years (17 on average) of prior work experience (Bruno and Tyebjee, 1984).

Figure 3–4 presents the distribution of work experience, showing that the bulk of technical entrepreneurs, 79 percent, had from three to 16 years

Figure 3–4
Work Experience of Technical Entrepreneurs Prior to
Starting their New Enterprises ($n = 111$)*

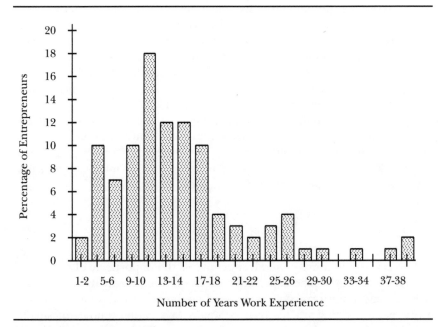

* Round-off error; adds to 103%

work experience before starting their new enterprises. Only 2 percent had less than three years and 22 percent had more than 16 years experience. (In contrast, the consumer-manufacturing entrepreneurs average 24 years of preenterprise work experience.)

How did the entrepreneurs spend this three to sixteen years of preentrepreneurial work experience before starting their new enterprises? As was the case with educational characteristics, the work experience of technical entrepreneurs differentiates them from other entrepreneurs. Collins and Moore characterize their Michigan entrepreneurs as lean on formal education but heavy on the education one gets in "the school for entrepreneurs" (Collins et al., p. 125). This perceived "school of hard knocks" features many job changes in which the prospective entrepreneur develops the skills destined to make him effective as an entrepreneur. The general entrepreneurs studied by Collins et al. clearly did not have one key work experience that provided the reason and basis for a new enterprise.

In contrast the technical entrepreneurs I studied did have one key work experience that gave them an opportunity to start a company that had a technological advantage, at least initially. Moreover, mainly through this technological edge, that key work experience enabled them to grow rapidly in sales and profits. In sharp distinction from life in California's Silicon Valley (i.e., 85 percent of technical entrepreneurs with prior work at five or more firms, according to Bruno and Tyebjee, 1984, and engineer-manager turnover rate of 30 percent per year, according to Rogers and Larsen, 1984), my data do not suggest a job-jumping phenomenon among the primarily Greater Boston entrepreneurs. In fact they have had relatively few employers with one, namely what I have been calling the technology source organization, being the most important. Silicon Valley seems to reflect a very different culture from Boston, but I do not know which one, if either, is representative of the rest of the United States.

The entrepreneur's total work experience can be broken down into three segments: his work if any prior to the technology source organization, his work at that "seeding" organization, and his work if any between the "source organizations" studied and the new enterprise. The mean number of years work experience before working for the source laboratory was 4.26 years (with variance about this mean of 5.59 years). More strikingly, for 41 percent of the technical entrepreneurs employment at the source organization was their first job. Fully 70 percent of the entrepreneurs had five or less years experience elsewhere when they went to work for one of these source laboratories.

What about the segment of the entrepreneur's work experience between the technology source organization and the founding of the new enterprise? Depending on the research sample chosen within the studies, between 50 and 65 percent went directly into their new businesses after (and sometimes even before!) they terminated employment at the source laboratory being assessed. The other 35 to 50 percent took another job after departing the so-called source lab, not yet starting up a new enterprise. About half of these new job-takers worked for five or less years between the source labs studied and the new enterprise. (In Chapter 4, I discuss the fact that the time lag between the source laboratory and the new enterprise has a strong effect on the degree to which technology is transferred; namely, the longer the time lag, the less the degree of technology transfer.) The average number of years between the laboratory and the new enterprise is 2.4 years. Figure 3–5 evidences the rate of new company formations after departure from the source organizations.

Where then do the technical entrepreneurs get their principal training, experience, and skills? This is answered more concretely with Chapter 4's evidences on technology transfer. Here the data primarily reflect the years spent in doing technical work. Some entrepreneurs cumulate most of their experience prior to the source laboratory and some gain their skills from work between the so-called source and the new enterprises. But the

Figure 3–5
New Enterprise Formation Related to the Years after
Termination of Employment at the Source ($n = 121$)

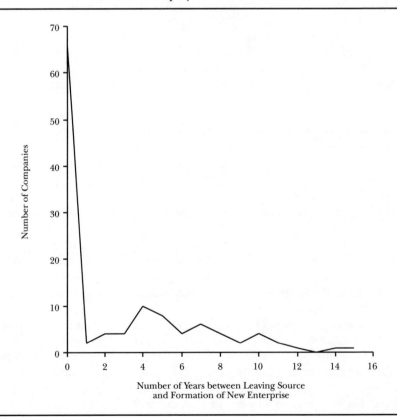

Number of Years between Leaving Source
and Formation of New Enterprise

vast majority of the technical entrepreneurs studied acquire their relevant experience and training during one key period of work experience, that at the technology source organization. The typical entrepreneur spent 7.4 years at the source organization (again biased upward by the inclusion of faculty entrepreneurs), longer by almost three and five years, respectively, than the average years worked before the laboratory and between the laboratory and the new enterprise. Fifty-three percent of the entrepreneurs spent five or more years at the source laboratory.

Almost 21 percent of the total sample of entrepreneurs worked only at what I have called a technology source laboratory. This is true not only for MIT spin-offs, but, for example, for those new enterprises formed out of the electronic systems company studied. Another 37 percent worked only before and at the lab, which makes up the total here of 58 percent who went directly from the source organization to the new enterprise. Twenty percent worked at the source and between it and the new enterprise. Only approximately 22 percent of the entrepreneurs worked before, at, and after the source laboratory before starting their new enterprises. These data are summarized in Table 3–9.

Productivity

Along with the notion that Greater Boston technical entrepreneurs usually have one key work experience, on the basis of which they start their new enterprises, must necessarily come some consideration of the elements of experience at that organization. Three measures help characterize the entrepreneurs' work experience at the source organizations: number of papers published and patents granted while at the source; percentage of time spent performing various types of work activities; and last, kind of technical work engaged in on a scale from basic research to development.

As indicated earlier in this chapter, I compare wherever possible the characteristics of the entrepreneurs' work experience in relation to those of the control group of employed scientists and engineers at those same

Table 3–9
Work Experience before Starting the New Enterprise

Work Experience	Technical Entrepreneurs (%)
Technology source organization only	20.6
Technology source organization plus employment between it and the new enterprise	19.6
Experience before the source organization and at it	37.4
Experience before, at and after the technology source organization	22.4

source organizations. Note that the typical entrepreneur spent 7.4 years at the technology source organization, while the typical nonentrepreneur staff member of these same technology source organizations had already been there 8.4 years when the data were collected, one full year longer than the entrepreneurs. (Obviously, the typical nonentrepreneur will be there even longer before eventually departing the source organization.) As a result when factors are considered such as papers published or patents granted during employment at the source, there is a bias of more years during which, for example, the nonentrepreneur staff member might have published more papers. Despite this bias, Table 3–10 indicates that the entrepreneurs, while employed by the technology source organizations, published almost three times as many papers per person as did the control group employees. In addition, almost twice as many entrepreneur employees as nonentrepreneurs published at least one paper while employed by these labs. Entrepreneurs are obviously unique individuals, but I know of no prior evidence that suggests they had been productive in the conventional ways by which engineers and scientists are measured in large organizations.

Examination of patents granted while at the source organizations reveals the same phenomenon. During employment by the technology source organization, the entrepreneur was granted 32 times as many patents on average (a clearer index of commercial entrepreneurial tendencies than papers) as his nonentrepreneur counterpart. In addition, 34 percent of the entrepreneurs were granted at least one patent while at these laboratories as opposed to only 5 percent of the control group. These data demonstrate a striking difference in terms of conventional technical productivity between the entrepreneurs and the control group. Even within the control group of employed scientists and engineers, those few who had previously been self-employed at one time or another had been granted far more

Table 3–10
Papers and Patents—Entrepreneurs vs. Control Group

	Technical Entrepreneurs	Employed Scientists and Engineers Control Group
Papers published per person while at the source laboratories	6.35	2.2
Percent who published at least one paper while at the labs	63%	38%
Patents granted per person while at the source labs	1.6	.05
Percent who were granted at least one patent while at the labs	34%	5%

patents prior to their employment by the source organization than their colleagues who had never been self-employed ($p=.001$). Apparently, not only do entrepreneurs start companies but they are among the most productive technical contributors while employed in research and development organizations. In further support, the spin-offs from the diversified technological corporation average 1.8 patents and 2.4 papers before they became entrepreneurs, and the biomedical founders average 1.2 patents and 8.1 papers, reflecting their more distinctly research-oriented backgrounds in hospitals and universities. But no control data exists for comparison with these two groups.

Why were the entrepreneurs so much more productive than their former lab colleagues? Their far higher productivity is obviously not explained by the similarity of the entrepreneurs' education to their former technical colleagues, and/or by their slightly briefer work experience. The causal influences at work here might now be hypothesized to be:

1. The higher productivity of certain technical persons, especially in regard to work that becomes the basis for patent awards, might influence (a) the exposure of these individuals to entrepreneurial opportunities, (b) their initial development of work that suggests possible outside applications, and/or (c) their own confidence in their likelihood of succeeding independently, any one of which might encourage entrepreneurial spin-off.

2. Whatever as yet unidentified set of "entrepreneurial behavior" influences that affect these individuals' technical productivity also affect their likelihood of leaving the labs to become a company founder. Unfortunately, the data do not permit clarification of these options.

Time Allocation

Another aspect of the entrepreneur's work experience at the technology source organization is the percentage of time spent on various activities such as report writing and development. The data presented in Table 3–11 show that on the average both the entrepreneurs and the control group spent about 50 percent of their time while at the source lab doing R&D work. Considering only the mean percentages of time spent at the various activities, the entrepreneurs appear to be much like the control group. However, the differences in the standard deviations about the means for the entrepreneurs and the control group indicate that on each activity the entrepreneurial group has much higher degree of variation. These data present a picture of individual entrepreneurs as working on particular activities for very different amounts of time, as opposed to the control group where percentages of time spent on the same activities do not differ much between individuals or from the average.

Entrepreneurs who had a longer employment at the industrial laboratory studied not unexpectedly spent a significantly greater amount of time on personnel supervision ($p=.01$), confirming an earlier MIT research study of R&D management at that same source organization (Rubin et al., 1965).

Table 3–11
Time Spent in Various Activities

Various Activities Engaged in while at the Laboratory	Technical Entrepreneurs		Scientists and Engineers Control Group	
	Mean %	Standard Deviation	Mean %	Standard Deviation
Report writing	10.61	15.88	9.59	0.40
Administrative duties	9.57	12.75	10.43	5.43
Meetings	8.09	8.62	10.51	5.51
Research	23.78	26.88	20.21	20.21
Development	28.57	27.83	29.99	9.99
Personnel supervision	11.10	13.03	11.11	6.11
Other	8.21	20.24	6.70	48.29
Totals*	99.93		98.54	

* Round-off errors.

Though exact job position is not uniformly recorded, many entrepreneurs had been technical supervisors in their source organizations. For example, 40 percent of the Instrumentation Lab entrepreneurs were technical supervisors (rather than just first line engineers) at the Lab before founding their own ventures. By contrast, only about 15 percent of the Lab staff members are supervisors. Several factors contribute likely explanations for the higher entrepreneurial defection of supervisors: (1) their personal characteristics that led them into supervisory roles; (2) their exposure to a wide variety of problems, as well as opportunities; and (3) the high degree of responsibility they typically are given as supervisors. They come into contact with suppliers and customers much more often than the average engineer, giving them a valuable source of market information and contacts.

Nature of Work

Beyond this simple inquiry into time allocation was a more serious attempt to get each entrepreneur (as well as each employed scientist and engineer in the control studies) to characterize the technical nature of the work he performed. The approach for doing this was borrowed from a classical U.S. Department of Defense (DOD) comprehensive study of sources of technical advances that were embodied in the Army's Bullpup missile project. The so-called Bullpup study was part of Project Hindsight, a detailed analysis of military R&D productivity (Office of the Director of Defense Research and Engineering, 1964). As such the nine classes of R&D work that follow, widely tested for reliability by the DOD, are referred to here as the individual's Bullpup Classification.

1. Investigations in pure and applied mathematics and theoretical studies concerning natural phenomena
2. Experimental validation of theory and accumulation of data concerning natural phenomena
3. Combined theoretical and experimental studies of new or unexplored fields of natural phenomena
4. Conception and/or demonstration of the capability of performing a specific and elementary function, using new or untried concepts, principles, techniques, and/or materials
5. Theoretical analysis and/or experimental measurement of the characteristics of behavior of materials and/or equipment, as required for design
6. Development of a new material necessary for the performance of a function
7. First demonstration of the capability of performing a specific and elementary function, using established concepts, principles and/or materials
8. Development of a new manufacturing, fabrication and materials processing technique
9. First development of a complete system, component, equipment or major element of such equipment, using established concepts, principles, materials—Prototype development

Since few people are interested in reading and/or remembering these classifications, suffice it to say that 1 is basic research, 9 is prototype development; the classifications become more developmental in orientation as they ascend in a continuum on the interval scale from 1 to 9.

1	2	3	4	5	6	7	8	9

Basic Research	$\rightarrow$	Increasingly Developmental	$\rightarrow$	Prototype Development

In some cases the assignment of classification by the coder was difficult due to the variety of work performed by the entrepreneur at the source technical organization. However, confidence in these ratings is bolstered by several statistical relationships. For example, the more highly educated staff members are heavily skewed to the basic research end of the Bullpup scale at Lincoln Laboratory (.001) and two other MIT labs (.08, .15) and among the electronic systems company spin-offs (.02). The assigned classification does correlate positively with the entrepreneurs' reporting of the percent of time spent on development work at MIT's Research Laboratory for Electronics (RLE) (.09) and at the Electronic Systems Lab (ESL) (.05); and it correlates negatively (as expected) with the entrepreneurs' reporting of the percent of time spent on research at RLE (.007), at ESL (.01) and among the electronic systems company spin-offs (.003). In addition these Bullpup classifications correlate with those assigned by the individuals' supervisors at two laboratories (.007, .13). These findings support adopting the "Bullpup" classification as a usable measure of the nature of work performed by the entrepreneur while at the source organization.

Figure 3–6 pictures the distribution of the entrepreneurs' technical work at the electronic systems company from which these data were collected on 35 spin-off enterprises. While at least one founder is classified in each Bullpup category, the distribution is skewed heavily toward the developmental end of the spectrum, as might be expected in an industrial systems-oriented organization.

In Table 3–12, a subset of ninety-four MIT-based entrepreneurs is categorized by the source organization from which they spun-off and by the type of work they performed at the source. The totals indicate that only two of the entrepreneurs performed work that was primarily basic research, while twenty-six did primarily prototype development work. This is understandable. Basic research does not present much technology that has immediate practical utility. It is not supposed to. Prototype development work on the other hand involves much technology of practical utility. That is its nature. Entrepreneurs should be expected to come to a greater degree from the more developmental types of work. Table 3–12 also provides further support for the credibility of the coding scheme used to categorize

Figure 3–6
Nature of Entrepreneurs' Laboratory Work in Electronic
Systems Company ($n = 35$)

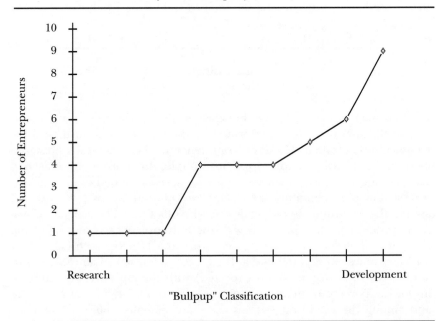

Table 3–12
Nature of Work of MIT Spin-off Entrepreneurs ($n = 94$)

Source	Work Classification ("Bullpup" scale)									Totals
	1	2	3	4	5	6	7	8	9	
Major Laboratories										
Electronic Systems Lab	—	—	—	—	3	—	1	2	4	10
Instrumentation Lab	—	—	—	—	1	—	—	1	7	9
Lincoln Lab	2	—	6	4	3	1	3	1	12	32
Research Laboratory for Electronics	—	—	2	1	3	—	4	—	2	12
Labs Totals	2	—	8	5	10	1	8	4	25	63
median = 7, mean = 6.6										
Academic Departments										
Aeronautics and Astronautics	—	1	3	8	5	—	—	—	—	17
Electrical Engineering	—	—	2	—	2	—	1	—	1	6
Materials Science	—	—	4	—	1	1	1	1	—	8
Departments Totals	—	1	9	8	8	1	2	1	1	31
median = 4, mean = 4.5										
MIT Totals	2	1	17	13	18	2	10	5	26	94

the entrepreneurs' prior work. The entrepreneurs from MIT academic departments, who include many faculty, did much more research-oriented work than those who originated from the major MIT laboratories ($p=.01$).

Table 3–13 shows the work ratings of randomly sampled MIT Lincoln Laboratory personnel and Instrumentation Laboratory personnel (from the control studies) along with their related spin-off entrepreneurs. In each comparison the work content of the entrepreneurs is more developmentally oriented than that of their laboratory counterparts ($p=.06$, $p=.13$). A subjective evaluation of the entrepreneurs' work done at two other MIT laboratories, the Electronic Systems Lab and the Research Laboratory for Electronics, draws the same conclusion (see also Chapter 4). The conclusion is clear: Those who work in development (rather than research) do have a greater tendency to become entrepreneurs. This may be due to the prospective entrepreneurs' inclinations to seek and to engage in work that has a practical direction, that translates ideas toward use, that is, development. Or, alternatively, an individual engaged in development rather than research may see more opportunities for potential commercialization, and thus become more likely to take an entrepreneurial path. Obviously, both influences may be at work.

Table 3–13
Nature of Work of Source Personnel and Spin-off
Entrepreneurs

Sample	Work Classification ("Bullpup" scale)									Median	Mean	Total
	1	2	3	4	5	6	7	8	9			
Lincoln Laboratory personnel*	9	15	22	16	16	4	11	1	56	5	5.75	150
Lincoln spin-off entrepreneurs	2	—	6	4	3	1	3	1	12	6	6.06	32
Instrumentation personnel†	3	3	2	18	14	1	6	6	81	9	7.33	134
Instrumentation spin-offs	—	—	—	—	1	—	—	1	7	9	8.44	9

*The difference in the work ratings of the Lincoln personnel and the Lincoln spin-offs is statistically significant at the probability level of .06 (Mann-Whitney U test).
† Significant at a level of .13 (Mann-Whitney U test).

The Source Organization Itself

More generally, the nature of the work being done at an organization may in itself affect the likelihood that an employee leaves to start his own firm. Finding that entrepreneurs individually tend to have more developmental (in contrast to basic or applied research) work experience suggests also that the more developmental the nature of the organization's work, the more likely that spin-offs occur. Each of the MIT labs was subjectively rated on the Bullpup scale (from basic research to development) by its director, with indications as to the change in this rating over the entire history of the lab. The ambiguous nature of this task makes the resulting ratings questionable. Nevertheless, the ratings do indicate that the bulk of subsequent entrepreneurs were employed during these labs' major development-oriented periods, even though the labs may have employed considerably more people at other times in their histories. However, a rank-ordering of the labs by overall developmental rating does not correlate with the frequency of new company spin-offs per 100 employees.

Examining a second lab characteristic, Cooper (1986) indicates mixed findings in regard to possible influence of the size of the source organization on the rate of spin-off formation, although his earlier studies suggest that smaller organizations are more prolific in generating technological entrepreneurs. In contrast, Feeser and Willard (1989) find larger incubators more prolific, especially of high-growth firms. To test this possible influence the MIT labs were arrayed in terms of funding size. The smaller MIT organizations do indeed tend to have higher frequency of spin-offs per million dollars of funding. Individuals may experience more independence when working in a smaller, more autonomous group and, therefore, become less apprehensive about venturing into the business world to form a new enterprise. Alternatively, the causal influences may be reversed. Entrepreneurial individuals may find more attractive and therefore seek work in the first instance in smaller technical organizations. When these persons finally gather the experience base needed for them to feel comfortable in setting up their own firms, their departures are then from a small organization, creating the possibly incorrect impression that smaller sources *generate* more entrepreneurs.

An individual's attitude toward work plays an important role in the learning process, as well as in the decision to stay or leave. A person who finds work challenging and enjoyable can be expected to learn more easily and develop skills more fully. With few exceptions the interviewed entrepreneurs speak very highly of their source technical organizations. For example, the MIT spin-offs typically say that their MIT lab work had been the most interesting work they had done, often at the leading edge or frontiers of science and technology. Eighty-nine percent indicate that their lab was a place in which they had learned substantial new technology, in contrast to primarily applying knowledge they had already possessed.

Eighty-four percent of the MIT-based entrepreneurs indicate that their work at their source laboratory or department organization was above average in challenge. Sixty-five percent indicate furthermore that their source work was at least as or more challenging than any other work experience they had before starting their new enterprises. Ninety-two percent of the entrepreneurs claim that they had derived above average enjoyment and personal satisfaction from their work at the source organization. Seventy-six percent felt that their laboratory experience was at least as or more enjoyable than any other employment before starting their new enterprises. Not surprisingly the challenge of the work and the satisfaction and enjoyment derived from doing the work are strongly related (Kendall tau=.60, p=.001). This relation between prior work challenge and satisfaction is true even for the consumer-manufacturing entrepreneurs (.04).

The entrepreneurs who came from the electronics systems company have similar attitudes. Only two of those individuals disliked the industrial laboratory and only one felt that it offered no challenge. These industrial spin-off entrepreneurs on the whole felt the challenge about equal to that of other companies they had worked for and say that they enjoyed the work at the source laboratory more than at other laboratories.

The dilemma arising from all of this contentment is then: Why do all these entrepreneurs leave? Obviously, the attractions of establishing their own firms finally overwhelm the attractions of remaining in their challenging and enjoyable source organizations. For many, as Ken Olsen indicated in Chapter 1, what had once been challenging had become routine. New heights needed to be conquered.

Occasionally an entrepreneur reports departing the source organization with unpleasant feelings. For example, of 23 Instrumentation Lab spin-offs providing this information, two felt their managerial talents were being underutilized (perhaps intentionally so, given Lab policies to encourage turnover), and a third entrepreneur had been asked to leave. But negative attitudes are uncommon. Similarly, only six of 53 medically oriented entrepreneurs in the San Diego area report that they had left their prior employ because of "dissonance" experienced there (Mitton, 1986). Comparative data are not available from the control study of scientists and engineers employed at the source labs that would permit labeling the entrepreneurs as more or less challenged by or satisfied with their employment.

As mentioned in Chapter 2, many entrepreneurs attest that when their primary projects were completed they felt that their work too at that organization was completed. They identified more strongly with the specific project than with the lab itself. The dead time that is inevitable before a new project gets underway is frustrating and gives the prospective entrepreneur time to rethink his objectives. Some entrepreneurs say that leaving to start a new firm was no more up-ending in anticipation than starting work on a new project. But the information required for a rigorous analysis of the effect of project completion on new company formation is not available.

Influence of Multiple Founders on Collective Work Experience

As shown earlier, most of the spin-off companies are formed by multiple founders, with a median of two. An obvious finding (from one study that collected details on each co-founder) is that the number of founders correlates strongly with the total prior years of work experience of the combined founding team (.01) as well as with the total founder hours worked per week in the new enterprise (.008). Each co-founder plainly contributes his years of work experience to the total of his colleagues along with his current effort, causing the startup enterprise to have an average of 21.7 years of founders' work experience.

More interesting are the findings from separating the total work experience of multiple co-founders into particular background fields. These data from a small sample of firms are shown in Tables 3–14 and 3–15. The number of founders correlates strongly with total years experience in sales work (.004) and in administrative work (.02), but not significantly with total years experience in technical work. Moreover, averaging the data into experience years per founder to get a measure of intensity, as founding teams get larger, the years in sales per founder also increase (.01) as do the years in administrative work per founder (.04). In compensation, the average years per founder in technical work decrease ($p=.04$) as founder teams increase in size. These indicate the more business-oriented nature of the

Table 3–14
Work Experience Composition as Function of Number of
Founders

Number of Founders	Number of Firms	Average Years/Firm in:		
		Sales	Technical	Finance and Administration
1	4	0.0	9.5	0.0
2	6	1.5	21.2	3.7
3	6	3.0	13.3	3.8
4	1	12.0	26.0	11.0
Average firm		2.3	16.1	3.3

larger teams, with sales and managerial experience gradually displacing technical experience among larger groups. The single founder enterprises had no work experience in sales or in finance and administration, and eight of ten single or dual-founder companies lacked sales experience. But in general, the experience mix changes as the founding team size gets larger. As is shown in later chapters, larger founding teams are also associated with more product manufacturing orientation in company direction and with larger initial capital funding (confirmed for a broad sample of enterprises by Cooper et al. (1989).

Thus, with respect to work experience, the average technical entrepreneur had 13 years of work, primarily at the key source organization studied from which he typically departed and directly founded his new enterprise. He worked primarily in applied development projects, not research, and often had supervisory responsibilities over other engineers. The entrepreneur dramatically out-produced his work colleagues along the conventionally measured lines of paper and patent output.

Table 3–15
Areas of Business Experience as Function of Number of
Founders

Number of Founders	Number of Firms with Founder Work Experience in:			
	Sales and Finance/ Administration	Sales Only	Only Finance and Administration	Neither
1	0	0	0	4
2	1	1	2	2
3	3	2	1	0
4	1	0	0	0

The key source organizations varied in their work content across the full spectrum from basic research to prototype development. But most entrepreneurs worked at those labs during the most development-focused periods of the lab's life. Smaller MIT labs have a higher frequency of new company spin-offs per million dollars of funding, possibly because they provide more autonomy and growth to their employees or inversely because more entrepreneurial individuals are attracted to initial work there. In most cases the entrepreneurs had found their prior source organization work to be both challenging and satisfying, and departed to meet new objectives, retaining good feelings toward their former employer.

Goal Orientation, Personality, and Motivation

Technical entrepreneurs cannot be described solely in terms of their fathers' careers or their own educational and work backgrounds. The goals, personality, and motivations of the entrepreneurs are a critical part of the entrepreneurial phenomenon. Any attempt to describe which individuals become entrepreneurs is incomplete without at least some concern for these issues, however "soft" the methods might be. Three different sources of information help define these entrepreneurial characteristics: two different sets of psychological tests, one aimed at specifying personality and behavioral preferences and the second focused on needs and motives; and structured interviewing of all the entrepreneurs, inquiring as to their reasons for starting their businesses.

Prior research suggests that technical entrepreneurs are expected to be extroverted individuals, with a high need for achievement and long-felt desires for independence, who are engaged in a continuing search of new challenges.

Psychological Types

The noted psychologist Jung believed that people are different in fundamental ways, reflected in how they prefer to function. Jung's four dimensions for psychological typing are: extroversion/introversion, sensation/intuition, thinking/feeling and judging/perceiving. The combinations of each possible preference along these dimensions create sixteen possible prototypes. The Myers-Briggs type indicator (MBTI) measures people in accord with Jung's typology, and decades of research by Educational Testing Services (using the MBTI) have amassed data on the psychological characteristics of individuals who perform various activities. Unfortunately, I could find no prior application of the Myers-Briggs test to technological entrepreneurs. A brief news item in *INC.* magazine (1988) did describe a mail questionnaire study, using the Myers-Briggs methodology, of 159 successful CEO-founders, including an unidentified number of technical firms. But the results charac-

terize the CEO-founder as very similar in personality to a "college professor", a finding rather hard to accept as credible.

Although subtle distinctions exist along each of the Jungian dimensions,
the archetypes are easy to define simply in terms of key preferences (Keirsey
and Bates, 1978). Extroverts are energized or "tuned up" by interacting
with other people. Introverts, on the other hand, draw their energy from
pursuing solitary activities, such as working quietly alone, reading or medi-
tating. Prior research indicates that 75 percent of the general U.S. population are extroverts (Bradway, 1964). Since entrepreneurship involves frequent interaction with colleagues, subordinates, customers, and others, I
expect to find entrepreneurs generally extroverted in their preferred functioning.

The "sensation"-preferring individual is "sensible", firmly grounded in
reality and facts. He or she focuses on past experience and actual past and
present happenings. The "intuition"-oriented person is innovative, future-
oriented, and imaginative, attracted by visions and inspirations. Despite the
fact that again 75 percent of the general U.S. population are oriented
toward sensation, I perceive founding entrepreneurs as more likely to prefer intuition as their mode of functioning.

Jung describes "thinking" individuals as preferring an impersonal basis
for choice—decision making based on logic, principles, law. "Feeling" individuals, in contrast, decide on a personal basis, subjectively considering
the effects of each choice on the decision maker and others. Because 60
percent of the thinkers are men and 60 percent of the feelers are women,
thinkers and feelers are distributed about equally in the general U.S.
population. Both thinking and feeling methods of decision making seem
reasonably applicable to entrepreneurs. But the dominance of highly educated men among technical entrepreneurs suggests that slightly more
thinking orientation should be expected among technical entrepreneurs.

The judging/perceiving dichotomy also divides the general U.S. population about 50-50. "Judging" persons prefer closure and the settling of
things, planning ahead and working to a deadline. They have a work ethic
such that work comes before all else. "Perceiving" individuals prefer to
keep options open and fluid, maintaining flexibility, and adapting to whatever happens. Entrepreneurship appears more akin to the judging preference.

My review of the Jung psychological dimensions thus concludes that I
expect technical entrepreneurs to be more extroverted (E), intuitive (N),
thinking oriented (T), and judging (J), identified as ENTJ types, or "the
field marshall" in the popular work by Keirsey and Bates (p. 73), perhaps
an apt label for some entrepreneurs!

The Technical Entrepreneur's Personality

To test the hypothesized personality profile, a shortened version (31 questions, about 20 minutes completion time) of the MBTI was distributed at

one meeting each of the MIT Enterprise Forum and the 128 Venture Group, monthly organizational groups in the Boston area that focus on technical entrepreneurship and venture capital activities. Data analyses were carried out on 73 usable responses, which include 48 people who had already founded one or more companies at some point in their lives (54 of the 90 companies they had founded are still active), and an additional six people with strongly expressed entrepreneurial desires (Cuming, 1984). Eighty-six percent of the respondents are men and 68 percent have received a master's or doctorate degree, making them comparable educationally to the previous descriptions of technical entrepreneurs.[1*]

The usable MBTI tests within the MIT Enterprise Forum and 128 Venture Group respondents were split into entrepreneurial and non-entrepreneurial groups. Among these 72 generally introverted technologists, those with stronger entrepreneurial tendencies are significantly more extroverted (E) $(X^2=10.60, p=.03)$. Relative to other engineers and scientists, technical entrepreneurs are characterized generally by more sociability, interaction, external orientation and interests, and a multiplicity of relationships. Also, even within this highly intuition-oriented group, the entrepreneurs tend to be slightly more intuitive (N) $(X^2=5.43, p=.25)$. They work on hunches, are speculative and future-oriented, use imagination and ingenuity in their actions. Similarly within this highly thinking-oriented group, entrepreneurs tend to show slightly further thinking preference (T) $(X^2=4.53, p=.34.)$ They try to be objective and impersonal, are analytically-oriented, behave with firmness in accord with standards.

Only with respect to Jung's judging/perceiving dimension are the results surprising. Despite a judging bias in the overall sample those with high entrepreneurial tendencies are significantly more perceiving-oriented than nonentrepreneurs (P) $(X^2=10.76, p=.005)$. Perhaps some entrepreneurs choose to "be their own boss" because they dislike the constant deadlines and pressures of the corporate world, reflecting their "perceiving preference".

Intriguingly Keirsey and Bates describe this ENTP personality profile, determined by the statistical findings shown earlier, as an "inventor", a title that seems apt to describe many technical entrepreneurs relative to other technically trained people. In fact, they say: "The ENTP can be an entrepreneur and cleverly makes do with whatever or whoever is at hand, counting on ingenuity to solve problems as they arise, rather than carefully generating a detailed blueprint in advance" (p. 186). Perry (1988) also categorizes one group of technical entrepreneurs as "inventors", but associates that group with motives to create and accumulate wealth.

The Technical Entrepreneur's Motives

As discussed earlier, McClelland and Collins and colleagues emphasize, to the exclusion of some other factors that I think are also important, the

---

[1*] Superscripts refer to Notes at the end of this chapter.

entrepreneur's motivations. This section describes some of the motivational characteristics of the entrepreneurs in the research samples. It is useful to discuss briefly the method used to measure motivation primarily because, even as McClelland says, "Human motivation has always been a topic of key interest to psychologists, but the lack of adequate methods for measuring it has seriously hampered the development of systematic knowledge of the subject" (Atkinson, 1958, p. 7). Following McClelland's lead I adopted the Thematic Apperception Test (T.A.T., codings based on verbal interpretations of fuzzy sketches) as the instrument for measuring motivation, utilizing McClelland's version, which concentrates on assessing three needs: need for achievement (n-Ach), need for power (n-Pow), and need for affiliation (n-Aff).

A summary of my previous comments about McClelland's views provides a definition of the need for achievement. Individuals high in need for achievement are presumed to take moderate as opposed to high or low risks. In addition they usually perform better at tasks that are moderate in risk. Neither low nor high risk situations are seen as achievement oriented because accomplishment is either too easy or impossible. On the other hand, moderate risk tasks can be accomplished in an innovative manner through the efforts of the achieving individual. High n-Ach individuals tend to choose and work hard at tasks in which they can achieve in personally determined novel or innovative ways. Their high n-Ach does not lead them to work hard and perform well at merely anything. The high n-Ach individual prefers concrete feedback on his actions and must perceive that the end results are at least in part due to his actions. The implication here is that the concrete feedback is external to the individual, such as profits from business rather than the personal satisfaction of being a good manager. As McClelland points out, however, money and profits do not have intrinsic interest for the high n-Ach individual. Instead, "Money, to them, was the measure of success. It gave them the concrete knowledge of the outcome of their efforts that their motivation demanded" (p. 237). Clearly inherent in the entrepreneurial situation are most of the characteristics that may satisfy individuals high in n-Ach.

Need for power is "that disposition directing behavior toward satisfactions contingent upon the control of the means of influencing another person... The means of control can be anything at all that can be used to manipulate another person" (Atkinson, 1958, p. 105).

Need for affiliation is concerned with the establishment, maintenance or restoration of positive relationships with other people. These relationships are most adequately described by the word "friendship". Statements of liking or of desire to be liked, accepted or forgiven are manifestations of this motive. Interpersonal feedback is implied in that the individual who is high in n-Aff has to have the internal feeling that he or she is liked or accepted. Note that this feedback is of a different sort than the feedback on performance or results necessary to an individual high in n-Ach.

Need for achievement has different behavioral manifestations than either need for power or need for affiliation in terms of the individual's relationships with people. Need for power and n-Aff are interpersonally oriented. Implicit in their definitions is the existence of other human beings with whom the individual can influence and control or be friends. Need for achievement, on the other hand, is more internalized. The individual may need other people to help him to satisfy his n-Ach. But his effectiveness with them is determined by other than n-Ach. Achievement need is a primary factor that influences behavior, other than interpersonal, affecting company performance (e.g., decision making characteristics, commitment to work, recognition of the need for plans generating tangible outcomes such as profits or sales volume). Need for power and n-Aff are primary factors that influence interpersonal behavior, affecting company performance (i.e., being concerned about people, being authoritarian, being paternal). Power need and affiliation need have strong implications as determinants of managerial style.

T.A.T. scores for the three needs were developed for a subset ($n = 51$) of technical entrepreneurs, with scoring done by the staff of the Harvard University Motivation Research Group so as to assure coding reliability. (Indeed, despite presumed "softness" of the underlying T.A.T. data the average intercoder reliability obtained is in the high 0.80 range.) The demographic characteristics of these 51 entrepreneurs, including family background, age, education, and work experience, matches closely with comparable data from the entire technological entrepreneurial group, lending support to the notion that the selected group is representative. Table 3–16 presents the needs data.

The most important finding from Table 3–16 is that the needs of technical entrepreneurs are not simply stated. Each measure varies over a wide range. For example, despite mythology and misstated empiricism to the contrary, all entrepreneurs do not have high n-Ach; only some do. Indeed, although McClelland focuses his descriptions of the n-Ach motive on entrepreneurial behavior, I cannot find in McClelland's writings any reference to studies of company founders. Evidently his claims regarding those he calls "entrepreneurs" rely more on data from businessmen and others. These data indicate that the median technical entrepreneur has

Table 3–16
Measurement of Technical Entrepreneurs' Needs
($n = 51$)

Need	Mean	Median	Range
n-Ach	5.9	5.0	–5 to 18
n-Pow	9.7	9.5	0 to 19
n-Aff	3.5	3.0	0 to 16

moderate n-Ach, moderate n-Pow, and low n-Aff. These can be interpreted to describe a typical technical entrepreneur as having some push to succeed, a willingness both to take charge and to share control with others, and little requirement for relationships with others. But a wide variety of motivational "types" is possible within the recorded data on the technological entrepreneurs.

These entrepreneurial needs are not independent of each other. Need for achievement is positively related to n-Pow ($p=.01$) and negatively related to n-Aff ($p=.01$); and higher n-Pow is reasonably linked to lower n-Aff ($p=.05$).

Since I hypothesize that these needs might well influence managerial behavior of the entrepreneur, I examine in Chapter 9 whether and how any of these drives affects performance of the newly created firm.

Why Do You Want to Start a Business?

Among the several hundred people examined in Edgar Schein's studies of "career anchors" he finds a small number of entrepreneurs. "These people discovered early in life that they had an overriding need to create a new business of their own by developing a new product or service, by building a new organization through financial manipulation, or by taking over an existing business and reshaping it in their own image" (Schein, 1987, p. 168). Many of the technical entrepreneurs in my studies had thought about starting or owning a business long before they formed their companies. A few had the thought and immediately went into business. Table 3–17 displays the number of years between the first occurrence of the idea to go into some business and the founding of the spin-off company for a subset of 62 entrepreneurs. Only 21 percent (13) of the entrepreneurs first thought about going into business within the same year they actually decided to form their companies. An additional 27 percent (17) thought about going into business more than one year but less than five years before forming their new enterprises. At the other extreme, 24 percent (15) of the entrepreneurs had thoughts of going into business for more than ten years, like Sam Morris in Chapter 1. For the entrepreneurs leaving the diversified technological company, as another data point, the median had been contemplating such a move for nine years.

Table 3–17
Years between Thinking of Going into Some Business and
Founding of Company ($n = 62$ entrepreneurs)

		Number of Years Between							
	0	1–2	3–4	5–6	7–8	9–10	11–15	16–20	>20
Number of entrepreneurs	13	7	10	8	5	4	7	7	1

Table 3–18
Years between Conceiving of Specific Spin-Off Company
and Its Founding ($n = 107$ entrepreneurs)

	Number of Years Between						
	0	1	2	3	4	5	>6
Number of entrepreneurs	56	28	10	4	4	4	1

The pattern is considerably different for the time between having the specific idea for the particular spin-off and the formation of the company. Table 3–18 displays those data for 107 technical entrepreneurs. Over half of the entrepreneurs more-or-less immediately formed their companies once the specific thought occurred. Nearly 80 percent of the firms were formed within less than two years after the specific thought for them occurred, and all except one in this group were formed less than six years after they were conceived.

It is difficult to find general explanations for why some entrepreneurs thought of going into business and immediately acted, while others did not. Each case is different. As to why the thoughts about starting the spin-offs materialize so rapidly into the formation of companies, explanations can only be given ex post facto (i.e., once the company materializes, we can conclude that the thoughts for it occurred fairly recently).

The reasons for going into any business depend on many factors. Traditionally, business "ownership" is a way in which the American goal of independence is evinced. From an economic theory perspective, business ownership is motivated by the profit that can be captured by the individual. Increasingly clear, as evidenced thus far in this book, is the fact that business initiation and whatever ownership results are really influenced by the complexity of factors such as the community environment, the family background of an individual, age, education, religion, and work experience.

Unfortunately, only one question in my research protocol relates directly to the entrepreneur's motivation for starting his own business:

At the time you started your new enterprise what feature of going into business for yourself did you consider most attractive? (Check all which apply, then rank those you have checked, 1, 2, ... with 1 being the most important.)

√   Rank
___   ___   Salary or wealth
___   ___   Being own boss—independence
___   ___   Challenge—do something others could not

		Challenge—taking on and meeting broader responsibilities
___	___	Freedom to explore new areas
___	___	See things through to completion
___	___	Other

Results from various samples of entrepreneurs indicate that the features of business initiation that appeal to them are conventional. Table 3–19 displays the features of going into business that were attractive to a subset of MIT spin-off entrepreneurs. The specific characteristic of independence, "being own boss", is the most appealing feature to 25 percent of those entrepreneurs. As a primary or secondary feature it appeals to over 40 percent of the entrepreneurs. The feature of independence, in one form or another, is of prime importance to 39 percent of the entrepreneurs. Occasionally the entrepreneurs present this need for being their own boss by evidencing their impossibilities with having anyone else as a boss. Alex d'Arbeloff, co-founder and CEO of Teradyne Inc., a half-billion dollar automatic test equipment company, says he had five jobs in his first ten years after graduation from MIT, being fired from three of them. "I had good ideas", Alex exclaims, "and I was very impatient about getting them implemented. People just didn't seem too interested, so I always pushed

Table 3–19
Attractiveness of Business Initiation to MIT Spin-Off
Entrepreneurs ($n = 72$)

Features of Going into Business[*]	Primary Feature	Secondary Feature
Monetary		
1. Salary or wealth	9	6
Challenge		
1. Do something that others could not	14	7
2. Taking on and meeting broader responsibilities	8	5
Independence		
1. Being own boss	18	11
2. Freedom to explore new areas	7	8
3. See things through to completion	3	6
Other[†]	13	6
Total responses	72	49

[*] This group includes 72 respondents to the relevant question; of these, 49 also indicate secondary features.

[†] The entrepreneurs who mention other attractions generally state features that are specific incidences of independence, challenge, or money.

harder and harder until someone told me to go." Starting Teradyne with Nick DeWolf, who by a fluke of alphabetical ordering was seated next to him throughout three years of ROTC classes, satisfied d'Arbeloff's search for independence.

Monetary appeal is stated to be of prime importance to less than 15 percent of the entrepreneurs and of primary or secondary importance to only 20 percent of the entrepreneurs in this cluster. Those who look for financial greed as the explaining drive of technical entrepreneurs will find that only a small fraction of the cases across-the-board fit that stereotype, by their own categorization. Of course, more entrepreneurs may well have been attracted from the outset by the potential of making money in a new enterprise than were willing to admit it. The drive for financial success has been questioned by many in our society, and this factor may have prejudiced the responses to this particular question.

The challenge that starting a new business affords is attractive to a sizable proportion of the entrepreneurs, despite the fact that most entrepreneurs had previously found challenge in their source organizations. Thirty percent list challenge (1) to do something that others could not or (2) to take on and meet broader responsibilities as the most appealing feature. Challenge as an attraction can easily be understood. These technical entrepreneurs are highly educated and highly trained. Employment either at their source laboratories or in industry was fairly well guaranteed for them. In these environments the talented individual is rather secure. Starting and trying to run and build a new enterprise, on the other hand, provides a more risky situation in which the individual's achievement is directly reflected in the success of his company. The business can succeed or fail on the accomplishments of the individual; and its success or failure is his responsibility. The challenge of business initiation lies in the setting wherein the individual can measure for himself his "true worth", a different and broader set of challenges than previously encountered. This is, or course, what frequently distinguishes these men as entrepreneurs. Schein's evidence (1978, p. 149) is supportive: "...these people seem to have an overarching need to build or create something that was entirely their own product. It was self-extension...a measure of their accomplishments."

An array of findings from several other samples from the research program presents much the same picture. As Table 3–20 demonstrates, even the consumer-oriented manufacturing entrepreneurs coincide in expressed drives with their technological brethren. In another sample of computer-related entrepreneurs, 16 of 22 said that "being my own boss" was the primary motive for getting started, agreeing also with 20 of 23 spinoffs from the large diversified technological firm. The biomedical entrepreneurs attest to similar motives.

Despite the consistency among the various subsamples, these three most frequent answers may nonetheless merely reflect the socially acceptable or "pat" responses. All three no doubt manifest other deeper motiva-

Table 3–20
Key Attractions of Going into Business (%)

Research Sample	Independence (Being own boss)	Financial (Salary or capital appreciation)	Challenge (Unique, broader)
MIT laboratory spin-offs	39	20	30
Spinoffs from electronic systems firm	38	23	25
Spinoffs from diversified technological firm	41	35	24
Searchers for venture capital	32	14	32
Computer-related enterprises	30	30	—
Consumer-oriented manufacturers	32	27	11

tions, rather than merely being motivations themselves. For example, an individual who indicates financial gains as his key motive for starting a new enterprise may really be indicating his n-Pow or higher social status. But unfortunately research probes in this direction do not reach deeper. In fact, statistical analyses of the entrepreneur's motivation in relation to other factors such as his religion, educational level, and his father's occupational status produce very little in explainable differences.

Don Valentine, successful Silicon Valley venture capitalist and investor in 175 companies including Apple, LSI Logic, Oracle, Tandem, and 3Com, has strongly held views as to entrepreneurial motives. "... the presumption is that employees of the big companies leave and go to venture companies to found startups to make more money. It's not that way. Andy Grove, Bob Noyce and others left Fairchild to found Intel, not to make more money. They left to make a product that Fairchild was either unable or unwilling to make or, for whatever reason, didn't get around to making. That's why ventures are started: from the lack of responsiveness in big companies. ... The press always talks to the successful entrepreneur about how much money he has now. But unless the entrepreneur is asked, he forgets to tell them that the real reason he wanted to found his company was because he had this idea, this vision, and his vision wouldn't work at Xerox PARC, or Hewlett-Packard, or IBM or wherever. He had to get to where he could make his vision work" (Sheff, 1990). Valentine's views agree with the dominant "being my own boss" motivation documented earlier. They are also anecdotally supported by the combination of anger and frustration displayed by several industrial spin-offs in the research samples who exclaimed that they would have stayed with the company if only their boss had let them pursue their pet idea.

The resulting findings in regard to goal orientation, personality and motivation affirm the hypothesized extroverted behavior, but characterize the personality traits of the typical technical entrepreneur as those of an inventor. Furthermore, the average technical entrepreneur does not have high need for achievement, but is more moderate in this regard as well as in his need for personal power. Goals sought by technical entrepreneurs include the long-desired independence of being their own boss and the challenge of doing something that others could not, these drives being far more prominent than a search for high income or wealth.

Summary and Implications

This chapter seeks to explain empirically the origins of the technology-based entrepreneur, comparing samples of entrepreneurs with appropriate control groups of scientists and engineers employed at key MIT entrepreneurial source organizations. The major conclusions are presented here and synopsized in Table 3–21.

Taking matters in the sequence presented in this chapter, the first set of conclusions relates to the influences of family background, a cluster of factors that the prospective entrepreneur cannot affect. Perhaps the most important research finding in regard to early breeding is what I have

Table 3–21
Characteristic Influences on Becoming a Technical
Entrepreneur

Family background
    "Entrepreneurial heritage"—son of self-employed father
    Some influenced by achievement-oriented religious background

Education and Age
    Master's degree, usually in engineering, more-or-less similar to their previous "source
        organization" technical colleagues
    Mid-30s age at founding, somewhat younger than their previous colleagues

Work experience
    Decade plus of work, slightly less than their earlier co-workers, dominated by
        experience in "source organization"
    Developmental (rather than research) work background
    Highly productive technologist, often with supervisory responsibilities
    Challenged and satisfied by "source organization" work

Goal orientation, personality and motivation
    "Inventor" personality
    "Moderate" needs for achievement and power, low need for affiliation
    Long-felt desire for own business
    Heavy orientation toward independence, as well as search to overcome challenges,
        less concern for financial rewards

labeled "the entrepreneurial heritage": a strong tendency for entrepreneurs to come from families in which the father was self-employed, providing a role model that presumably stimulated his son's desire for independence. This phenomenon characterizes 50 to 65 percent of the technical entrepreneurs, is at least twice what should be expected from a random sample of the U.S. population, and is significantly more than what is found in "control" populations of scientists and engineers.

For those who do not come from entrepreneurial homes, aspirations inculcated through the family's religious background do affect the incidence of technical entrepreneurship. In line with McClelland's writings about achievement motivation being linked to certain religious and ethnic family backgrounds, relatively more Jews and fewer Catholics establish technology-oriented firms.

But careful analyses dispel the myth of the first-born son as being significantly related to technical entrepreneurship. Relative to their siblings proportionately as many first-borns are likely to remain as employed scientists and engineers as are likely to spin-off to form their own firms.

In regard to "growing up", in contrast with prior studies of "general" entrepreneurs, the research here demonstrates the technical entrepreneur to be well-educated, his median education being about a master's degree, more typically in engineering than in science, and only infrequently educated in all other disciplines including management. These advanced educational characteristics, however, do not differentiate the entrepreneur from his prior technical colleagues at the "incubating" labs.

The mid-30s is the dominant age range of technical founders, with MIT laboratory spin-offs being on the younger side and faculty entrepreneurs being older exceptions to the age "rule" in several subsamples of entrepreneurs. The average technological entrepreneur is typically younger by two or three years than his average prior work colleague.

With respect to these aspects of family background and upbringing that are beyond the control of the prospective entrepreneur, that person should gain comfort from what is essentially the "flip side" of these dominant relationships. For example, to take the strongest finding, as many as two-thirds of the technical entrepreneurs did have self-employed fathers. But one-third to one-half in each of the research samples did not. Even larger "minorities" exist in regard to each of the other family and "growing up" factors. The data demonstrate a wide pathway for those wanting to form their own firms.

In contrast to his earlier years, the aspirant entrepreneur does have greater control over the duration and nature of his professional work experience. The typical technical entrepreneur had 13 years of work experience, more than half of it at the incubator technical organizations labeled the source organizations. Close to two-thirds of the entrepreneurs went directly from these dominant technical work experience sites into their own companies.

At the source organizations the entrepreneurs significantly outproduced their technical colleagues along the conventional output measures of papers and patents. Many had already risen into technical supervisory roles. Indeed, starting a company might be just another avenue for the productive energies and knowledge of these outstanding people.

Their work backgrounds evidence that not only are entrepreneurs more likely to come from engineering rather than science, but especially from the developmental (not research) end of the R&D work spectrum. Translating technology into use is more likely to spawn entrepreneurs than is the more basic creation of new technical knowledge. Going from a challenging and satisfying work environment in search of still further challenge is typical of most technical entrepreneurs.

Technical professionals who are considering starting a company should also gain comfort from the wide diversity of goals, personalities and motives included among actual technical entrepreneurs. From a personality perspective, technical entrepreneurs are likely to be more extroverted than their rather introverted technical colleagues. As a group they represent extremes in orientation to both intuitive and analytic thought processes, both dimensions already strong among engineers and scientists. The technical entrepreneur is also perceiving oriented, generating a personality profile that Keirsey and Bates have rather aptly (for many) labeled the inventor.

Motivational studies show wide ranges of basic needs within the technical entrepreneur population. Despite the fact that all those studied are indeed entrepreneurs, they do not all have high need for achievement, although of course some do. The median technical entrepreneur has moderate need for achievement, moderate need for power, and low need for affiliation.

Most technical entrepreneurs seem to be fulfilling a long felt need (or at least ambition) in starting their companies, reflecting at least several years of prior general contemplation about going into their own businesses. But when asked to state why, these technical entrepreneurs reveal primarily a heavy orientation toward independence, being their own boss, some reflection of a continuing search for new and bolder challenges, and considerably less focus on financial gains than might be expected by cynical observers of entrepreneurs. More specifics about precipitating events in forming the business are discussed in the next chapter.

Notes

1. The psychological profiles of the 54 entrepreneurs in this sample were first compared with data on the U.S. general population. On a statistically significant basis the group of technical entrepreneurs is more introverted (I) ($p=.10$), more intuitive (N) ($p=.05$), and more thinking-oriented (T) ($p=.05$) than the general population, and not different in judging preference (J). Indeed

Keirsey and Bates describe the INTJ personality that was found as a "scientist" (p. 72), a bias that might have been expected for this overall sample of well-educated technologists, relative to the U.S. population as a whole. These results add confidence to use of the shortened version of the MBTI tests.

References

John W. Atkinson. *Motives in Fantasy, Action and Society* (Princeton: D. Van Nostrand Co., 1958).

P. J. Bearse. "An Econometric Analysis of Black Entrepreneurship", in J. A. Hornaday et al. (editors), *Frontiers of Entrepreneurship Research, 1984* (Wellesley, MA: Babson College, 1984), 212–231.

Katherine Bradway. "Jung's Psychological Types", *Journal of Analytical Psychology*, 9 (1964), 129–135.

A. V. Bruno & T. T. Tyebjee. "The Entrepreneur's Search for Capital", in J. A. Hornaday et al. (editors), *Frontiers of Entrepreneurship Research, 1984*, (Wellesley, MA: Babson College, 1984), 18–31.

O. F. Collins, D. B. Moore with D. B. Unwalla. *The Enterprising Man* (East Lansing, MI: Michigan State University Press, 1964).

A. C. Cooper. *The Founding of Technologically-Based Firms* (Milwaukee: The Center for Venture Management, 1971).

A. C. Cooper. "Spin-offs and Technical Entrepreneurship", *IEEE Transactions on Engineering Management*, EM-18,1 (1971), 2–6.

A. C. Cooper. "Entrepreneurship and High Technology", in D. L. Sexton & R. W. Smilor (editors), *The Art and Science of Entrepreneurship* (Cambridge, MA: Ballinger Publishing, 1986), 153–167.

A. C. Cooper & A. V. Bruno. "Success Among High Technology Firms", *Business Horizons*, 20, 2 (1977),16–22.

A. C. Cooper, C. Y. Woo, & W. C. Dunkelberg. "Entrepreneurship and the Initial Size of Firm", *Journal of Business Venturing*, 4, 5 (1989), 317–332.

John W. Cuming. *An Investigation of Entrepreneurial Characteristics Based on the Myers-Briggs Type Indicator* . Unpublished S.M. thesis (Cambridge, MA: MIT Sloan School of Management, 1984).

J. Doutriaux. "Evolution of the Characteristics of (High-tech) Entrepreneurial Firms", in J. A. Hornaday et al. (editors), *Frontiers of Entrepreneurship Research, 1984* (Wellesley, MA: Babson College, 1984), 368–386.

Henry R. Feeser & Gary E. Willard. "Incubators and Performance: A Comparison of High- and Low-Growth High-Tech Firms", *Journal of Business Venturing*, 4, 6 (1989), 429–442.

William B. Gartner, Terence R. Mitchell & Karl H. Vesper. "A Taxonomy of New Business Ventures", *Journal of Business Venturing*, 4, 3 (1989), 169–186.

Robert D. Hisrich. "Entrepreneurship/Intrapreneurship", *American Psychologist*, February 1990, 209–222.

INC. "The Entrepreneurial Personality", August 1988, 18.

David Keirsey & Marilyn Bates. *Please Understand Me: An Essay on Temperament Styles* (Del Mar, CA: Prometheus Nemesis Books, 1978).

D. G. Marquis & I.M. Rubin. "Management Factors in Project Performance" Unpublished paper (Cambridge, MA: MIT Sloan School of Management, 1966).

David C. McClelland. *The Achieving Society* (Princeton: D. Van Nostrand Co., 1961).

D. H. McQueen & J. T. Wallmark. "Innovation Output and Academic Performance", in J. A. Hornaday et al. (editors), *Frontiers of Entrepreneurship Research, 1984* (Wellesley, MA: Babson College, 1984), 175–191.

D. G. Mitton. "The Begatting Begins: Incubation Patterns in the Developing Health Science and Biomedical Industry in the San Diego Area", in R. Ronstadt et al. (editors), *Frontiers of Entrepreneurship Research, 1986* (Wellesley, MA: Babson College, 1986), 509–525.

Office of the Director of Defense Research and Engineering. *A Trial Study of the Research and Exploratory-Development Origins of a Weapon System ... Bullpup.* (Washington, D.C.: 1964).

C. Olofsson, G. Reitberger, P. Tovman, & C. Wahlbin. "Technology-based New Ventures from Swedish Universities: A Survey", in N. C. Churchill et al. (editors), *Frontiers of Entrepreneurship Research, 1987* (Wellesley, MA: Babson College, 1987), 605–616.

L. Perry. "The Capital Connection: How Relationships Between Founders and Venture Capitalists Affect Innovation in New Ventures", *Academy of Management Executive*, 11, 3, (1988) 205–212.

D. M. Ray & D. V. Turpin. "Factors Influencing Entrepreneurial Events in Japanese High Technology Venture Business', in N. C. Churchill et al. (editors), *Frontiers of Entrepreneurship Research, 1987* (Wellesley, MA: Babson College, 1987), 557–572.

Edward B. Roberts. "Entrepreneurship and Technology: A Basic Study of Innovators", *Research Management*, 11, 4 (1968), 249–266.

E. B. Roberts & H. A. Wainer. "Some Characteristics of Technical Entrepreneurs", *IEEE Transactions on Engineering Management*, EM-18, 3 (1971), 100–109.

Everett M. Rogers & Judith K. Larsen. *Silicon Valley Fever* (New York: Basic Books, 1984).

I. M. Rubin, A. C. Stedry, & R. D. Willits. "Influences Related to Time Allocation of R&D Supervisors", *IEEE Transactions on Engineering Management*, EM-12, 3 (September 1965).

David Sheff. "Don Valentine Interview, Part Two", *Upside*, 2, 4 (June 1990), 48–54; 72–74.

Edgar H. Schein. *Career Dynamics* (Reading, MA: Addison-Wesley, 1978).

Edgar H. Schein. "Individuals and Careers", in J. W. Lorsch (editor), *Handbook of Organizational Behavior* (Englewood Cliffs, NJ: Prentice-Hall, 1987).

Joseph A. Schumpeter. *Capitalism, Socialism, and Democracy* (New York: Harper & Row, 1966).

D. L. Sexton & R. W. Smilor (editors). *The Art and Science of Entrepreneurship* (Cambridge, MA: Ballinger Publishing, 1986).

R. W. Smilor, D. V. Gibson & G. B. Dietrich. "University Spin-out Companies: Technology Start-ups from UT-Austin", in *Proceedings of Vancouver Conference* (College on Innovation Management and Entrepreneurship, The Institute of Management Science, Vancouver, BC: May 1989).

N. R. Smith, J. S. Bracker, & J. B. Miner. "Correlates of Firm and Entrepreneur Success in Technologically Innovative Companies", in N. C. Churchill et al. (Editors), *Frontiers of Entrepreneurship Research, 1987* (Wellesley, MA: Babson College, 1987), 337–353.

R. D. Teach, F. A. Tarpley, Jr. & R. G. Schwartz. "Who are the Microcomputer Software Entrepreneurs?", in J. A. Hornaday et al. (Editors), *Frontiers of Entrepreneurship Research, 1985* (Wellesley, MA: Babson College, 1985), 435–451.

T. T. Tyebjee & A. V. Bruno. "A Comparative Analysis of California Startups from 1978 to 1980", in K. H. Vesper (Editor), *Frontiers of Entrepreneurship Research, 1982* (Wellesley, MA: Babson College, 1982), 163–176.

J. M. Utterback, M. Meyer, E. Roberts, & G. Reitberger. "Technology and Industrial Innovation in Sweden: A Study of Technology-based Firms Formed Between 1965 and 1980", *Research Policy*, 17, 1 (1988), 15–26.

A. H. Van de Ven, R. Hudson, & D. M. Schroeder. "Designing New Business Startups: Entrepreneurial, Organizational, and Ecological Considerations", *Journal of Management*, 10 (1984), 87–107.

W. Lloyd Warner & James C. Abegglen. *Big Business Leaders in America* (New York: Harper & Brothers, 1955).

Max Weber. *The Protestant Ethic and the Spirit of Capitalism* (New York: Scribner, 1956).

CHAPTER 4

The Technological Base of the New Enterprise

The forty plus years since World War II have defined technological advance as an integral part of our society and a key basis for both military and industrial competition. Among other vital contributions, technology provides the raison d'etre for the formation and growth of myriad new firms. The opportunities presented by advancing technology, however, have not been seized by all potential entrepreneurs. As shown in Chapter 3 the unique companies that are technology-based new enterprises are for the most part founded by individuals highly trained in existing technology. The new technical ventures have originated from similar technology-based organizational environments or sources.

All entrepreneurs must possess some sort of entrepreneurial spirit. In addition, however, high-technology companies at their formation depend on the transfer of technology or technological know-how learned or created by the entrepreneur from a variety of incubating sources. Increasing numbers of studies are examining that incubation process (Cooper, 1984; Cooper and Bruno, 1977; Doutriaux, 1984; *The Economist*, 1983; Feeser and Willard, 1990; Roberts, 1968; Smilor and Gill, 1986). Fewer focus on detailed aspects of the technical base of the firm at its time of origin (Meyer and Roberts, 1988; Olleros, 1986; Roberts and Hauptman, 1986). Consequently, the theoretical framework presented below rests more on logic and surmise from experience than from prior academic research. This chapter presents that framework and empirical evidence on the determinants of the initial technological basis for these new enterprises.

A Theoretical Perspective

Figure 4–1 presents a schematic (or descriptive flow model) of what might be involved in the personal generation and/or use of technology, here applied as the basis for a new firm. The question underlying the model is: Under what circumstances does a founder of a new firm use know-how he or she was exposed to and possibly acquired as an employee of a prior "technology source organization"? Solid lines in the diagram represent the

100

Figure 4–1
The Personal Transfer of Technology

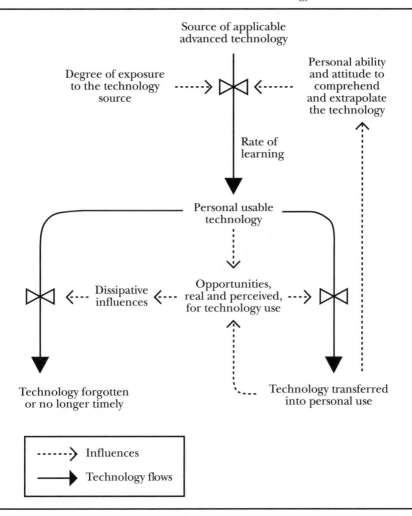

hypothesized flows of technology from one state to another; dotted lines indicate influences on those flows. The diagram is a closed-system representation of the factors affecting this special form of technology transfer. External forces, such as economic climate or specific market opportunities, are intentionally omitted from this attempt to focus on the organization-to-organization transfer of know-how and skills that is being carried out through personal entrepreneurial movement.

First, the diagram assumes there must necessarily be some technological know-how worthy of transfer. This "source of applicable advanced technology", including both skills and knowledge, might derive from several different cognitive realms in combination or separately (e.g., work experience, educational programs, the contents of a journal, knowledge of a friend). It may be new or old information, capabilities developed in bench work or theories underlying processes, narrowly specified or widely applicable. Every further step in the diagrammed process leading to personally transferred technology into a new firm builds from this first platform. Chapter 3 shows that entrepreneurs come more often from development rather than research backgrounds. This suggests a testable hypothesis: namely, that development work at the source organization provides a higher degree of technology transfer to new enterprises than does research work.

Next, to become transferable an individual must personally absorb the information that is available at the source of technology. To learn such technology an individual must have at least some modest "degree of exposure to the technology source", and also have the "personal ability and attitude to comprehend and extrapolate" as well as perhaps interpret and apply the information. I hypothesize that the greater the exposure and the more advanced the personal ability, the more the individual learns from the technology source. Length of exposure to a technology source is only one of several dimensions that potentially affect learning. The kind of work being done at the source and perhaps the individual's role in carrying out the work are other potential influences on exposure. The breadth of exposure, to knowledge creation as well as use, may also influence the rate of possible learning. Chapter 3 also alerts us to a special condition in a number of these start-ups: The entrepreneur often simultaneously works at both his old and his new employers, engaging in so-called moonlighting activities or part-time entrepreneurship. This provides concurrent exposure to both potential source as well as potential use of advancing technologies, which I hypothesize should produce increased degree of technology transferred to the new firm.

Personal ability is assumed here to include educational preparedness for learning and to play a role in this transfer process. The layman has difficulties in acquiring knowledge of theoretical physics, even when opportunely exposed to excellent sources of such knowledge, without a prior grounding in physics or advanced mathematics principles; and attitudinal preparedness, a willingness and an interest in absorbing new skills and tools, is also hypothesized as required. Within the same organization, some delight in its stimuli and challenge; others are bored, and cannot wait to get home at the end of a day, ignoring opportunities the organization may proffer.

These two influences, the degree of exposure to a source of potential new technology and the personal ability and attitude to comprehend the knowledge, combine in a not-well understood "gating function" to affect the individual scientist's or engineer's "rate of learning" of new scientific or technological information. The learning adds to the individual's storehouse of potentially "usable technology". But the mere existence of personally accessible information is not sufficient for useful technology transfer to occur. "Dissipative influences" and lack of opportunity can block effective exploitation of potentially useful information. For example, continued lack of exposure to opportunities, or lack of ability to recognize opportunities, for application of his or her newly acquired know-how no doubt causes the individual eventually to forget the technical know-how. Or while not forgotten, the technology may cease to be as useful because problems for which it was once applicable no longer exist (or their priority has lessened). The technology becomes obsolete. Thus, logically, in time the pool of usable knowledge gradually shrinks unless replenished by new learning.

The final flow in this process, the one of principal interest here, is the "technology transferred into personal use". The prime influence on this final stage of transfer appears a priori to be the "opportunities for technology use", such as in the formation of a new enterprise. But second order influences are probably at work too. For example, the greater the base of usable technology the greater the likelihood of some matching between know-how and application opportunity. But someone who has achieved useful technology transfer in the past is more likely to be accustomed to sensing new opportunities as they arise. The use of the technology feeds back, as in Figure 4–1, to strengthen the individual's ability to recognize and take advantage of usable technology. That person is not only ready and waiting when opportunity knocks; he or she knows where to wait! Moreover, an individual who perceives an opportunity for technological exploitation can move forward with intent, reducing what might otherwise be a longer delay in attempting technology use. This perception of opportunity, therefore, reduces the possible dissipative effects of time, leaving more personal technology still useful for the new enterprise formation.

The Importance of Initial Technology Transfer

The process of individual transfer of technology, as outlined in Figure 4–1, presumably occurs to some extent whenever any new enterprise is begun. What is not so clear is the degree to which this transfer takes place and the role that specific organizations play in this process. The first issue faced in gaining more insight into the process is the need for a simple measure of the degree of importance of initial technology transfer. Each company in the samples is rated on a four-point scale representing the extent to which it depended initially on technologies developed at the source laboratory under study. The rating is based on observation of the circumstances surrounding formation of the enterprise, consideration of the nature of the company's products, and statements by the entrepreneur.

As an initial trial of this way of assessing technological dependence, this measure was determined for 125 MIT-based spin-off companies. To assure reliability of these ratings, two different coders evaluated a subset of the companies, with almost no disagreement between their classifications. In addition, the entrepreneurs' former laboratory supervisors were asked to rank the companies on this variable. Their classifications correlate significantly with those assigned by the coders ($p = .007$). The intercoder reliability and the supervisor concurrence in this trial provide confidence in the use of this measure throughout the spin-off studies.

The degree of importance of technology transferred from the source organization on the formation of the spin-off firms is divided into: direct, partial, vague, and none.

Direct: The company in its present form would not have been started without the source-learned technologies. It now utilizes or utilized at the beginning mostly what the founder(s) learned at the source.

Partial: An important aspect of the company's work originated with source-learned technologies. The individual who transferred the technology might have supplemented the source-learned technologies at other employment between the laboratory and the new enterprise.

Vague: Nothing specific was transferred. However, general technical background and know-how learned at the source were very important. In this case, the company might have been started without the source experience.

None: Essentially nothing that the company does is related to source technologies. The individual who started the company may have learned an extensive amount at the source but is not utilizing this (from a technological standpoint) in the new enterprise.

The definitions used in the scale are specific to the source organization under study; therefore, a rating on importance of technology transfer for a spin-off company from MIT's Lincoln Laboratory relates only to the degree of dependence on Lincoln technologies. In addition, note that the terms partial and vague somewhat understate the importance of the source-learned technology in the formation of these new enterprises. In each of these categories, learned technology is unquestionably important; the difference is only in degree. Only the category of "none" dismisses the importance of the learned technology.

As shown on the left in Table 4–1, 32 new companies are directly affected by the learned technology, 34 partially, 43 vaguely, and 16 not at all. In all, then, 109 companies of these 125 felt the source-learned technology important to their creation. Even stronger dependency on source technology is uncovered in my separate sample of 29 biomedical products companies. Fifty percent of that group indicate that their firm could not have been started without technology from their source organization, with an additional 13 percent testifying that an important aspect of the company's work originated at the previous employer. This dependency is comparable

Table 4-1
Degree of Importance of Technology Transfer
($n = 125$ MIT spin-offs)

| | Distribution | | Impact | | |
	No. of Companies	% of Companies	Weight	Weighted Score	% of Weighted Score Total
Direct	32	26	3	96	46
Partial	34	27	2	68	33
Vague	43	34	1	43	21
None	16	13	0	0	0
Totals	125	100		207	100

to the 14 Canadian high-technology firms where 13 had initial technology similar to that used in the last prior job of the founder (Doutriaux, 1984, p. 374). Furthermore, *The Economist* (1983) indicates that a study of 182 companies created in Silicon Valley between 1977 and 1982 shows that 75 percent of their founders had already been working on the technologies that form the core of the firm's knowledge base and 54 percent on similar products. In addition Cooper and Bruno (1977) demonstrate that the high growth companies among 250 high-technology firms founded in the San Francisco peninsula during the 1960s utilize the same technology and serve the same markets as their prior organizations. This finding is reaffirmed by Feeser and Willard (1990) in their comparison of *INC. 100* computer firms with a matched set of low growth companies.

Table 4–1 indicates that only 13 percent of the MIT spin-off companies have no technical dependence whatsoever on the spawning source laboratory, whereas 53 percent are critically or importantly dependent on laboratory-learned technologies. Some of the cases of "no transfer" arise from the tight definition of "spin-off" employed in the research, which forces inclusion in the samples of some companies in which the individual from the source organization studied is only a weak member of the founding group. Other cases are more straightforward in that the company is working in areas totally disassociated from work carried on in the so-called source organization. For example, 30 percent of the spin-offs from the diversified technological corporation list earlier work experience elsewhere as the primary technology source for their new enterprises; another 18 percent identify education, hobbies and other sources of the initial base technology. Due to lack of information on the many co-founders who did not work previously in one of the source organizations studied, the measure of technology transfer explicitly ignores any role of those co-founders in transferring technology from their prior employers. It, therefore, understates the over-all dependence of new firms on technologies transferred from some founder's source organization.

To provide some feeling for the relative impact of the source tech-nologies on these 125 companies as a whole, my assistants and I devised a weighting scheme that social scientists would point out is not rigorous. Nevertheless, as indicated in the right half of Table 4–1, weights ranging from zero to three are assigned (somewhat arbitrarily) to correspond with the indicated degrees of technology transfer from none to direct. If all 125 companies are directly dependent on source-learned technologies, the overall impact rating would be 375 (3×125 companies). As indicated in the "Weighted Score" column, when each company is scored according to the importance to it of its original source-derived technology transfer, the overall impact rating for all companies is 207, or 55 percent of the maximum possible dependence score. This measure, while by no means conclusive, does suggest that the effect the spawning labs have had on the technologi-cal basis of all new enterprises formed by their alumni is rather significant. This measure also suggests, as tabulated in the last column, that nearly half of the measurable technology transferred is manifested in only one-quarter of the companies—not surprisingly, those companies most directly techni-cally dependent on the several source organizations as their original tech-nological bases.

The importance of transferred technology from industrial source or-ganizations is more varied than from the MIT labs and departments. Direct technological dependence accounts for only 16 percent of the spin-offs examined from a large electronic systems company, with partial transfer representing an additional 24 percent, according to the entrepreneurs' own estimates. These estimates may be prejudiced downward, however, due to these entrepreneurs' concern about possible legal actions by that corporate source organization. The Legal Department of that electronics company told me of relationships with 12 of the identified 39 spin-offs, including lawsuits, licensing agreements, and waivers of rights to technology. Most businesses educate large numbers of people in developing and using complex technology. But these firms usually do not want these people transferring outside the walls of the company the knowledge they have learned inside. The typical firm wishes to protect its huge investments by preventing this knowledge from leaving the company. Most business is fearful of competitors acquiring knowledge of vital inside operations and proprietary technology. Many companies are just as concerned about the people who leave and utilize the technology they have learned to form profitable enterprises in competing or even noncompeting areas.

In contrast, the large diversified technological corporation also studied was apparently less concerned about the commercialization of its spun-off technology. Indeed, 48 percent of its spin-off founders indicate that his company could not have started in its present form without technology from the parent (direct transfer), while 17 percent indicate partial technological dependence. Perhaps inappropriate for a profit-making firm, the diversified technological firm's attitudes are quite similar to MIT's strongly positive historic orientation toward its scientific and technological advances being transferred out to industry and society. The MIT policies are obviously the ones of most direct impact on the 125 firms in the primary sample.

This discussion does not address the question of the financial value of this transferred technology. Nor does it treat what impact this initial technology might have on success or failure of the new firm. The latter issue is addressed in Chapter 9. In my study of new biomedical companies, however, the intensity of technology transfer from prior source organizations does correlate positively and significantly with the new firm's products embodying a new technology or being first of a kind ($p < 0.08$), both representing clear potential competitive advantages to the young high-technology firm.

Sources of Advanced Technology

Having established that transferred technology serves as a critical base for many new enterprises, I begin here to test the model presented in Figure 4–1 and to present the results. Data do not exist to verify all elements of that implicit set of hypotheses, but much of the figure can be examined with information gathered from the research.

Development Work

The source organizations studied all engage in a broad spectrum of research and development work. But Chapter 3 demonstrates that technical entrepreneurs are especially biased in their prior orientation to developmental activities, and that they tend to reflect the more advanced development aspects of even the more research-oriented labs.

Now, as hypothesized earlier, I move the argument one step further to claim that development work, not research, is the much more fertile bed of immediately applicable advanced technology that potentially can become the basis of a new company. Only in the rare circumstances of a major breakthrough, such as the transistor, does research become the immediate basis for a product-oriented firm. Development, in contrast, takes research results and new technical knowledge and advances them toward and into application and use. Indeed, the recent formation of numerous new firms in the biotechnology and genetic engineering field reinforces in a rather unique manner the evidence for this developmental dependency. The biotech firms, most of them founded by Ph.D. researchers, have all had to go through a prolonged and extensive R&D stage, after formation of the firm, to bring their prior knowledge base to a state of development adequate for product development and release.

To support this argument I examine first those entrepreneurs who had no delay between leaving the source labs and starting their own companies. Table 4–2 displays the level of technology transfer associated with the nature of work that the entrepreneur had performed at the source organization for fifty-one entrepreneurs who went immediately into their new enterprises. The nature of prior work was measured by using the "Bullpup ratings", defined in the classic Project Hindsight study by Sherwin and Isenson (1967). Those ratings go from 1, basic research, to 9, prototype development, with carefully defined gradations along the scale (see Chapter 3 for further detail). As the entrepreneur's prior work becomes more developmental, an increasing proportion of direct technology transfer occurs (4 of 13 = 31 percent, 7 of 11 = 39 percent, and 11 of 20 = 55 percent, in order). This parallels an increasing proportion of summed direct and partial technology transfer as well (54 percent, 67 percent, and 85 percent, respectively). The type of work done at the source organization, the base for potential applicable technology, significantly affects the level of technology transfer into the new firm ($p = .025$). Thus, a double filter is at work: More development-oriented individuals become entrepreneurs; and more development-oriented entrepreneurs establish companies that depend more directly on technology transferred immediately into their companies.

Viewed slightly differently, research-oriented people who establish new companies have less work-derived technology to apply immediately in their new settings. They must draw from a broader array of other, mostly older, research results to generate a base for a new firm. Alternatively, they must have longer time available, either before starting a firm or once the firm is underway, for their more recent research to be brought into development and application. This corresponds in direction but not degree to the Project Hindsight findings that little Department of Defense nonmission research had found its way into operational military use within a twenty-year period of the research being carried out (Sherwin and Isenson, 1967).

Table 4–2
Nature of Source Work and Technology Transfer for
Immediate Company Founders ($n = 51$)

Nature of Prior Work* (Bullpup ratings)		Technology Transfer*				
		Direct	Partial	Vague	None	Total
Research	1 – 3	4	3	4	2	13
	4 – 6	7	5	5	1	18
Development	7 – 9	11	6	3	0	20

* Kendall tau = .25, $p = .025$.

Table 4–3
Nature of Source Work and Technology Transfer for
Delayed Company Founders ($n = 41$)

Nature of Prior Work	Technology Transfer		Total
	Direct–Partial	Vague–None	
Research (1–3)	4	3	7
Development (4–9)	12	22	34

Another cut at the data bolsters this argument. Table 4–3 shows the nature of source work versus degree of technology transfer for those entrepreneurs who did delay after leaving the source organization before setting up their new firms. Researchers who wait have *relatively* more apparent opportunity for meaningful transfer than do development types who wait. Research results, being more basic, "age" less rapidly than development; if anything, the research results may become more relevant and applicable with the passage of time. Tables 4–2 and 4-3 document that of those who did research, 54 percent (7 of 13) who immediately set up companies experienced direct or partial technology transfer and 57 percent (4 of 7) of those who waited had similarly high transfer. But development-stage work becomes more obsolescent and less applicable to the new firm, if the developer delays. Of those who had engaged in development 76 percent (29 of 38) who immediately set up companies experienced direct or partial transfer, but only 35 percent (12 of 34) of those who waited had similarly high transfer.

Degree of Exposure

A number of different factors might affect the degree of exposure by a "would-be" entrepreneur to potentially applicable technology. The nature of the work itself (such as indicated in the previous discussion on research versus development), the person's responsibility for that work, the duration of relationship to the work setting all combine to produce exposure to advanced technology.

Years at the Source

Due to data limitations the only "exposure-related" assumption tested is that the amount of time the potential entrepreneur works at a source organization relates to the degree to which he learns and probably transfers that organization's technology into the new enterprise. Remember that the measure of degree of technology transfer really measures technical dependence. However, individual differences might produce differing patterns of learning over time. Some individuals are rapid learners, quickly absorbing most of what is known within their environs, reflected by the top curve in Figure 4–2. Others tend to have a more steady pace of knowledge absorption or even a pattern of learning that starts slowly and gradually accelerates, indicated by the two lower curves in Figure 4–2.

Analysis of the relationship between the number of years a MIT spin-off entrepreneur spent at the source laboratory and the degree to which he transferred its technology produces statistically significant results (Kendall tau = 0.17, p = .03). Moreover, tests of several possible curvilinear relationships, approximating the top and bottom curves in Figure 4–2, fail to generate significant fits with the data. Similarly, those electronic systems company spin-offs who spent longer in that company report that their work at the lab caused them to learn more technology as opposed to just applying their prior knowledge (.03). Those longer employed entrepreneurs also established new enterprises based on more direct technology transfer than did their shorter employed colleagues (.03).

Part-Time Founding

Another aspect of exposure might be its timing relative to the time of possible use in the new enterprise. Over half of the companies, perhaps a surprising number, were started on a part-time basis, with the entrepreneur "moonlighting" in the new firm while still working "full time" in the source organization. "Part-timers" have concurrent exposure to the needs of the new firm and the solutions from the prior (i.e., also still current) organization. Part-time entrepreneurs from MIT labs engaged in more direct technology transfer to their new operations than full-time entrepreneurs (.03). Furthermore, in the biomedical firms I studied the duration of part-time

Figure 4–2
Possible Patterns of Individual Learning

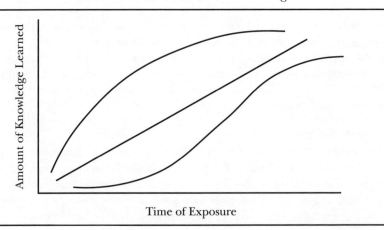

Amount of Knowledge Learned

Time of Exposure

commitment was longer on average, 30 months, than in any other of my entrepreneurial samples. In those medical firms too, part-time founded companies had higher degrees of advanced technology coming from university and other source organizations (.05). Logically, based on exposure to market needs, or selfishly, based on conservation of effort, the part-time entrepreneur who is still working at a source laboratory should be trying to gain an advantage by making his efforts work for double purposes. He is more likely to transfer into his company technology related to lab projects on which he is currently working and other ideas that are current in his laboratory. His double-hatted role facilitates the part-time entrepreneur becoming an effective agent of technology transfer.

Personal Ability and Attitude

One factor no doubt affecting learning rate is the individual's ability to perceive, comprehend, and extrapolate the relevant technology. While ability might be measured in many ways, looking at the entrepreneur's education as a possible surrogate for technology transfer ability does seem reasonable. Furthermore, possible effects of aging on professional ability to acquire and move technology should also be assessed.

Formal Education

The measure of degree of technology transfer used here does not differentiate on relative sophistication of the technology, only on the receiving company's relative technological dependence. In any event formal educational attainment is not necessarily a valid representation of an individual's capacity to learn or of his acquired fund of knowledge. For example, in at least one new enterprise a technician with no formal education beyond high school supplied the technical basis for a sophisticated product in a new technological area for his company.

Despite these qualifications the data were tested for the association between educational level and technology transfer and a weak positive relationship was found (Kendall tau = 0.13, $p = 0.07$). But in Chapter 3 I show that technical entrepreneurs cluster around the master's degree educational level. If that clustering also affects the degree of source technology exploited through transfer to the new firm, a quadratic relationship should be expected, as pictured in Figure 4–3.

This hypothesis led to regrouping the data to separate moderately educated technical entrepreneurs from those with either lower or higher education, as indicated in Table 4–4, clustering the groups to equalize as closely as possible the totals around the margins. Statistical testing (i.e., the F-test) supports the notion that entrepreneurs with Master's degrees (more or less) transferred the most technology ($p = .05$), consistent with the general shape of the relation pictured in Figure 4–3.

Figure 4–3
Possible Relationship between Educational Level and
Technology Transfer

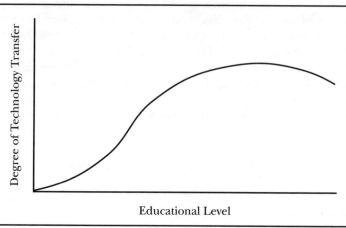

Age

Just as education (up to a point) is supportive of an individual's ability to comprehend and his likelihood of transferring new technology, age may well suggest limitations on these abilities. Despite its controversial nature, much evidence indicates that a person's technical abilities tend to peak between 30 and 40 years of age and begin to wane after that. In any event Chapter 3 shows that the ages for entrepreneurial behavior have similar characteristics.

Table 4–4
Relationship between Grouped Educational Levels and
Technology Transfer *

Educational Level	Degree of Technology Transfer		Total
	Direct–Partial	Vague–None	
Moderate education (B.S. plus courses, M.S., and M.S. plus courses)	33	24	57
Lower and higher education (college without degree, and B.S., plus professional engineering degree and Ph.D.)	19	23	42
Totals	52	47	99

*Group boundaries established to equalize as closely as possible the marginal totals.

Younger people, being closer to new technological developments, might transfer more technology than older entrepreneurs. On the other hand, older people have experienced more technology over their professional lives and, therefore, presumably have greater mental "storehouses" of technology, as well as perhaps more mature perspectives on market opportunities for applying (i.e., transferring) those technologies. One hundred-fifteen entrepreneurs are grouped in Table 4–5 according to their age at the founding of their companies and their level of technology transfer. The greatest proportion of direct and partial transfer occurs in the age range of 26 to 30 years; the lowest is in the range of 46 to 50. Direct and partial technology transfer statistically decrease as age increases ($p = .03$). Closer scrutiny of Table 4–5 indicates that this result is dominated by the strong differences among those under 30 years, with essentially no difference between direct–partial and vague–none in the over-30 categories.

Still further analyses of the data show that this technology transfer effect arises in part from a reluctance of the older men who do eventually become entrepreneurs to leave the source lab and immediately set up their own companies. Those who leave the lab, take another job, and later start their own firm are older by the time they become entrepreneurs. Even within that group of delaying company founders (illustrated in Table 4–3), age negatively and significantly correlates with degree of technology transfer (tau = minus .26, $p = .01$). Aging does appear to have some effect overall on the extent of transferred technology, but apparently a restricted one.

Attitude

Beyond issues of ability, an individual's attitude toward his or her work is hypothesized to play an important role in the learning process. A person who finds work challenging and enjoyable can be expected to learn more

Table 4–5
Age of Entrepreneur and Technology Transfer ($n = 115$)

Age (yr)*	Technology Transfer*			
	Direct	Partial	Vague	None
21–25	—	—	1	—
26–30	9	10	4	1
31–35	10	10	13	6
36–40	6	7	7	3
41–45	4	2	5	2
46–50	—	2	4	1
>50	2	2	2	2

*Kendall tau = 0.15, $p = 0.03$

easily and develop skills more readily than a person who finds work boring or painful. In addition, a person who likes what he or she is doing probably wants to continue that type of work in future employment or self-employment.

When applying this consideration to technology transfer, I expect that those entrepreneurs who found their source work challenging and enjoyable would reflect their source technology more in their new companies than those who did not. Table 4–6 shows how ninety-four entrepreneurs subjectively rate the challenge of their source work. The sole entrepreneur who felt his work was not challenging transferred none of the source's technology. As the rating of challenge increases, the proportion of direct and partial transfer of technology generally increases into the new enterprise ($p = .001$).

This same relationship is found between technology transfer and satisfaction with work. Table 4–7 displays data for 97 entrepreneurs. As their ratings of their earlier source lab work experience rise from hate and dissatisfaction to much satisfaction and very enjoyable, an increasing proportion of the entrepreneurs transfer source technology directly or partially into their new companies ($p = .015$).

The challenge of the work and the satisfaction and enjoyment in doing the work do strongly relate (tau = .60, $p < .001$). Consequently, the entrepreneur's overall attitude toward his work has much to do with the level of technology he transfers into his new company. Partial correlation analyses indicate that the link between challenge and transfer (0.19) is far stronger than that between satisfaction and transfer (0.04). It is worth noting that the causal relationship between technology transfer and perceived work satisfaction might be in the opposite direction. An entrepreneur who, upon

Table 4–6
Challenge of Source Work and Technology Transfer
($n = 94$)

Challenge*		Technology Transfer*				
		Direct	Partial	Vague	None	Total
No challenge	1	—	—	—	1	1
	2	—	1	2	3	6
	3	1	—	—	—	1
	4	3	3	1	—	7
	5	4	6	11	3	24
	6	9	7	13	3	31
Very challenging	7	11	8	4	—	24
Total						94

* Tau = 0.30, $p = 0.001$.

Table 4–7
Satisfaction and Enjoyment of Source Work and
Technology Transfer ($n = 97$)

Satisfaction and Enjoyment[*]		Technology Transfer[*]				
		Direct	Partial	Vague	None	Total
Much dissatisfaction and hate of work	1	—	—	—	—	0
	2	—	1	—	—	1
	3	—	—	1	—	1
Neutral	4	2	1	1	3	7
	5	2	4	2	6	14
	6	13	8	17	3	41
Much satisfaction	7	11	11	10	1	33
Total						97

[*] tau = .21, p = .015

retrospection, senses important transfer of technology from his previous job to his new enterprise might well more appreciate that prior job and express strong satisfaction with it.

Dissipative Influences

Thus far I have examined a number of forces hypothesized to contribute a base and learning to help establish the potential entrepreneur's usable store of technology. But technology does not remain ripe for plucking. Some of it decays in utility by becoming less timely; other ideas or skills are just forgotten or lost over time. This section discusses what one colleague has labeled "the half-life of technical information" and its impact on the technological base of the new enterprise.

Years Between Source and New Enterprise

A common business practice is to have employees sign restrictive contracts that, among other things, usually forbid the employee, upon termination of work with the company, to work for a competing company or to enter into any business in which he can utilize the methods and knowledge he has acquired. This restriction generally holds for one to three years after termination of employment. While many reasons are given for having this restrictive period, one of the prime reasons is that a company does not want an employee to use its own technology to compete against it. Companies feel that a period of exclusion considerably reduces the employee's capability to effectively transfer company technology.

The information collected on the spin-off entrepreneurs strongly supports this contention, even though many of the entrepreneurs come from university laboratories that do not follow these restrictive practices. Figure 4–4 (a repeat of Figure 3-5) shows the incidence of new enterprise formation over time, following departure from a source organization. Sixty-three of the 121 spin-offs in this grouping were formed immediately after the entrepreneur left the source or while he was still at the source. Eighty-five were formed within four years after leaving. I will comment on the causes of this time delay in the next section.

In Table 4–8, 118 of these companies are grouped by the time of company formation, after termination of source employment, and by their degree of technology transfer. The data displayed need little interpretation. Of the thirty-two companies that directly utilize source technology, thirty (94 percent) were formed within four years after their founder left

Figure 4–4
Rate of New Company Formation Related to the
Number of Years after Termination of Employment at the
Source Organization

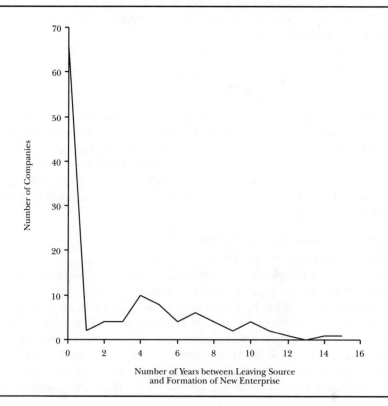

Number of Years between Leaving Source
and Formation of New Enterprise

the source. Of the thirty-three that partially utilize source technology, twenty-eight (85 percent) were formed within four years after source employment was terminated. Viewing this in another way, 36 percent of the companies formed within four years (30 of 84) directly utilize source technology and 33 percent of them (28 of 84) partially utilize that technology. In comparison, only 6 percent of the companies formed after four years directly utilize source technology and only 12 percent partially utilize that technology. The relationship between time after leaving and degree of transfer of technology is very strong (tau = .37, $p < 0.001$). As time elapses, the technological advantage that a scientist or engineer possesses at a frontier R&D organization is in part lost. The individual's technological knowledge from lab employment becomes less relevant, and the degree of transfer necessarily decreases as time away from the source lab increases. (In the next section of this chapter I consider the possibility that the direction of causality assumed here, that time lag affects technology transfer, might indeed be reversed.)

Another perspective on this same issue is provided by data from 23 product-oriented firms spun-off from the diversified technological corporation. As shown in Table 4–9, as each next major product is released, the passage of time leads to the product's decreased dependence upon source organization technology, with presumably (but not documented) increased dependence upon technologies generated by the new firm itself post-founding.

The potential for losing technological competitive advantage due to increased delay should specifically reflect the relative advancedness of the source organization. The research data indicate that the prospective entre-

Table 4–8

Years between Source and New Enterprise vs. Technology
Transfer ($n = 118$)

Number of Years[*]		Technology Transfer[*]			
		Direct	Partial	Vague	None
Immediately	0	25	19	13	5
	1–2	2	3	2	—
	3–4	3	6	4	2
	5–6	1	3	6	2
	7–8	—	1	9	1
	9–10	1	1	1	3
	> 10	—	—	3	2
Totals		32	33	38	15

[*] tau = .37, $p < 0.001$

Table 4–9
Product Dependency on Source Organization

	Product 1	Product 2	Product 3
Number of firms	23	13	6
Years until launch	0.7	5.3	7.0
Source technology used	57%	35%	4%

preneur may reinforce rather than dissipate the knowledge he gained at a university or other advanced laboratory if he next works for another university or nonprofit government-sponsored laboratory, rather than for a commercial concern. The degree of technology transfer is strongly but inversely related to the number of years of commercial experience between the source laboratory and the new enterprise. Practically no relationship exists between the degree of technology transfer and the number of years of noncommercial experience between the laboratory and the new enterprise. Indirectly, this notion supports the hypothesis that commercial experience has more of a decaying effect on the degree of source technology transfer to the new firm than does noncommercial experience. Unfortunately, no data were collected on whether important other technologies are transferred from the intervening organization(s) to the new enterprise.

My separate study of biomedical firms, however, finds that ongoing contacts with clinical environments, after the formation of the new firm, link to the development of products that incorporate newer technologies ($p = .02$) and/or special specifications (.03). But those analyses also determine that the initial spin-off transfer by the founders is far more important to the biomedical companies than the later continuing technology transfer.

Opportunities for Technology Use

Nearly half of the spin-off entrepreneurs engaged in other activities between the time they left their source work and the time they formed their companies. Reasonably this group of entrepreneurs had different motives for leaving the sources and possibly were attracted to business ownership for different reasons than the group that left and immediately formed their companies. Surprisingly each group, the "immediate" entrepreneurs and the "delaying" entrepreneurs, mentions nearly identical reasons for leaving their source work, and both groups give similar reasons for being attracted to business ownership.

Technology as a Factor in Leaving

The only difference found between these two groups of entrepreneurs is the importance of the source technology as a factor in their leaving the source organizations. Table 4–10 displays those data. Over half of the entrepreneurs (27 of 52) who left the sources and immediately formed their companies state that the utilization of technology was the prime reason for their leaving. In comparison only 20 percent of the entrepreneurs (9 of 44) who formed their companies later state that technology utilization was an important factor. When asked if they would leave without the technology, 70 percent of the entrepreneurs who did not immediately form their companies say "yes" while only 36 percent of those who immediately entered their new ventures respond affirmatively.

As expected, those entrepreneurs who state that the use of laboratory technology was their prime reason for leaving formed companies with higher degrees of technology transfer from the source organizations ($p = .001$). Those persons also had longer laboratory employment (.02) and a shorter average time lag between leaving the lab and setting up their own company (.002).

These findings are reflected nearly across the board in all the spin-off samples. Among the entrepreneurs who left the electronic systems company, many had received patents for advancing laboratory technology. Those patent holders transferred technology to a significantly greater extent to their new firms ($p = .04$), sometimes with licenses from the source corporation, an explicit form of recognition of opportunities inherent in the new technology. Another example of identifying opportunities for technology use is provided by 60 percent of the spin-offs that had started directly from the diversified technological corporation, where the founder's decision to start a company depends on his knowledge of a product or service that he felt was not being adequately developed or commercialized by the source. In these cases, each founder usually took some or all of the product's technology with him to the new business, usually by agreement with the

Table 4–10
Technology as a Factor in Leaving Source Employment
($n = 96$)

| Question | Went Into Business | | | |
| | Immediately | | Later | |
	Yes	No	Yes	No
Use technology as prime reason for leaving source*	27	25	9	35
Would leave source without technology †	18	31	26	11

* Differences in groups: $X^2 = 8.77$, $p = .005$
† Differences in groups: $X^2 = 8.20$, $p = .005$

source corporation. The evidenced importance of untapped new product opportunities in company formation agrees with venture capitalist Don Valentine's Silicon Valley experiences that are cited in Chapter 3.

At times the need for specialized equipment or materials to conduct lab R&D work has led the technologist to develop such materials for himself. Eric von Hippel (1988) has documented this as leading often to user-based innovation. It also generates an understanding of an opportunity and the means to satisfy that opportunity that can produce an entrepreneurial start-up. David Kosowsky required quartz crystal filters for his research in the area of statistical communication theory at the MIT Research Laboratory for Electronics. In seeking to satisfy his requirement he developed a means of producing these filters that led directly to the business he subsequently formed, Damon Corporation. And "Doc" Edgerton's creation of stroboscopic photography techniques to carry out his dissertation led directly to the formation of EG&G, as described in Chapter 1.

The significance of technological opportunity as a motivator of the potential entrepreneurs' behavior, as shown in Table 4–10, provoked a challenge of the presumed directions of causality in the hypothesized flow model of Figure 4–1. Does seeing a technological opportunity change the behavior of prospective entrepreneurs? Most of the influences on technology transfer that are assumed in that diagram must remain unmodified following this reexamination. Indeed the perceived current use for technology cannot cause the entrepreneur to perform development rather than research work at the source lab some several years earlier, as shown in Table 4–2. Nor does immediate opportunity cause him to shorten his overall duration of lab work, which might be hypothesized. In fact, the opposite is true; those with longer lab experience transfer more technology *and* also saw exploitation of lab technology as their prime motivator for setting up a new firm (as indicated earlier). Nor can the current desire to use lab technology have had a prior effect on the entrepreneur's education or age, both shown previously as influences on technology transferred to the start-up firm.

Sensing an entrepreneurial opportunity arising from a source lab technology might have one important effect. It might cause a potential entrepreneur not to seek additional work experience upon departure from the lab, but rather to establish his own enterprise immediately. With this perspective the data contained in Table 4–8 are reinterpreted in Table 4–11 to indicate how the extent of perceived possible dependency of a new enterprise on source lab technology might cause shortened delay in setting up a new firm. In this format the data clearly indicate (rank correlations = 1) the potentially strong influence that perceived opportunity for technology exploitation might have on the timing of setting up a firm. And of course the more immediately the firm is established after the individual's departure from the source organization, the less possibility for technology dissipation to occur, as suggested by the lower left dotted arrow in Figure 4–1.

Table 4–11
"Perceived" Technological Dependency of
New Firm on Source Technology and Delay
in Starting Enterprise ($n = 118$)

Technological Importance	Average Delay (yr)	Median Delay (yr)
Direct	0.7	0
Partial	1.8	0
Vague	4.1	4.5
None	4.9	5.5

Summary and Implications

This chapter attempts to explain empirically the initial technological base for the spin-off new enterprise. A conceptual model is presented of a variety of hypothesized influences on the flow of technology from an advanced R&D source organization into a newly founded company. The measure used to assess the final technology transferred is a four-point scale of importance of the technology to the new firm: direct, partial, vague, and none. Evaluating the spin-offs with this measure generates a broad distribution of outcomes, but with perhaps half of the manifested technology transferred into about one-quarter of the new companies.

The results of the tests of the implicit hypotheses embodied in Figure 4–1 are synopsized in Table 4–12. Of the influences listed, only the negative effect of aging seems relatively weak in its impact.

Development-oriented work at the source organization, not research work, is the primary origin for most of the transferred technology. A double filter is in effect: More development-oriented individuals become entrepreneurs, and more development-oriented entrepreneurs transfer technology immediately into their own firms.

Greater exposure to the technological source through longer years of service at the labs leads more frequently to significant technology transfer to the new company. This effect is amplified if the entrepreneur starts his company on a moonlighting basis, working part-time in the new enterprise while continuing full-time in the source lab.

Personal ability to perceive, understand, and apply advanced technology, reflected in an advanced formal educational level, is supportive of entrepreneurial technology transfer, although some negative effects seem to be provided by Ph.D.-level education. Somewhat countering the positive influence of education is the weak but negative influence of an

Table 4–12
Characteristic Influences on Degree of Technology
Transfer to the New Enterprise

Source of Advanced Technology
  Development-oriented work at source organization

Exposure to the Technology
  Longer service at source
  Part-time founding

Personal Ability and Attitude
  Moderate educational level
  Negative effect of aging
  Sense of challenge and satisfaction with source

Dissipative Influence
  Years between source and new enterprise

Opportunities for Technology Use
  Technology as prime reason for departure, with possible influence as well on time
  lag in setting up the new enterprise

---

entrepreneur's aging on his transferring source-learned technology to the
new company.

Individual attitudes reinforce abilities. Those who see their former
source organization's work as challenging and satisfying more often find
and transfer technology to their own start-ups. This interpretation is somewhat clouded by the fact that obvious success in technology transfer might
enhance one's feelings of satisfaction with a prior employer.

Far more significant than personal aging, the principal dissipative influence on technology transfer is a delay between terminating employment
at a source organization and establishing the new enterprise. The decaying
effect on the technological basis of the company is strong and nearly immediate, essentially full dissipation of transferability occurring within four
years after departure. As noted earlier part-time entrepreneurial work obviously minimizes the effective delay in technology transfer from a source
organization.

Finally, those who identify utilization of some technology as the prime
reason for leaving the source organization make highly important transfers
to their new firms. They sense and immediately attempt to apply commercially an advanced technology. In fact, if the perceived opportunity to
exploit lab technology affects the timing of new enterprise establishment,
then the dissipative effect of time lag described above is caused primarily
by the lack of apparent opportunity.

Prospective entrepreneurs who value the potential benefits of transferred technology must heed the importance of extensive participation in advanced development projects as the most likely seed beds of their firms. When combined with a sensitivity to market opportunities to exploit the technology, the entrepreneurial candidate should be able to move the technology rapidly into commercial application, avoiding the eroding effects of delay.

It is critical to point out in closing this chapter that few of the entrepreneurs left their sources with a product that had actually been developed at work. The technology transferred is advanced knowledge that they had learned and then applied in the creation of their new enterprises. The effect of this technology transfer on the success of the new firm is discussed in Chapter 9.

References

A. C. Cooper. "Contrasts in the Role of Incubator Organizations in the Founding of Growth-oriented Firms", in J. A. Hornaday et al. (editors), *Frontiers of Entrepreneurship Research, 1984* (Wellesley, MA: Babson College, 1984), 159–174.

A. C. Cooper & A. V. Bruno. "Success among High-Technology Firms", *Business Horizons*, April 1977, 16–22.

Jerome Doutriaux. "Evolution of the Characteristics of (High-Tech) Entrepreneurial Firms", in J. A. Hornaday et al. (editors), *Frontiers of Entrepreneurship Research, 1984* (Wellesley, MA: Babson College, 1984), 368–386.

The Economist. "The New Entrepreneurs", December 24, 1983, 61–73.

H. R. Feeser & G. E. Willard. "Founding Strategy and Performance: A Comparison of High and Low Growth High Tech Firms", *Strategic Management Journal*, 11(1990), 87–98.

Marc H. Meyer & Edward B. Roberts. "Focusing Product Technology for Corporate Growth", *Sloan Management Review*, 29(1988), 7–16.

F. J. Olleros. "Emerging Industries and the Burnout of Pioneers", *Journal of Product Innovation Management*, 3, 1 (1986), 5–18.

Edward B. Roberts. "Entrepreneurship and Technology: A Basic Study of Innovators", *Research Management*, 11, 4 (July, 1968), 249–266.

Edward B. Roberts & Oscar Hauptman. "The Process of Technology Transfer to the New Biomedical and Pharmaceutical Firm", *Research Policy*, 15 (1986), 107–119.

Chalmers W. Sherwin & Raymond S. Isenson. "Project Hindsight", *Science*, 156, 23 (June 1967), 1571–1577.

R. W. Smilor & M. D. Gill, Jr. *The New Business Incubator* (Lexington, MA: Lexington Books, 1986).

Eric von Hippel. *Sources of Innovation* (New York: Oxford University Press, 1988).

CHAPTER 5

The Financial Base of the New Enterprise

Entrepreneurial people provide the initiative, the energy, and the vision to launch a new company. Advanced technology often provides the unique competitive advantage over existing companies or the basis for creating a new market. But money provides "the grease", the wherewithal to make it happen, even for the high-technology firm.

As Rex Morey and his partner were pulling their downward-pointed radar rig along the back country road, they were not expecting to meet any prospective investors for their brand new start-up. But Bernie and Ned Shannon pulled their Shannon Brothers Contractors truck along side and asked if they needed any help. Morey said he was doing fine and, in answer to the Shannons' curiosity, explained that he was attempting to map the subsurface conditions under the roadway, using a high pulse radar derived from earlier research at MIT Lincoln Lab, further developed at EG&G. Quick to sense an opportunity, the Shannons eventually invested the first money in Geophysical Survey Systems, Inc. (GSSI), hoping to score big by being able to improve bidding estimates on excavation jobs through advance knowledge of the mix of gravel and rocks that would be uncovered. They also brought in as a substantial co-investor their friend Jim Grimes, owner of one of the largest gravel pits in New Hampshire. The first meeting of the new Board was held in my office at MIT, as I had been persuaded several months earlier by Morey's enthusiasm to help him as an initial investor and director. The next several years of association provided quite a roller coaster experience, as the company approached bankruptcy on several occasions. Four years after GSSI's founding, the original investors rescued it once more, the new financing occurring at only one-tenth the initial valuation. After years of technological and market frustration, marked by numerous conflicts and eventual resignation of the original founders, GSSI was finally sold in January 1990 to OYO Geospace, a Japanese corporation wanting to expand on GSSI's established and growing position in selling its radar systems for use in resource exploration. The early outside investors were rewarded for their patience with a compound annual rate of return of about 14 percent over nearly two decades.

In recent years numerous books and articles have been published on how to manage venture capital investing along with even more publications on how to raise capital for the new enterprise. The related research literature has been growing at a rapid pace, as indicated by the fact that 11 percent of the papers presented at the annual Entrepreneurship Research Conference since 1981 have focused on venture capital and an additional 6 percent on other aspects of financing (Hornaday and Churchill, 1987). A reasonable number of these papers have looked at technology-based firms and several contain data regarding initial financing of these companies.

Utilizing the research literature and my own experience, this chapter establishes a general background for understanding the financing of a technological firm by first discussing the several stages of a company's financial development and the variety of relevant potential financial sources. This discussion then leads to a focus on expectations as to the initial sources of capital for the technological enterprise. My research studies provide detailed data to verify these expectations and indicate the actual sources and extent of initial capitalization. Finally, I discuss factors that influence the initial financing and also test those hypotheses with the research data. Subsequent capital acquisition by the technical enterprise, including the search processes employed by the entrepreneurs in seeking venture capital, the role of business plans, the nature of venture capitalist decisions, and the decision to go public, is treated in detail in Chapters 7 and 8.

Stages of Financial Development

The new technology-based firm evolves through a succession of several stages of corporate growth and parallel development of its financial needs. The time during which a company can be classified in a particular phase varies widely among firms and the dividing line between phases is at best fuzzy. Yet, the relative stage of evolution does strongly influence the type and amount both of capital required and especially of capital available. Ruhnka and Young (1987) surveyed executives of 73 venture capital companies and found that in general they perceive five different stages of new company development, from seed to "exit" (going public or selling out), each stage characterized by distinctive features. Of most importance these potential investors estimate sharp decline in their risk of loss of investment, depending on the stage at which they invest, from 66 percent at the seed stage down to 20 percent at the exit stage (p. 181). Dean and Giglierano (1989) find similar results among the 38 California venture capitalists they surveyed, with estimated "complete losses" going from 33 percent of "founders round" investments to only 7.5 percent of financings at the mezzanine stage. Wetzel (1983) finds comparable changes in the perception of investment risk by private individual investors, with their expected failures ranging from 70 percent of their investments in "inventors" to only 20 percent of their investments in "established firms" (p. 29). To clarify the relationship between company evolution and financing I shall first examine the general characteristics of the firm at each stage, which in turn imply the nature of its likely financial backers. My empirical research is later used to illuminate these financing relationships.

Traditionally the technical firm has been visualized as going through a usually precompany R&D stage, followed by three phases of corporate development: (1) start-up, (2) initial growth, and (3) sustained growth. (Van de Ven et al. [1989] label the comparable stages of new business creation as initiation, start-up and take-off.) The R&D phase often takes place in the laboratory of some other source organization, as demonstrated in the preceding chapter, or the basement of a founder's home, often while the founders are still employed "full time" for another organization. It involves experimental verification of product principles and may include attempts to determine commercial applicability. Few resources other than the founders' time are generally employed at this "pre-venture stage". In recent years the financial community has become more involved in this precompany stage, often with university laboratories or their direct spin-offs, in the long-term funding of ambitious R&D programs with hoped-for commercial outcomes. Sometimes the mechanism of an "R&D limited partnership" is employed during this stage, especially in recent years with biotechnology companies, but the R&D stage (especially research) remains a largely precorporate or at least unfunded aspect of most new companies' formation and development. The R&D stage does sometimes overlap with the "zero stage" of start-up firms, as is discussed next.

Phase 1: Start-up

The start-up phase begins with the founding of the company and ends, more-or-less, when the company has experienced significant sales (at least a few hundred thousand dollars per year) and has developed one or more products or services that exhibit growth potential. In recent years the start-up phase has been subdivided conceptually into the "seed stage" or "zero stage" and the "first stage". During the so-called zero stage, the new company works out its basic technology, formulates its initial strategy, and rounds out the start-up team.

At the outset of its seed phase the company often lacks an operating prototype of its intended product and even has little in the way of a formal business plan. Many companies carry what I described above as a precompany R&D stage into this seed phase, continuing to solve key product development issues and moving toward an operating demonstration prototype of their initial product. Following the seeding activities is the more conventional first stage, during which the company generally has produced a reasonably well-defined business plan, an emerging organizational structure built up around several key committed people, and a product for which at least some level of commercial applicability has been demonstrated.

During the entire start-up phase the new technological firm typically devotes considerable time to product development. It is dealing with only a few customers but is actively seeking new marketing/sales channels. The firm is housed in modest facilities, using barely adequate equipment. It has little or no available financial collateral. Typically, few of the people in the company have substantial management experience; a large portion of the founders and their early employees are technical people by education and work experience. The company is able to react quickly when opportunities arise. However, the company is usually losing money.

The financial needs of the firm during the start-up phase are many. It needs capital to finance product development, primarily to support salaries for the technical personnel, despite the fact that many or all of them are being paid lower salaries than in their previous jobs, earning by their financial sacrifice the so-called "sweat equity" ownership in their company. Some capital is also needed for equipment. Working capital may be required if the company is already producing products for sale. As the company is losing money, the entrepreneur must turn outside the firm for capital.

But what type of investor is willing to supply initial capital to such a new company? Since the investment appears to be so risky, the potential payoff must be high to outweigh the perceived probability of failure. The capital source must be patient, willing to wait for five to ten years for a return. The investor must trust unproven management to develop, produce, and sell a product or service that often does not yet exist. Such an investment is viewed by many as analogous to putting several hundred thousand dollars or more on the Daily Double!

What many prospective entrepreneurs and investors do not realize is that few technology-based firms actually fail, that is, declare bankruptcy or close their doors under desperate conditions. Timmons and Bygrave (1986, pp. 163-164) list seven different studies that demonstrate complete failure to range from 14.7 percent to 35 percent in total over the first five to ten years of company life, with the mean being about 20 percent, dramatically different from the characteristically high and early failure rate of non-technical start-ups. But large numbers of these firms merely survive—staying in business, generating neither significant year-to-year growth nor decline, producing little profit. Under these more usual circumstances, the investor is not able to get his money back or any return on it. No one is interested in buying the company, the firm is not able to go public, and it has neither the motivation nor the cash to buy out its initial investors. This scenario produces what investors in new enterprises call "the living dead", that group of going no place survivors that constitute a large part of many investors' portfolios.

Phase 2: Initial Growth

For those firms that emerge from start-up, the initial growth phase can be said to begin when the company has completed the development of a product line and has sufficient sales to justify an expectation of rapid growth. The phase may be regarded to end when the company has lived up to such expectations and demonstrated a capability to operate profitably and grow quickly. During this phase the company matures somewhat. It begins to work on product quality and on lowering unit costs. Although it is gaining new customers, it also begins to face some competition from other small firms and sometimes from large corporations as well, giving the young company strong incentives to develop new products. The company is operating profitably, but the resultant cash flows are typically insufficient to supply the needed growth capital.

The problems that the firm faces are also changing. Plant and equipment are needed. Working capital needs are expanding with the growth in sales. Key management personnel are needed as production, sales and marketing, and R&D become important functional areas. Management and operations control become important to keep the company operating efficiently.

The type of financial backers that the firm attracts tends to change with the company's characteristics. The risk and uncertainty associated with the company decrease. The young company still offers the opportunity for a large payoff, and the probability of absolute failure has decreased further. The investment need not be locked in for more than two to three years, if the founders are willing and the financial market permits the company to go public or to be sold to a larger firm during these next few years. The company no longer needs a gambler to supply capital, but phase 2 investors must still be speculators at least over the near term. As Chapter 10 illustrates, however, many firms encounter difficulties in developing their second and later products, even following a successful initial product launching of the company, and are unable to transition to continuing growth of revenues and profits.

Phase 3: Sustained Growth

If it solves its initial start-up and early growth problems, the successful company emerges as a growth business. It has annual sales in the millions of dollars and employment numbers in the hundreds. The enterprise begins to face many of the problems of the large corporation but on a smaller scale. The firm serves many customers with a variety of products and services and faces strong competition. Profits and cash flows are sufficient to meet the majority of its capital requirements, but new growth possibilities are continually being presented. Indeed, growth rate of the company may be the source of its most serious challenges, including financing the growth.

The major problems facing the entrepreneur change significantly during phase 3; he is now required to think about overall corporate direction, development of multiple product lines, employee morale, communications, and long-range planning. Potential merger or acquisition candidates present themselves; and the company itself is courted by larger corporations. Tax and legal considerations loom increasingly large. The entrepreneur may find himself no longer the central figure of the company and he may wish to sell his interest and retire or start over again. The company has ceased to be a new enterprise and has become a growth business, perhaps the IBM of the future!

Despite, indeed perhaps because of, its speculative future prospects, the company has undoubtedly become attractive to the public. If it had not previously issued stock publicly, it can now turn to the public financial markets with some degree of confidence. Long-term loans are also available since the company has sufficient assets to serve as collateral. The technological enterprise, through its ingenuity, efforts, persistence, and good luck, has stood the test of time and established itself as a going concern. Of course, lurking in the background are many of the earlier issues: Will the next product offering succeed? Might a competitor "blindside" the firm with a leapfrog technology? Will a group of key employees quit and start a new firm with the next generation product? Any of these threats could return the company to a less attractive stage of performance or at least induce stagnation of further growth prospects.

Financial Sources

A wide variety of financial sources are potentially available to fund the technology-based company's capital requirements through the successive stages of its growth and development. Each source theoretically considers the unknowable risks of failure, presumably a function of the stage of development of the enterprise, against the also unknowable prospects for financial return, and makes its investment decisions. Embryonic technological enterprises are quite different from most other new firms: Not only are they marked by high technical, market, and managerial uncertainty, they also lack tangible resources, as well as tangible measures for their early-stage performance. Many research- or technology-based companies start out with little more resources than an oscilloscope and a soldering gun or a magnifying glass. Many begin with only the entrepreneur's intelligence and drive as inventory. With little else for collateral, the entrepreneur's searches for funds from banks and other formal financial institutions are often fruitless. Alternatively, the term venture capital often comes to mind when thinking of the initial financing of new enterprises. And yet, it has long been true (and still is) that the bulk of financiers known as venture capitalists do not support the earliest stage of capital acquisition for the vast majority of technology-based enterprises. As I later show, both in this chapter and in Chapter 7, venture capitalists generally prefer later-stage investments in growing enterprises, not early-stage investments in technological start-ups (see Rubenstein, 1958, for historical evidence).

Despite all the attention paid to them, venture capital firms of all sorts (including "wealthy family funds", conventional venture capital firms, seed capital funds, Small Business Investment Companies, as well as financial and non-financial corporations operating their own funds), still account for only a small fraction of initial financing of new firms. At the extreme, Dunkelberg and Cooper (1983, p. 370) found less than 0.5 percent of their sampled companies were funded by venture capital firms. But this is in part due to their sample including "few of the growth-oriented, high-technology companies which are of particular interest to venture capital firms." Bygrave and Timmons found that the 1978 decrease in capital gains tax was followed by a sharp increase "in the proportion of ... first round (venture capital) investments in highly innovative technological companies ($p<.0001$) The proportions of seed-stage companies (also) showed an increasing trend ... ($p<.002$)" (1985, p. 115). More recent data on the entire venture capital industry (*Venture*, 1989, p. 55) show that only 2 percent of their 1987 investments went into so-called seed-stage firms and only 11 percent went into so-called start-ups, a significant decline from the 22 percent startup funding in 1981. Electronic Trend Publications indicates the percentage for start-ups declined even further in 1988 to 7 percent (Welles, 1990). In fact most of the receiving firms are well past the situation of initial funding. Fully three-fourths of the venture capital industry's investments now go for second-stage and later-stage financings or leveraged buyouts of established companies.

If not the venture capitalist, then to whom does the entrepreneur turn for funds to finance his dreams? Although initial requirements may be low, who is willing to gamble on the start-up's success? And once the future begins to look promising, where can the entrepreneur find several hundred thousand to a few million dollars of growth capital? The many classes of potential financiers for new technical enterprises are examined here in an effort to determine their resources, attitudes toward risk, selection criteria, preferred investment terms, and postinvestment relationships with the young technical firm. I consider them in the order of their general likelihood of being an initial investor in the new firm.

Personal Savings

Undoubtedly the most available source of capital to the entrepreneur is his personal savings. Indeed Dunkelberg and Cooper (1983) found personal savings, either alone or in combination with other sources, to be the primary source of financing for 59 percent of the 890 owner-started firms they studied in a wide variety of mostly nontechnical industries. Tyebjee and Bruno (1982) indicate similar dominance of personal savings in the funding of 185 California technology-oriented companies. And Smilor, Gibson and Dietrich (1989) confirm that founders used their personal funds to start 74 percent of the 23 University of Texas spin-out companies they studied. However, those savings are typically quite limited and the average individual scientist or engineer in his early 30s has difficulty in raising more than $25,000 to $50,000 on the strength of his savings account, his signature, and his available collateral. The entrepreneur must realize though that he may be required by other investors to gamble much of his own assets on his company as a sign of good faith. It is especially important that he and his co-founders own the bulk of the company initially, as later dilution of their ownership necessarily follows from the required raising of increasing amounts of outside capital. The entrepreneur should recognize that his potential capital gain is phenomenal if the company proves successful and that he should be risking much of his own "wealth" if the future looks bright.

Personal savings, therefore, are the foundation of initial capital. Usually additional funds are not needed for several months up to close to a year or more, depending on the scale of initial efforts. The entrepreneur can make many nonmonetary forms of investment in the company in the form of patents, developed products, and free labor, previously referred to as sweat equity. However, the assets of the entrepreneur are all too soon exhausted and he must then turn to outsiders for capital. If the entrepreneur is personally wealthy, from birth or from previous entrepreneurial success, the need for outsider investments may be delayed significantly.

Family and Friends

Next to personal savings the assets of an entrepreneur's relatives and friends are probably most available. The Dunkelberg and Cooper study (1983) finds that friends and relatives were the most important source of capital for starting 13 percent of the businesses in their survey of members of the National Federation of Independent Businesses. Such investments often take the form of short-term loans, although the loans may later be changed into "equity" investments at the insistence of subsequent investors. The main advantage of such funds is that they are relatively easy to get. The investors know the entrepreneur and have assessed his or her capabilities. Sometimes the entrepreneur, unsure of whether his venture will succeed, properly feels reluctant to "take advantage" of such close personal relationships to raise money. The major disadvantage if friends and relatives do invest is that they often feel that the investment gives them the right to advise or actively interfere with management. Therefore, although such "naive" money is relatively easy to obtain, problems may result from its acceptance.

One colleague reported that a venture capitalist friend in Palo Alto claimed he could raise start-up capital for a high-technology firm just by riding his bicycle around the neighborhood. This may be a function of what kind of friends and neighbors one has. Indeed Stephen Albano did raise the $30,000 needed to start Offtech, now Ricoh copier's major New England full-service distributor, by waving to his neighbor Bob McCray during a spring walk through their neighborhood in a small New Hampshire town (Logan, 1986). McCray turns out to be quite the opposite of a naive investor, falling clearly into the "angel" category discussed next.

Private Individual Investors, or "Angels"

The great majority of initial investing in high-technology firms by outside investors has traditionally been undertaken by wealthy individuals. In the jargon of the investment community, these private venture capitalists have been referred to as "informal risk capital investors" and often as "angels". Gordon Baty long ago characterized the traditional private investor as having a tax bracket favoring capital gains. Furthermore, being "accountable only to himself for his actions, he can afford the inevitable loss and he often has motivations for investing which are not strictly economic" (Baty, 1964). Noneconomic motivations include a sense of gambling, participation in an exciting growth company, especially the involvement with young bright people, and sometimes satisfying his sense of social responsibility, perhaps related to his wealth. Unfortunately the current lack of tax differences between regular income and capital gains may well affect this individual's motives and actions.

The private investor seldom seeks out investments in new enterprises. Instead he learns of opportunities from contacts within the financial community of which he is often a member. Investment bankers, commercial bankers, and brokers (plus his lawyer, accountant and countless friends and relatives) all refer companies to him. Occasionally the prospective individual investor participates in local groups like the MIT Enterprise Forum, where early-stage entrepreneurs present their aspirations and problems.

William Wetzel (1983, 1986, 1987) and his followers (Aram, 1989; Neiswander, 1985; Tynes and Krasner, 1983) have carefully analyzed this individual, whom Wetzel calls a business angel, and the market in which the angels operate. The private investor's or angel's resources are considerable, with their venture investment portfolios aggregating in the neighborhood of $50 billion according to Wetzel (1987, p. 412) and a study funded by the Small Business Administration (SBA) (IC^2 Institute, 1989, p. 40). Acting alone or through a syndicate of friends and acquaintances he can raise as much as $1 million for a given deal, although he seldom does. The several studies cited find that a large fraction of the initial deals are for $50,000 to $300,000, typically involving an angel and one or more of his friends, each putting up $25,000 to $100,000. Freear and Wetzel (1989) find that "nonmanagement private investors" previously provide equity capital for 35 percent of the companies that later receive venture capital funding. In those special cases their median investments amount to $225,000 if at the seed stage and more than double that if made at the later start-up or first round stage. These investors usually do not seek a controlling interest or management position in the company, but most prefer to be consulted on major management decisions.

Such investors rely heavily on the advice of their friends and other backers when making investment decisions. Few make a detailed analysis of the situation, evaluating the company primarily on the basis of its management, although Aram (1989) finds that angels in the Great Lakes region are frequently entrepreneurs themselves who tend to invest in companies related to their own expertise. The investments are usually straight equity (Conlin, 1989). Thus, the wealthy individual investor tends to qualify as the type needed in the company's initial phase. The entrepreneur needs only to find the right angel for his company. This is not easy, despite the computerized "matching network" created by Wetzel for private investors in the New England area. For a small fee entrepreneurs enter brief descriptions of their ventures and their financial needs into the computer data base, and are referred to the two or three closest "matches" of potential individual investors who have also listed their preferences in the computer program. This matching service has now been replicated in many other parts of the United States (IC^2 Institute, 1989, 1990; Logan, 1986). Patterns of private investor behavior seem comparable in California and New England (Tynes and Krasner, 1983) as well as in the midwest (Aram, 1989; Neiswander, 1985).

Wealthy Family Venture Capital Groups

More-or-less next in line, at least historically, in likelihood of investing at the outset of a technological enterprise is the formal private venture capital investment group established by a wealthy family. Shortly after World War II several wealthy families in search of capital gains created such organizations to invest family resources in young businesses, especially those based on advanced technologies. The largest of these groups, led by such people as Laurance Rockefeller, J. H. "Jock" Whitney, and Payson and Trask, became both well-known within the investing community and instrumental in funding numerous technological enterprises. Rather than invest informally and as individuals (as the angels above) those families usually funded an autonomous investing organ (corporation or partnership), managed by a staff of full-time employees who analyze incoming investment proposals, make the investment decisions (usually without family participation in the decisions), and work with the investee companies during the postinvestment period. Venrock, founded by the Rockefeller family, is probably the best known of the current survivors of these organizations.

As these family groups developed they evolved a certain style of operations that became the basis for today's U.S. venture capital industry, with resulting advantages and disadvantages to the entrepreneur who deals with them. The advantages to an entrepreneur who gets funds from such family groups are many. Other investors look more favorably at the new company because these larger family groups have a reputation of choosing only the best companies. This, of course, makes it easier to obtain additional capital later. Their resources are essentially unlimited, making it possible for the entrepreneur to come back later for more capital. The staffs of such family groups have had top quality reputations, with both business and technical expertise. The final advantage is that they are patient investors, willing to wait five or ten years for their returns, and they do not have to answer to stockholders or outside investors for their performance.

Some disadvantages are also identifiable with seeking investments from the organized family groups. They have been very discriminating in choosing their investments, investing typically in less than 1 percent of the proposals they receive; the entrepreneur must submit a detailed proposal (called a "business plan" and discussed in depth in Chapter 7) to be considered. The investors demand one or more positions on the board of directors of the company and detailed ongoing reports of operations. They may insist on placing a staff member in an operating position in the company if growth does not materialize or, even worse from the entrepreneur's perspective, they may step in and replace the founding entrepreneurial head of the company. They are also rather slow in reaching a decision, so the entrepreneur must approach them several months before he needs the money.

In evaluating a young company the aspect that seemed most important to these family groups during their early and formative years was the quality of the management, followed by the market for the product. They also consider the state of product development and the underlying technology. Investments are usually made in the form of convertible debentures, debt that can be converted into stock, providing some modicum of investor protection in the event of company liquidation. The size of their investments generally ranged in their early years between $300,000 and $500,000 and of course have grown in magnitude over the past forty years, but not by more than about a factor of two. The family venture capitalists often avoided initial financing, but now are willing to put in small sums as early-stage investments, especially in companies headed by entrepreneurs with whom they have had prior experience.

Venture Capital Companies

The family venture capital groups were the models for the formation of specialized closed-end investment companies that focused on venture capital (VC). The first of these was American Research and Development Corporation (ARD), organized in Boston in 1946, in large part through the efforts of the then Chairman of MIT, Karl T. Compton, and a number of prominent alumni and friends of MIT, to move research and technological ideas forward into the market. The heads of MIT's Departments of Chemical Engineering and Aeronautical Engineering acted as advisors, and the treasurer of MIT served as treasurer of ARD. ARD was funded initially with $3.4 million in investments primarily from several Boston-area insurance companies, joined as investors by MIT, Harvard, Rice Institute, and the University of Rochester. ARD later went public, sold out still later to become a division of Textron Corporation, and finally was sold by Textron to a member of the Mellon family.

Initially and for several years ARD invested primarily in ideas promoted by senior MIT faculty, housing those start-up companies in MIT buildings with a unique cost-sharing arrangement that only today is beginning to be replicated at other universities. This approach led to ARD's participation both in the formation of and investment in such early MIT spin-off companies as High Voltage Engineering and Ionics. Gradually ARD's approach changed under the guidance of Georges Doriot, a professor at the Harvard Business School who served as president of ARD from its beginning, who moved ARD's attitudes and policies toward company management into imitation of the larger family groups in almost every respect. A full-time staff of professionals annually screened hundreds of incoming proposals, giving careful consideration to perhaps 10 percent of them, and eventually investing in 2 to 3 percent of the companies. In its early days ARD usually took dominant stock ownership position in a company through an investment of $100,000 to $500,000 in the form of convertible debentures, with $200,000 buying 80 percent of High Voltage in 1946 and $100,000 gaining 75 percent of Ionics in 1948.

ARD's principal success by far is the $70,000 start-up investment, in 1957, that purchased 78 percent of Digital Equipment Corporation (DEC), that success dwarfing all other actions ever taken by ARD, accounting for 86 percent of ARD's total value distributed to its stockholders (Stevenson, Muzyka and Timmons, 1986, p. 383). Importantly, the uniqueness of this DEC investment performance is in character with distributions of investment returns in other venture capital portfolios, as evidenced by Huntsman and Hoban (1980): "This risk-return tradeoff is consistent with venture capital *folklore*, which suggests that about one in ten investments will be home runs and that these relatively few highly successful ventures will, in effect, carry the portfolio." Scherer (1965, p. 1098) has identified similar skewed outcomes for the profitability of patented inventions and Ravenscraft and Scherer (1987, p. 43) indicate comparable distribution of performance of go-go conglomerates. These results are far more skewed than the general Pareto notion of an 80-20 rule, that is, that 20 percent of the activities will generate 80 percent of the results. If also true of high-technology companies, then most new enterprise investors gain nearly all their returns from very few of their investments.

Following the lead of pioneers like ARD other professionally managed venture capital funds were formed, usually raising their money privately from wealthy individuals, banks, pension funds, and corporations. ARD itself acted as an incubator for many other Boston-based venture capital firms, providing the experience and contacts that led many of its alumni to form other venture capital organizations. Whereas ARD and its former employees had a strong bias toward companies located in the Greater Boston area, funds were formed in other parts of the United States with tendencies toward their own regional biases, such as that managed by Arthur Rock in the Silicon Valley area. Florida and Kenney (1988) trace the evolution of these regional venture capital complexes and their current behaviors as investors, especially in regard to high technology firms. Gradually these professional funds proliferated and came to dominate the venture capital sector, becoming far larger in magnitude of total funds managed and invested than the earlier-formed wealthy family funds. Venture capital companies, such as Hambrecht and Quist, Kleiner Perkins, Morgenthaler and Associates, and TA Associates became well-known for technology-oriented investments in particular. Recent analysis demonstrates "a core group of highly skilled and experienced venture capital firms accounting for a disproportionate share of highly innovative technological venture (HITV) investing. The 21 venture capital firms that were most active in HITV investing represent less than 5 percent of the 464 firms in our data base, yet they were involved in nearly 25 percent of all the investments in HITVs" (Timmons and Bygrave, 1986, p. 168).

In general the professionally managed funds follow patterns of investment analysis, decision making, and management similar to those practiced first by the family funds and then by ARD. Careful screening and selectivity characterize their investments, only 2 percent of the proposals received get a favorable response (Maier and Walker, 1987, p. 208). Venture capital firms aspire toward high rates of potential return to compensate their investors for the presumably high risks being taken, the returns sought rising significantly as they invest in earlier stages of the new firm (Dean and Giglierano, 1989). They devote considerable postinvestment effort to monitoring and actively assisting their portfolio companies. Indeed, founders of many high-technology companies "reported that they actively seek out those venture capitalists with noteworthy reputations for their nonmonetary, high value-added contributions to fledgling firms" (Timmons and Bygrave, 1986, p. 169). In recent years this initially U.S. phenomenon has spread globally and venture capital funds interested in investing in technology-based firms now exist throughout western Europe and Asia, albeit sometimes with very different operating styles (Maital, 1989; Smith and Ayukawa, 1989; E.V.C.A./Peat Marwick McLintock, 1990). New venture capital firms recently formed in Japan bring to over 100 the number operating there (Nikkei, 1990).

Among the hundreds of venture capital funds is a small group of so-called seed funds, like the Zero Stage Capital Equity Funds that I co-founded, that focus on investments primarily in the initial and early stages of technology-based firms. These seed or zero stage funds follow in the tradition of the earliest activities of the wealthy families and of ARD in helping to put together the start-up enterprises, working very closely with the company founders to round out their team, more sharply define their business objectives, help develop a completed business plan. Seed funds often provide more value in advice and "sleeves rolled up" assistance than in the capital itself. Responding to this need for a rather different set of skills, the original co-founders of Zero Stage Capital had backgrounds that were quite different from the usual VC financiers: Gordon Baty, Arthur Obermeyer, and I had significant multi-company experiences as founders, managers, and counselors of start-up high-technology companies; Paul Kelley and Joseph Lombard had long prior experience on the financing and advising side of new and young firms. With the same kind of pain and suffering faced by other pioneering entrepreneurs, we eventually raised $4.5 million from a group of Limited Partners in 1981 and began to invest not more than $150,000 of our seed funds in each new company we selected. Gradually we found that even "seedlings" need somewhat larger investments. More important was the need for filling their "experience gap". The resulting Venture Advisor concept that we evolved utilizes "an experienced business professional who joins the founding team of a new venture on a part-time basis, sharing his (her) knowledge, experience, and contacts to promote rapid development of the firm" (Baty, 1988). This is in the spirit of, but tends to go even farther than, the help provided by most other VCs.

Today, seed funds typically invest from $200,000 to $500,000 at the initial stage of a new company, with perhaps matching funds available for participation in a later second round of financing. Seed funds seldom have "deep pockets" (e.g., Zero Stage now manages $40 million in four pooled funds) and seek kindred spirits as investing partners to share the initial investment so as to ease the difficulties that might be experienced later in securing the hopefully much larger requirements of growth financing (Allen et al., 1989). One unique seed fund that has helped many Massachusetts start-ups is the Massachusetts Technology Development Corporation (MTDC), state chartered and funded during a period of low public availability of venture capital. It works very closely and effectively with other Boston area seed funds to help initiate and enhance early growth of local technical firms. Other state governments and regional development organizations have helped to establish seed capital funds to support the start-up and expansion of emerging firms, such as the Ben Franklin Partnership's role in encouraging the formation of Zero Stage Capital of Pennsylvania. One survey finds at least twenty-five states engaged in related activities (IC^2 Institute, 1990, p. 20). The fragile conditions of high-technology new enterprises at this seed stage cause Paul Kelley, the originator of Zero Stage Capital, to refer to the firms as "stomach burners" from the venture capitalist's perspective. He exclaims that the anguish and aggravation generated is such that "you have to contribute a little bit of your stomach on each one of these deals". Seed-stage investing is clearly not for the faint hearted.

Small Business Investment Companies

A special form of venture capital fund that was especially important in the United States during the early 1960s is the Small Business Investment Company (SBIC), enacted by Congress in the 1958 Small Business Investment Act. Private capital was given tax incentives and low interest leveraged loans from the U.S. government to invest in small business. Several hundred SBICs were chartered with combined assets of nearly $1 billion, but only 50 had assets greater than $1 million. The resulting generally small financial organizations invested heavily in real estate and the trade sector, with some of the larger ones investing in new technical companies. In this early period about 15 to 20 percent of the SBIC investments were made in early-stage companies that were less than one year old. Probably less than 10 percent of the SBIC capital was invested in technologically oriented companies. Overall, however, the SBICs did have significant impact due mainly to the fact that their funds were available during a time that was otherwise relatively dry of small business investing resources.

A recent analysis of SBIC activity shows that "the number of SBIC financings to firms 1 year or younger exceeded the combined financings of 3 year old and 2 year old firms" (Feigen and Arrington, 1986), indicating that they still serve an important role for young companies. However, the 66 independent SBICs control less 1 percent of the capital in the venture capital industry, restricting the magnitude of their overall impact (*Venture Capital Journal*, 1990, p. 12). Data are not available on how many of them focus on technologically oriented investments although personal experience supports that some large SBICs, such as Bank of Boston Ventures, have been active and important participants in investments in early stage technology-based firms.

Nonfinancial Corporations

Beginning in the early 1960s and increasing significantly only in the 1980s, major manufacturing firms have become interested in supplying venture capital to young technological companies. Many of them are seeking to supplement their in-house R&D efforts by backing entrepreneurs in hopes of gaining access both to technology and engineering talent. Initially, companies such as DuPont, Ford, Texas Instruments, and Union Carbide, experimented with this approach of direct venture capital investment in new or early stage companies. Later, Exxon, Inco, Lubrizol, and Monsanto demonstrated active and effective programs of investment that encouraged widespread participation by many Fortune 500 corporations. Today, 92 U.S. industrial corporations are managing $2.6 billion in dedicated venture capital pools, in addition to many more corporations that invest in pooled funds managed by professional venture capitalists (*Venture Capital Journal*, 1990, p. 12). The 3M Company recently revealed its program of extensive investments in 27 venture capital funds on a global basis, with $75 million committed toward gaining "windows on new technology" and providing bases for later alliances with emerging companies (Hegg, 1990). Still more corporations occasionally make a strategic investment in a new company but do not have ongoing venture investment activities.

Nonfinancial corporations differ significantly from the previously discussed venture capitalists in regard to their motivations, selection criteria, and attitudes toward the technological enterprise. Their prime consideration is usually technology. Most investing firms choose only a few technical fields in which to invest, sometimes related or complementary to their current lines of business, at other times wholly unrelated, depending on the corporation's present strategy of concentration or diversification. The quality of the entrepreneurial team is usually the second most important decision criterion. Corporations have tended to avoid providing initial capital, often because they do not see the opportunities soon enough or because they cannot act fast enough, instead preferring somewhat later growth financing. In recent years this tendency has changed somewhat, especially in areas of medical technology and advanced materials, where a number of nonfinancial corporations have developed close ties to venture capitalists that allow the corporations to see and participate in early stage financings. Japanese companies have aggressively begun to participate as investors in U.S. high-technology firms, especially in electronics and bio-technology, but these investments usually come after the start-up stage (Sun, 1989; Welles, 1990).

The nonfinancial firms often are willing to provide technical, marketing, and managerial assistance to the companies in which they invest, potentially more valuable than the funds themselves if these services can be accessed and utilized effectively by the investee. This assistance, which in earlier writings I have labeled "venture nurturing" (Roberts, 1980), plus very "deep pockets", may be primary advantages provided by the corporate venture capitalist, but combine with some potential disadvantages. The corporation may have a tendency to interfere more in the day-to-day operations of the young firm than the entrepreneurs find desirable. Furthermore, the corporate investor may oppose the firm "going public", preferring to merge it eventually into its own operations. Young entrepreneurs often think that going public is the ultimate measure of and route to personal glory and financial success, but the facts are that far more technical companies eventually sell out to larger companies than go public. Thus, the entrepreneur frequently is leery of corporate funding at early stages of the firm but becomes less naive and resistant to their funds and help as his company moves forward.

Commercial Banks

In some areas of the United States, commercial banks have taken an active role in supplying capital to new technical enterprises, even though the bank itself is restricted by regulations as to how it can invest. During the early years of a company the more venturesome banks supply short-term loans secured by projected accounts receivable based on contracts or orders received by the firm. These sometimes can get converted effectively into intermediate or even long term loans through constant renewals and renegotiation. Bruno (1986, p. 113) cites Ashton-Tate, the large software company, as having refused venture capital funds offered to it so as to avoid having to dilute the founders' ownership percentage, securing instead a $6 million line of credit from Bankers Trust. Banks can also help through mortgage financing of a company's real estate and long-term lease financing of its laboratory or manufacturing equipment. Bank-owned SBICs, discussed earlier, can of course become direct investors and the SBICs and/or bank commercial lending officers can assist in establishing relationships with conventional venture capital funds. Bank-affiliated venture capital funds account for almost $2 billion, approximately 6 percent of the venture capital industry's total resources (*Venture Capital Journal*, 1989, p. 11).

The bank's motives for its lending, investing, and referral activities are primarily future profits to be generated through regular banking business with a growing corporation. By helping to finance the firm when it is young the bank hopes to retain the company's conventional banking business when the company becomes large and successful. Thus, the bank's attitudes and patience may well differ from other potential investors. Banks are strongly influenced by the Small Business Administration's (SBA) loan guaranty program, which substantially lessens a bank's risks of lending to a small company. In the last decade SBA guaranty approvals have declined from about 24,000 per year to nearly 15,000 (*Venture*, 1989, p.55), but I have no information available on the extent to which these affect the initial funding of technological firms.

Public Stock Issues

During several short periods since the early 1960s the start-up entrepreneur could turn even initially to the public market in the United States for very early-stage capital, especially for a high-technology or otherwise "glamorous" company. Although few people active today in venture financing will remember, 1969 was the peak year to date in numbers of initial public offerings (IPOs) by early-stage companies. More recent IPO peaks occurred in 1983 and then again in 1986. But those speculative times are usually short-lived. In contrast when a more conservative mood prevails, especially in "bear market" conditions, it becomes very difficult, certainly very costly, for even the successful growing new enterprise to raise public funds.

A young technical company has many reasons to go public, discussed thoroughly in Chapter 8. The entrepreneur and his venture capital backers may wish to realize capital gains; the entrepreneur may want a public market to insure that his holdings are liquid when he dies; the new enterprise may want the prestige of being listed on the financial pages of the newspaper. Or, relevant to this chapter's discussion of the financial base of the company, the company may find that the public market will supply the least expensive or otherwise most attractive funds for its further growth and development.

Regardless of motivations the entrepreneur needs expert advice from the financial community before attempting a public stock issue. The U.S. Securities and Exchange Commission (SEC) has extensive and complex requirements that affect the process of going public, as do many state regulatory bodies. A U.S. firm has several different ways to raise public money, including both underwritten and nonunderwritten methods, and in recent years including the possibility of going public in Britain. Underwriters vary greatly in criteria and effectiveness, and need to be carefully evaluated by the entrepreneur. Large investment banking houses, for example, seldom underwrite issues of less than $10 million and then usually only when the firm meets other performance criteria. Thus, early-stage entrepreneurs need to deal with the smaller underwriters, with whom greater caution is recommended. But despite these cautions the data in Chapter 8 reveal that the public markets serve the technical firm's earlier requirements as often as their growth capital financial needs.

The Initial Capital Base: Amount and Source

This review, coupled with my own experience in new venture financing, leads to two related expectations: (1) initial capital is usually small in amount, typically under $50,000, and rarely more than a few hundred thousand dollars; (2) initial capital is supplied most frequently by the entrepreneurs themselves from their own savings, second by their families and friends, and third by private investors, all these being sources of capital outside of the formal channels. More substantial but still initial funding from wealthy family funds, special seed funds, and more conventional venture capital funds are expected to be the primary complements of the informal sources.

Many entrepreneurs begin their companies with a minimal amount of initial capital and often find their operations hampered by a shortage of capital. Other entrepreneurs, perhaps wiser or just monetarily more fortunate, raise substantial funds before beginning their ventures and have their operations proceed relatively free of financial constraints. Figure 5–1 presents the distribution of initial capital of 113 new technology-based companies spun-off from MIT departments and laboratories. Twenty-three percent of these companies (26) began with funds of less than $1,000. Almost half started with less than $10,000. Only 22 percent (25) had initial funds equal to or in excess of $50,000, of which the vast majority (20 of 25) began operations on a full-time basis.

The comparative studies listed in the Appendix (tracks 2 and 3) show similarly small initial funding. One sample, 38 spin-off firms from a large electronic systems company, includes 18 percent with less than $1000; 42 percent with less than $10,000; only 18 percent with more than $50,000. Twenty-three spin-offs from a large technologically diversified corporation had somewhat higher but still small average start-up equity of $67,000 and additional average initial loans of $48,000. The seventeen manufacturing

firms that provide new energy conversion or conservation systems had
initial financing between $8,000 and $700,000 with a median of $50,000.
Twenty-five biomedical start-ups range in initial equity from 0 to $850,000,
but average only $75,000. Their initial loans range from 0 to $450,000. The
total initial financing of the biomedical group, equity plus loans, goes from
$1000 to $1 million, and average $130,000. Rather remarkably, clusters of
companies incorporated ten years apart experienced the same distribution
of initial capital, with a median of $15,000. The most recently formed
group of technological firms studied had initial financing ranging from
$3,000 to $300,000, half of them starting with under $10,000. The consumer-
oriented manufacturing firms in the research also had modest beginnings,
over half of them starting with less than $10,000. These results mirror
findings by Teach, Tarpley, and Schwartz (1985) that 49.7 percent of the
microcomputer software entrepreneurs they studied had initial capitalization
of $10,000 or less.

The precise amounts of initial capital for the 154 companies in the
data samples being examined in depth in this chapter range from zero
dollars for several firms to one company's $900,000. Close to half of these
firms started on a part-time basis. Of the 52 firms begun on a part-time
basis that provided financing data, 58 percent started with less than $10,000
while only 38 percent of the full-time operations began with so little.

Figure 5–1
Amount of Initial Capital (113 MIT spin-off companies,
separated into full-time and part-time founders)

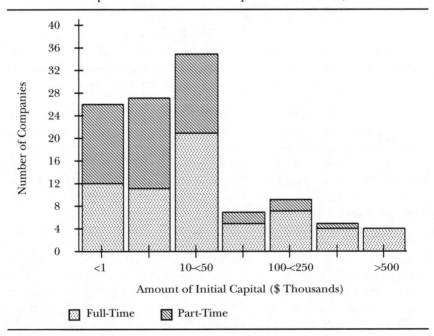

As reported in Table 5–1 personal funds of the founders are the primary sources used to finance the start of more that 70 percent of these companies, and family and friends are key contributors to the start of an additional 5 percent. These percentages are consistent across all subgroups of MIT spin-off companies, as well as those from a large electronic systems company and from a sample of entrepreneurial firms whose early years were carefully assessed. Similar personal or "close" sources funded 20 of 23 companies spun-off from a large diversified corporation, as well as 80 percent of the consumer-oriented manufacturers. The other companies were begun through funds obtained primarily from private investors (angels), venture capital firms, or nonfinancial corporations at which the founders worked, with a few funded by the public stock market. These same two sources, the founders themselves and private investors, are also the dominant initial financiers of a sample of 21 companies that were carefully evaluated for later funding by one venture capital firm studied. An independent study of software entrepreneurs affirms that 76.3 percent had raised essentially all of their initial capital from personal funds (Teach, Tarpley, and Schwartz, 1986).

The nonfinancial corporations listed here can conceptually be expanded to include all forms of the entrepreneur's first customers, a source often used by custom developers and some other start-ups. One entrepreneur who, in addition to his personal savings, was funded twice in this way is Patrick J. McGovern, whose International Data Group (IDG) "is today a $500 million international empire with 125 magazines in 40 countries" (French, 1990, p. 36). Pat followed his MIT undergraduate editorship of *The Tech* into a remarkable career in the computer-related publishing field. When he started his first company, International Data Corporation, now the market research arm of IDG, he persuaded 12 companies to pay him $10,000 each, entirely up-front, for a survey of computer use. Later, when McGovern launched" the weekly newspaper *Computerworld* ... he invested his

Table 5–1
Primary Source of Initial Capital (*n* = 154 companies)

Source	Number of Companies	%
Personal savings	114	74
Family and friends	8	5
Private individual investors	11	7
Venture capital funds	8	5
Nonfinancial corporations	9	6
Commercial banks	0	0
Public stock issues	4	3
Totals	154	100

entire net worth, about $50,000, to develop a prototype", and sold 30,000 prepaid subscriptions at a trade show again to gain his needed working capital from his intended customers." He has never raised money from investors or taken a bank loan" (p. 37).

Those starting on a part-time basis are even more likely to use their own personal funds to finance the early years of the company. Does this reflect less need for funds by part-time founders, not yet fully committed to their firms? Their "burn rate" of cash expenditures is likely to be smaller than the full-time companies. Or does it reflect "savvy entrepreneurs," moving ahead appropriately slowly on their own funds, holding on to equity ownership until further progress is achieved? Is the direction of causality here really the other way? Does lack of outside capital support force the entrepreneur to utilize his own limited personal resources and thus restrict him to starting on a part-time basis only? Unfortunately the data do not permit answers to these questions.

As anticipated in the discussion above no equity capital was supplied by commercial banks, but bank credit came early and frequently into these companies. Many of the companies had early sales by contract to government or large industrial organizations, and the banks often granted loans to these firms, attaching the contract payments as security.

In Table 5–2 the amounts of initial capital and their sources are shown in detail for 110 new enterprises. The specific amounts of money provided by the various categories of investors are obviously incidental to the specific time periods at which these companies were incorporated and to some extent to the specific industries in which they are involved. New biotechnology companies, not included in this sample, typically generate far more initial capital than new software firms, of which several are in this sample. Regardless of timing and industry specifics, my experience indicates that the relative distributions of which financial sources are actively involved at

Table 5–2
Amount of Capital by Source ($n = 110$ companies)

Source	Amount of Initial Capital ($ thousands)							
	<1	1<10	10<50	50<100	100<250	250<500	≥500	Total
Personal savings	22	27	27	1	3	0	0	80
Family and friends	1	0	3	1	0	0	0	5
Private investors	0	0	2	2	3	1	0	8
Venture capital funds	0	0	0	2	2	3	1	8
Nonfinancial corporations	1	0	2	1	1	1	2	8
Commercial banks	0	0	0	0	0	0	0	0
Public stock issues	0	0	0	0	0	0	1	1

the outset of new technical firms, and which ones provide more rather than less amounts of initial capital, are strongly persistent.

The numbers in Table 5–2 demonstrate empirically what might have been assumed beforehand: In the relatively few cases where money is obtained through "outside" forms of financing (those sources other than the founders or their families or friends), those sources provide far greater average amounts ($p = .001$). Of twenty-six firms begun with funds from outside sources, twenty (77 percent) had initial capital equal to or in excess of $50,000. Of eighty-five companies funded by personal or "close" money, only five (6 percent) were begun with comparable amounts. Similar patterns are found in each research cluster. For example, among the enterprises being evaluated by a venture capital firm for "step-up" funding, those that had initially been self-financed had started with considerably less initial capital (an average of $90,000) than the companies funded by private investors (an average initially of $215,000) (0.02).

The primary reason for this difference is understandable. The amount of money that the founders and their associates have is limited by the fact that these are personal funds. Indeed some of the founders do have a healthy personal stake from the sale of previous ventures, such as five of 21 self-funded spin-offs from the MIT Instrumentation Lab, to cite an extreme case. And when Jerome Goldstein started BioClinical Group, now renamed and publicly traded as Advanced Magnetics, Inc., he risked several hundred thousand dollars of his gains from the sale of Clinical Assays to Baxter-Travenol. Yet most entrepreneurs use only accumulated savings from past earnings, not from sales of prior companies. The "outside" financial sources by their nature have a much greater supply of money available for investment in a technological entrepreneurial start-up.

Three examples from Chapter 1 illustrate the tails of the distributions shown previously in terms both of amounts and sources of initial funding. Digital Equipment Company received its initial funds from a venture capital company, albeit only $70,000; Tyco Laboratories raised over $1 million in a public offering; and Medical Information Technology got its half-million-dollar stake from an industrial corporation. Each is an exception, as the overall data clearly demonstrate.

The more basic question of why some entrepreneurs seek out and receive funds from outside sources and why other entrepreneurs either do not seek or do not receive initial outside capital cannot be answered simply. Some more insight is provided in the Chapter 7 discussion of venture capitalist decision making. Three possible answers to this question are apparent and all somewhat applicable: (1) The need for outside funds does not exist; (2) the desire for outside funds does not exist; (3) the entrepreneurs are unable to obtain outside funds, despite existing need and desire.

Analyses discussed later demonstrate that the need for initial funds varies significantly among new enterprises as a function of their industry, type of business, and indeed size of founding group, among other influences. Clearly, many firms do not need outside financing. At the opposite end of the spectrum, the general interviews and several specific studies of venture financing reveal many failed attempts at raising capital by entrepreneurs who end up using only personal or family and friends funding. Some entrepreneurs do not know how to go about seeking outside funding and use their own monies as a default.

However, other entrepreneurs know of the more formal sources. In my judgment they might well be successful in gaining outside commitments; but they choose not to. Some entrepreneurs want little or no equity financing at the outset because they wish to retain a maximum amount of ownership and control. They often seek primarily debt from outside sources, resulting usually in relatively small loans because of the founders' limited net worth. Then, to cope with the constraints of their limited funds, the entrepreneurs gear their operations to reduce their need for funds (e.g., they render a service instead of producing hardware or they engage in custom-oriented development and production that can be contracted with larger firms or government agencies). The contracts provide advances and/or progress payments that minimize additional financial requirements. One study provides statistical support for this explanation, demonstrating that the entrepreneurs who initially prefer debt to equity tend to finance the companies themselves ($p = .015$) and have lower initial capitalization ($p = .11$). This stated intent of first starting a contract R&D or consulting firm, in hopes of generating the capital base to move into product development and manufacture, is discussed in the next chapter. The strategy frequently fails.

Several of the self-financing entrepreneurs seem rather less rational and more emotional in their emphatic opposition to sharing the profits of their labors and their ideas with others "who do nothing more than provide money". Not understanding that initial capital for a high-technology company is a very risky investment, such entrepreneurs repeatedly call venture capitalists "vultures" who want something for nothing. Underlying this rather naive and often angry position, and also involved with many other aspects of financing, are a complexity of motivations that I cannot even attempt to explain. One study of 20 entrepreneurs who were seeking capital compared those who had initially supplied more than 50 percent of their equity from personal funds with those who had obtained that much from outside investors. The self-funded entrepreneurs have significantly higher evaluations of the importance of independence of action ($p = .025$). In addition those who initially engaged primarily in self-financing include five of the seven entrepreneurs in this cluster who indicate that independence of action is the most important reason for starting a company. Some call this "different strokes for different folks". It is important to note that need for independence is not at all the same as n-Ach (see Chapter 3), and may well lead to different entrepreneurial outcomes.

On occasion, lack of understanding by technical entrepreneurs about sources of finance produces unusual problems, as in the case of one firm financed by a group of "bookies" who had funds available due to a temporary crackdown on bookmaking activities. The bookmakers wanted their money back three years later, just when the enterprise really needed the funds to finance expansion. As a result this company literally was forced out of business.

Influences on Initial Capitalization

The earlier chapters have produced portraits of technical entrepreneurs whose motives and preparedness for company formation have varied widely. These personal and situational differences contribute to major differences in, among other things, the initial capital base of the enterprise. The hypotheses to be tested here are that several symptoms of seriousness of intent and readiness to launch an ambitious new company are the primary influences upon the amount of initial capitalization of the firm. Three elements are examined, each of which is expected both to require and lead to larger initial capital base: (1) the number of founders who have come together for the venture; (2) their preparation of specific plans for the undertaking; and (3) the orientation of the new business toward product development and manufacture, in contrast with contract R&D, consulting, or other custom services.

Number of Founders

The number of founders can influence the amount of initial capital both directly and indirectly. Multiple founders are likely to reflect a more substantial intended undertaking, such as product development and manufacture rather than just research or consulting. This implied need for greater funds both generates and justifies its supply. And the larger team is itself likely to be more impressive to outside sources, partly explaining the research finding that outside sources are more willing to invest in multifounder companies. Furthermore, as the number of founders increases more personal funds are available from which to draw money. This has a direct effect in that over 70 percent of the companies studied are financed initially by personal funds. Indirectly, the more founders there are, the greater possibility that one of them knows a receptive "outside" source.

Table 5–3 presents the initial capital amounts associated with the number of founders of 109 companies. The largest proportion of companies that began with less than $10,000 (62.5 percent) is in the group of one-founder companies. In general, the larger the number of founders, the less the occurrence of financing under $10,000 and the greater the occurrence of funding in excess of $50,000.

The companies begun primarily with the entrepreneurs' own funds and those started with others' funds were analyzed separately. Nearly half

Table 5–3
Amount of Initial Capital by Number of Founders
($n = 109$ companies)

Number of Founders[*]	Amount of Initial Capital ($ thousands)						
	<1	1<10	10<50	50<100	100<250	250<500	≥500
1	17	8	11	1	3	—	—
2	2	10	9	2	1	2	—
3	3	5	10	—	1	1	2
4	1	4	1	1	1	—	—
5	1	—	2	1	1	1	—
6	2	—	—	1	—	1	—
7	—	—	—	—	1	—	1
8	—	—	—	—	—	—	—
9	—	—	—	—	—	—	1

[*] Kendall tau = 0.25, $p = .01$

of the firms founded by a single individual using his own money began with less than $1000. In the founder-funded firms greater amounts of initial capital are provided as the number of founders increased (tau = .19, $p = .03$). (Indeed, when Jack Pugh and I co-founded our consulting firm, Pugh-Roberts Associates, Inc., we *each* invested $1,000!) Looking across the entire sample of companies, at each size of founding group the average amount of funds supplied by others is greater than the average supplied by the founders themselves. All but six of twenty-six companies that obtained funds primarily from others were started by multifounder teams. And for outsider-funded firms, the same finding holds that the number of founders and the amount of initial capital received are positively related (tau = .23, $p = .08$).

These results are wholly supported by other studies of venture capital decision processes. Tyebjee and Bruno find (1984, p. 1060) that "lack of managerial capabilities significantly increases the perceived risk ($p < .05$)" of investing, those capabilities reflecting multifunctional skills only achievable with a multi-founder team. Their earlier work (Bruno and Tyebjee, 1982, p. 290) also shows that deficiencies in the venture's management team explains funding rejections in a third of the cases. Goslin and Barge (1986, p. 366) agree that "the significant factor leading to [venture capitalist] funding is the management team", which also coincides with Dean and Giglierano's more recent (1989) findings. MacMillan, Siegel, and Subba Narasimha (1985, p. 125) conclude that "just under one-half [42 percent] of venture capitalists will not even consider a venture that does not have a balanced team." Multiple founders obviously increase the likelihood that a stronger and more balanced management team has been assembled, as demonstrated in Chapter 3. And Teach, Tarpley, and Schwartz (1986, p. 553) concur, finding "the larger the [software] venture team the larger the amount of initial capital ($X^2 = 22.47$, $p = .001$)". In a sample of 1903 start-up ventures, not limited to technical companies, Cooper, Woo, and Dunkelberg (1989) also find that larger initial size firms have significantly larger initial capitalization.

Most personal background characteristics of the technical entrepreneurs, including age, education, religion and motivation do not relate statistically to the amount or source of initial capital. Other researchers (MacMillan et al., 1985, p. 122) have found that entrepreneur personality characteristics are strong influences on venture capitalist investment decisions, but my data are not adequate to validate these findings. Even prior patents by the founders do not generally relate to larger initial capitalization, the sole exception in my research samples being the firms spunoff from the electronic systems corporation ($p = .02$). However, the data analyses do reveal that those individuals with the greatest amount of commercial work experience start their companies with more initial capital financing (.08). These individuals, by virtue of their more extensive familiarity with the industrial and financial community, are probably more aware of venture capital sources and how to approach them successfully. Their greater industrial experience no doubt also provides some modicum of comfort to the investors. Not necessarily in conflict with this finding is that a significant fraction of entrepreneurs coming out of MIT labs and departments felt their previous association with MIT had aided their capital seeking efforts.

Specific Plans

Not all the entrepreneurs have specific plans for their companies when they decide to start them. Twenty-four of fifty-three entrepreneurs (45 percent) who responded to questioning indicate that they had neither specific short-term nor long-term plans at the beginning of their companies. With no specific plan considerable investment is not necessary. The *Alice in Wonderland* adage applies here: If you don't care where you are going, any path will get you there. Nor is an investment likely to be attracted from an outside professionally managed financial source when the nature of the future work is so uncertain.

In Table 5–4 the amounts of initial capital for 29 firms started with specific plans are compared with the amounts for 23 firms started without specific plans. Seventy-four percent of those without specific plans started with less than $10,000, whereas only 24 percent of the companies with specific plans were formed with so little funding. Furthermore, 38 percent of the companies begun by founders with specific plans received funding in excess of $50,000 while only 9 percent of the companies lacking specific plans had so much initial capital. Clearly entrepreneurs with specific plans raise more initial capital than those without plans ($p = .001$).

Table 5–4
Specific Plans for the Company and Amount of
Initial Capital ($n = 52$ companies)

Initial Capital ($ thousands)	Specific Plans	
	Yes[*]	No[*]
< 1	4	7
1 < 10	3	10
10 < 50	11	4
50 < 100	2	0
100 < 250	3	2
250 < 500	3	0
≥ 500	3	0
Totals	29	23

[*] Mann-Whitney U, $p = .001$.

As might be expected from the discussion thus far the more institutional sources of financing are much more inclined to support ventures that have a specifically planned future. Table 5–5 shows that ten of twelve companies that received other than personal or "close" funding had specific operational plans at the start. Clearly most investors see firms with plans as better bets. In addition the entrepreneurs who prepare detailed plans no doubt foresee needs for greater capital and go out to get it. Of course not all prospective entrepreneurs who plan get funded. MacMillan and Subba Narasimha (1986, p. 409) find that "excessively optimistic fore-

Table 5–5
Specific Plans for the Company and Source
of Initial Capital ($n = 49$ companies)

Source	Specific Plans	
	Yes	No
Personal savings	15	19
Family and friends	2	1
Private investors	3	1
Venture capital funds	3	0
Nonfinancial corporations	3	0
Commercial banks	0	0
Public stock issues	1	1
Totals	27	22

casts of performance can create a fatal credibility problem", which might explain in part why only 10 of the 27 planners in the Table 5–5 firms received outside funding. Rea (1989) also finds that "a business plan in which too many things could go wrong" is the most important reason for rejection by venture capitalists.

The two companies in this cluster that received outside funds despite lack of specific plans are unique exceptions. One was founded by several MIT employees who were about to be left without work when MIT decided to abandon its atomic energy research. These founders along with nearly ninety other MIT employees engaged in the same work formed a company without specific goals, but backed strongly by private investors, an investment that is easily understood. The other situation involves a new company formed from the division of a larger corporation, spun-off in its entirety due to rising costs. The venture was backed by a public stock issue generated by the parent company. Excluding these two unusual cases only companies with specific plans obtained money from the more sophisticated sources of financing.

Product Orientation

The type of business being started obviously has major influence on the amount and intended use of initial capital. It may also affect the firm's

Table 5–6
Ranked Needs for Initial Capital ($n = 107$ companies)

	Type of Business				
	Aggregate ($n = 107$)	Hardware ($n = 33$)	Software ($n = 10$)	Contract R&D ($n = 22$)	Consulting ($n = 20$)
Rank					
1	Product development	Product development	Other	Lab equipment	Technical personnel
2	Lab equipment	Production facilities	Technical personnel	Product development	Lab equipment
3	Technical personnel	Inventory	Lab equipment	Accounts receivable	Accounts receivable
4	Accounts receivable	Other	Accounts receivable	Production facilities	Development
5	Production facilities	Accounts receivable	Development	Technical personnel	Inventory

ability to attract capital from outside investors. Table 5–6 presents responses from 107 entrepreneurs who rank their needs for capital. The wide variances reflect the types of business entered. In hardware production capital is first needed for product development, then for production facilities and working capital. Software companies need working capital for their technical personnel payroll and to finance their accounts receivable (A/R), but they also need funds for computer equipment and for product development work. Firms performing contract R&D exhibit needs for lab equipment, product development, working capital and production facilities. Even individual consultants need funds for lab equipment and to fund development work.

None of the groups finds that either marketing expenses or production and clerical workers place a stress on their capital needs. As is shown in the next chapter the typical technological enterprise at its time of founding displays unfortunately little emphasis on marketing. Also the Boston area labor market, relevant to most of the firms studied, has been especially efficient until recently in terms of a new company's ability to find both skilled and semiskilled hourly workers.

Given the differences in specific needs how do the capital requirements vary in amount by type of business? The consultants and the software houses require the least capital; nearly 80 percent of them were capitalized initially at less than $10,000. Indeed, one software entrepreneur started his company on $700 he received from selling his automobile. Van de Ven et al. (1989, p. 232) document a comparable case in which the two founders of Medformatics, in December 1980, "made personal investments totaling $1,000 and obtained a $37,000 line of credit from a local bank" for their software start-up. At the opposite extreme are the hardware production firms, but even here 84 percent were capitalized at under $50,000. This relatively modest figure is explained in part by the fact that 60 percent of those companies were started on a part-time basis.

Information gathered from 110 firms indicates that forty-seven (43 percent) of them are based on specific products that had already been developed or that the entrepreneurs planned to develop immediately. A firm dependent on a product needs capital, whether for product development or production facilities or market launch. Such a firm would have difficulties getting operations underway without substantial capital. Since the sixty three other firms in this grouping do not have a product or immediate product objectives, they need considerably less initial capital to get going.

Table 5–7 displays the amount of initial capital for 43 companies that had a product or specific product plans initially and for 59 companies that did not. The group with initial products was initially financed to a greater extent ($p = .02$). This situation is driven statistically by the fact that 21 of these firms without product began with less than $1000, while only three companies with products had similarly small initial funding.

Table 5–7
Product Initially and Amount of Initial
Capital (n = 102 companies)

Initial Capital ($ thousands)	Product Initially	
	Yes[*]	No[*]
< 1	3	21
1 – < 10	12	13
10 – < 50	15	15
50 – < 100	2	5
100 – < 250	5	3
250 – < 500	4	0
≥ 500	2	2
Totals	43	59

[*] Mann-Whitney U, $p = .02$.

A product-oriented company's capital requirements do vary according to the nature of the product, its stage of development, development requirements, its production process, as well as the demand for the product. Among 21 firms in one sample those with a "proprietary" product (one with some degree of "protectability" from patents or secrecy or not easily duplicated technological uniqueness, the combination perhaps of "specific plans" and a product orientation) had significantly higher initial capitalization ($200,000 on average) than those without a proprietary product ($129,000) ($p = .07$). But it is doubtful that capital required would ever be much less than $1,000. Indeed each of the three product-oriented firms listed in Table 5–7 that began with less than $1,000 was in the process of developing its first product, making its meager funding slightly more understandable.

The initial capitalization of the firms that began without a product is readily explained. Many were initially engaged in activities such as technical consulting or computer programming. Little or no financing is needed to start them. Others were involved with work such as systems design and development, which requires capital primarily to support technical personnel and equipment. Here capital needs vary, depending on the size of the project to be undertaken. There is a bit of the chicken-versus-egg issue here. In some of these cases companies that had problems in raising initial outside capital had already abandoned their earlier intentions and started to do things that were not capital intensive. Thus, lack of available initial capital often influences the apparent lack of "initial" product orientation.

Some differences are observable in the sources of capital for both groups. Thirty companies (70 percent) formed around a product were financed by founders or close associates, while thirteen generated other funds. Fifty-one companies (82 percent, slightly more than the above group) without an initial product focus were funded by founders or family or friends, while eleven received funds elsewhere. Tyebjee and Bruno (1984, p. 1057) demonstrate a strong bias by venture capitalists (over 90 percent of their deals) in favor of product manufacturing companies, but the venture capitalists' reluctance to get involved at ground zero no doubt prevents their product preference from being reflected in the data shown here.

Most companies find that their initial funds are insufficient to support their growth during their early years. Sixty percent of the companies studied sought capital a second time and nearly half sought funds a third time. The additional financing of high-technology companies is discussed in more detail in Chapter 7. Their experiences in going public are evaluated in Chapter 8.

Summary and Implications

This chapter presents an overall assessment of the capital market for technology-based firms, focusing on the links between the stages of evolution of a firm and the investment preferences of various capital sources. Figure 5–2 portrays the relationship of the stage of evolution of the technology-based firm to the likely availability of capital from the various investment sources. The diagram is inexact and is meant to convey the average tendencies of each class of investor during the three stages of a company's development. As should be expected, the investment behavior of each group contains considerable variance, as evidenced by the presentation of empirical findings in the chapter.

This Figure 5–2 synopsis of capital sources and the earlier discussion in this chapter have not included pension funds or insurance companies, both major participants in U.S. capital markets. Until recently neither type of institution directly participated in initial or even early round financing of high-technology firms. Both sources are major investors in the pooled funds managed by venture capital firms, and a growing number of pension funds and insurance companies have initiated programs of direct venture investing, especially in later corporate growth financing. Corporate pension funds have been major players since 1981 and public employee retirement funds have become active since then. In recent years over 20 states have become involved in venture capital fund investments, often with the combined motives of economic stimulus of their own regions as well as increased returns on their investment portfolios (Maier and Walker, 1987, p. 210). Perhaps surprising, even union pension funds are beginning to go beyond investing in pooled venture capital funds to engage in selective direct investments in young firms, but seldom in technological start-ups (Spragins, 1989).

Figure 5–2
Primary Investment Preferences of Capital Sources

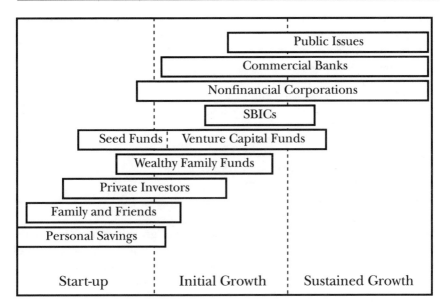

Stage of Company Evolution

The review of financing alternatives produced expectations that the initial capital base of high-technology firms is generally small in magnitude, and provided primarily by the entrepreneurs themselves, their families and friends, and private investors (angels). More substantial amounts of initial funding, when obtained, were expected to come from wealthy family funds, seed funds, and more conventional venture capital companies.

Data from the studies of technological firms support these expectations, providing evidence of the usual small initial capitalization (almost half with less than $10,000) and the dominance of personal savings as the principal source of initial capital (74 percent of the companies). Prospective technological entrepreneurs need to realize that most of their predecessors got started on a shoestring. The effect of initial funding on entrepreneurial success is discussed in Chapter 9.

"Outside" sources of capital are responsible for the larger initial investments when they occur. Outside sources do vary greatly, and private investors are far more likely to provide initial funds than are most venture capital companies. Prospective entrepreneurs should seek out these private investors, perhaps taking advantage of the computerized contact mechanisms described in this chapter.

As listed in the Table 5–8 summary, larger amounts of initial capital are associated with several cues provided by the founding entrepreneurs of

Table 5–8
Initial Capital for the Technological Enterprise

Primary Sources of Capital
    Personal savings
    Family and friends
    Private individual investors (angels)

Larger Initial Capital Associated with
    Outside initial investors
    Larger co-founding team
    Full-time, rather than part-time, commitment
    Specific plans for business development
    Hardware focus, rather than software or consulting
    Initial product available or targeted

---

their seriousness of intent to build a successful growth company. For example, more funds are both contributed and raised by the larger groups of co-founders, especially when they are involved in the companies from the outset on a full-time basis. Specific plans for the company lead to greater initial capitalization, as well as to raising outside capital. The needs for initial capital vary enormously by amount and intended use as a function of the type of business being started, with consulting firms and software companies requiring far less than hardware developers and producers. This is reflected in the positive effect of product orientation upon initial capital. Prospective entrepreneurs need to express, in actions and in writing, their aspirations, the rationale for potential attractiveness of the contemplated businesses, their strategies for achieving corporate growth and success, and their personal commitments to those objectives. Larger initial capital base is at least one likely outcome of this behavior.

References

D. N. Allen, D. R. Costello, & J. P. Danford. "Seed Venture Capital and University Research Commercialization: The Zero Stage-Penn State Connection". Unpublished paper (State College, PA: The Pennsylvania State University, 1989).

J. D. Aram. "Attitudes and Behaviors of Informal Investors Toward Early-Stage Investments, Technology-Based Ventures, and Coinvestors", *Journal of Business Venturing*, 4, 5 (1989), 333–347.

G. Baty. *Initial Financing of the New Research-Based Enterprises in New England* (Boston: Federal Reserve Bank of Boston, Research Report No. 25, 1964).

G. B. Baty. "The Role of the Venture Advisor in High-Tech Startups", *Review of Business* (New York: Business Research Institute, St. John's University), 10, 1 (Spring/Summer 1988), 12–15.

A. V. Bruno. "A Structural Analysis of the Venture Capital Industry", in D. L. Sexton and R. W. Smilor (editors), *The Art and Science of Entrepreneurship* (Cambridge, MA: Ballinger Publishing, 1986), 109-117.

A. V. Bruno & T. T. Tyebjee. "The One that Got Away: A Study of Ventures Rejected by Venture Capitalists", in J. A. Hornaday et al. (editors), *Frontiers of Entrepreneurship Research, 1982* (Wellesley, MA: Babson College, 1982), 289–306.

W. D. Bygrave & J. A. Timmons. "An Empirical Model for the Flows of Venture Capital", in J. A. Hornaday et al. (editors), *Frontiers of Entrepreneurship Research, 1985* (Wellesley, MA: Babson College, 1985), 105–117.

E. Conlin. "Adventure Capital", *INC.*, September 1989, 32–48.

A. C. Cooper, C. Y. Woo, and W. C. Dunkelberg. "Entrepreneurship and the Initial Size of Firm", *Journal of Business Venturing*, 4, 5 (1989), 317–332.

B. Dean & J. J. Giglierano. "Patterns in Multi-Stage Financing in Silicon Valley", in *Proceedings of Vancouver Conference* (Vancouver, BC: College on Innovation Management and Entrepreneurship, The Institute of Management Science, May 1989).

W. C. Dunkelberg & A. C. Cooper. "Financing the Start of a Small Enterprise", in J. A. Hornaday et al. (editors), *Frontiers of Entrepreneurship Research, 1983* (Wellesley, MA: Babson College, 1983), 369–381.

European Venture Capital Association/Peat Marwick McLintock. *Venture Capital in Europe, 1990 EVCA Handbook* (London: E.V.C.A./Peat Marwick McLintock, 1990).

G. L. Feigen & L. M. Arrington. "The Historic Role of SBICs in Financing the Young and Growing Company", in R. Ronstadt et al. (editors), *Frontiers of Entrepreneurship Research, 1986* (Wellesley, MA: Babson College, 1986), 457–458.

R. Florida & M. Kenney. "Venture Capital and High Technology Entrepreneurship", *Journal of Business Venturing*, 3(1988), 301–319.

J. Freear & W. E. Wetzel, Jr. "Equity Capital for Entrepreneurs", in *Proceedings of Vancouver Conference* (Vancouver, BC: College on Innovation Management and Entrepreneurship, The Institute of Management Science, May 1989).

K. French. "Patrick J. McGovern, '59: Chronicler of the Information Age", *Technology Review*, 93, 6 (August/September, 1990), MIT 36–37.

L. N. Goslin & B. Barge. "Entrepreneurial Qualities Considered in Venture Capital Support", in R. Ronstadt et al. (editors), *Frontiers of Entrepreneurship Research, 1986* (Wellesley, MA: Babson College, 1986), 366–379.

G. L. Hegg. "A Corporate View of Venture Capital", in *Managing R&D Technology: Building the Necessary Bridges* (The Conference Board, Research Report #938, 1990), 28–30.

J. A. Hornaday & N. C. Churchill. "Current Trends in Entrepreneurial Research", in N. C. Churchill et al. (editors), *Frontiers of Entrepreneurship Research, 1987* (Wellesley, MA.: Babson College, 1987), 1–21.

B. Huntsman & J. P. Hoban, Jr. "Investment in New Enterprise: Some Empirical Observations on Risk, Return and Market Structure", *Financial Management*, Summer (1980), 44–51.

IC² Institute. "Risk Capital Networks", in *Technological Alliances for Competitiveness* (Austin, TX.: The University of Texas, 1989), 39–42.

IC² Institute. *Technological Alliances for Competitiveness* (Austin, TX: The University of Texas, Summer 1990).

William Bryant Logan. "Finding Your Angel", *Venture*, March (1986), 38–44.

I. C. MacMillan, R. Siegel, & P. N. Subba Narasimha. "Criteria Used by Venture Capitalists to Evaluate New Venture Proposals", *Journal of Business Venturing*, 1 (1985), 119–128.

I. C. MacMillan & P. N. Subba Narasimha. "Characteristics Distinguishing Funded from Unfunded Business Plans Evaluated by Venture Capitalists", in R. Ronstadt et al. (editors), *Frontiers of Entrepreneurship Research, 1986* (Wellesley, MA: Babson College, 1986), 404–413.

J. B. Maier, II & D. A. Walker. "The Role of Venture Capital in Financing Small Business", *Journal of Business Venturing*, 2 (1987), 207–214.

S. Maital. "Europe's Venture Capital Boom", *Across The Board*, June (1989), 5–6, 64.

D. K. Neiswander. "Informal Seed Stage Investors", in J. A. Hornaday et al. (editors), *Frontiers of Entrepreneurship Research, 1985* (Wellesley, MA: Babson College, 1985), 142–154.

Nikkei Financial Daily. "Venture Capital Business Finally Emerging", *Japan Economic Journal*, July 21, 1990, 22.

D. J. Ravenscraft & F. M. Scherer. *Mergers, Sell-Offs, and Economic Efficiency* (Washington, D.C.: The Brookings Institution, 1987).

R. H. Rea. "Factors Affecting Success and Failure of Seed Capital/Start-Up Negotiations", *Journal of Business Venturing*, 4, 2 (1989), 149–158.

E. B. Roberts, "New Ventures for Corporate Growth", *Harvard Business Review* 58, 4 (July–August 1980), 134–142.

A. H. Rubenstein. *Problems of Financing and Managing New Research-Based Enterprises in New England* (Boston: Federal Reserve Bank of Boston, April 1958).

J. C. Ruhnka & J. E. Young. "A Venture Capital Model of the Development Process for New Ventures", *Journal of Business Venturing*, 2 (1987), 167–184.

F. M. Scherer. "Firm Size, Market Structure, Opportunity, and the Output of Patented Inventions", *American Economic Review* 1965, 1097–1125.

R. W. Smilor, D. V. Gibson, & G. B. Dietrich. "University Spin-out Companies: Technology Start-ups from UT-Austin", in *Proceedings of Vancouver Conference* (College on Innovation Management and Entrepreneurship, The Institute of Management Science, May 1989).

C. Smith & Y. Ayukawa. "Japan Venture Capital: A Different Game than U.S.", *Venture Japan*, 1, 4 (1989).

E. E. Spragins. "Working for the Union", *INC.*, April 1989, 173–174.

H. H. Stevenson, D. F. Muzyka, & J. A. Timmons. "Venture Capital in a New Era: A Simulation of the Impact of Change in Investment Patterns", in R. Ronstadt et al. (editors), *Frontiers of Entrepreneurship Research, 1986* (Wellesley, MA: Babson College, 1986), 380–403.

M. Sun. "Investors' Yen for U.S. Technology", *Science*, 246 (8 December 1989), 1238–1241.

R. D. Teach, F. A. Tarpley, Jr. & R. G. Schwartz. "Who are the Microcomputer Software Entrepreneurs?", in J. A. Hornaday et al. (editors), *Frontiers of Entrepreneurship Research, 1985* (Wellesley, MA: Babson College, 1985), 435–451.

R. D. Teach, F. A. Tarpley, Jr. & R. G. Schwartz. "Software Venture Teams", in R. Ronstadt et al. (editors), *Frontiers of Entrepreneurship Research, 1986* (Wellesley, MA: Babson College, 1986), 546–562.

J. A. Timmons & W. D. Bygrave. "Venture Capital's Role in Financing Innovation for Economic Growth", *Journal of Business Venturing*, 1 (1986), 161–176.

T. T. Tyebjee & A. V. Bruno. "A Comparative Analysis of California Startups from 1978 to 1980", in K. H. Vesper (editor), *Frontiers of Entrepreneurship Research, 1982* (Wellesley, MA: Babson College, 1982), 163–176.

T. T. Tyebjee & A. V. Bruno. "A Model of Venture Capitalist Investment Activity", *Management Science*, 30, 9 (September 1984), 1051–1066.

E. R. Tynes & O. J. Krasner. "Informal Risk Capital in California", in J. A. Hornaday et al. (editors), *Frontiers of Entrepreneurship Research, 1983* (Wellesley, MA: Babson College, 1983), 347–368.

A. H. Van de Ven, S. Venkataraman, D. Polley, & R. Garud. "Processes of New Business Creation in Different Organizational Settings", in A. H. Van de Ven, H. L. Angle, and M. S. Paule (editors), *Research in the Management of Innovation: The Minnesota Studies* (New York: Ballinger/Harper & Row, 1989), 221–297.

Venture. "Looking Back", May 1989, 54-56.

Venture Capital Journal. "Special Report", March 1989, 9–17.

Venture Capital Journal. "Special Report", April 1990, 11–18.

E. O. Welles. "The Tokyo Connection", *INC.*, February 1990, 52–65.

W. E. Wetzel, Jr. "Angels and Informal Risk Capital", *Sloan Management Review*, Summer (1983), 23–34.

W. E. Wetzel, Jr. "Informal Risk Capital: Knowns and Unknowns", in D. L. Sexton & R. W. Smilor (editors), *The Art and Science of Entrepreneurship* (Cambridge, MA: Ballinger Publishing, 1986), 85–108.

W. E. Wetzel, Jr. "The Informal Venture Capital Market: Aspects of Scale and Market Efficiency", in N. C. Churchill et al. (editors), *Frontiers of Entrepreneurship Research, 1987* (Wellesley, MA: Babson College, 1987), 412–431.

CHAPTER 6

Evolving Toward Product and Market-Orientation

A Still Evolving Tale

The dormitory's vendor of pinball machines and arcade games decided to pull out. He just could not make enough profit to justify coping with all the aggravation of servicing the machines in an MIT undergraduate house. Douglas Macrae, sophomore vice president of MIT's MacGregor House, thought this might be a good chance to make some spending money while serving the dorm's needs. With an agreed 50-50 split of revenues between himself and the dorm, Doug took over the operation, his first "business" venture. Sales boomed and problems from house residents eased off, now that one of their own was taking care of things and the dorm was generating a profit that went for house events. Growing student "addiction" to game-playing during the next two years led Doug to invest his profits into more machines, installing them in other MIT dormitories, requiring that his friend and classmate ('81) Kevin Curran join Doug to handle the increasing demands on their time. Doug says, "We did some wild things to promote business. For example, we used to give five quarters as change for a dollar. What did we care? All of it would soon be deposited in our machines."

A senior year co-op assignment at Computervision (founded by MIT alumnus Phil Villers) gave Doug new skills in chip design. He and Kevin developed an "enhancement board" that could be wired into Atari's "Missile Command", one of the arcade video games in the dorm, providing it with new features that appealed to their customers. The enthusiastic market feedback encouraged them to get more serious about their activities, and they began to sell the board through game hobbyist magazines for $295, a reasonable markup they felt over their $30 out-of-pocket costs. Assembly and shipping was done from Doug's off-campus apartment living room, with John Tylko, a two-year-older MIT grad joining as the third partner to help this growing enterprise. They incorporated as General Computer Company, investing $25,000 in profits from their prior activities and borrowing an additional $25,000 from Doug's mother to expand more rapidly.

The trio started working on their second enhancement board, intended for the popular PacMan arcade game, when suddenly the roof fell in. Atari sued General Computer and each of its principals for $5 million each, claiming patent infringements and other violations, getting the federal court to issue an injunction against all their operations. Within a few hectic weeks a settlement was reached, much to everyone's apparent delight, and General Computer received a major game development contract from Atari, now moving them primarily into the software side of creating new games. The PacMan enhancement, still underway, was licensed to Bally-Midway as Ms. PacMan, becoming the largest selling video arcade (and later Atari home video) game, generating over a half billion dollars of revenues to Bally and Atari and huge royalties to General Computer. With arcade and home video game development activities rapidly moving to fever pitch, the threesome dropped out of their MIT classes, with Doug and Kevin not completing their bachelor's degrees, and the company moved into expanding space in the nearby Atheneum Building, a converted mill behind MIT. Doug remembers the 1983 year with particular pride, as General Computer hired more graduating MIT seniors than either IBM or Hewlett-Packard, attracting the best with the lure of this unusual start-up venture that, among other benefits, provided employees the opportunity of game playing 24 hours a day. The frequent new product introductions were celebrated by shooting off a brass cannon into the adjacent Charles River canal, part of the boisterous environment the founders created. Ten percent of each product's royalty revenues were divided among that product's development team, based on the team members' consensus votes of who deserved what portion of these gains, making several 22-year olds the thrilled recipients of six-figure incomes.

External crisis marked the beginning of the next stage of General Computer's evolution, when Atari suddenly announced massive quarterly losses on its games empire. By the end of that same day John Tylko had assembled a group of employees into a new ventures unit and began developing software and games for Apple's Macintosh computer. They chose the Macintosh because the graphics orientation of that machine had much in common with the company's experiences with gaming. But they soon began to experience problems with the limited internal memory and the consequent slow loading of software programs on the Macintosh's initial versions. Needing faster inputting speed they opened up their Macintoshes (thereby automatically voiding the warranty on the computers) and rigged up high speed disk memories to get the desired performance. This led to quick recognition that such a product would be a very desirable add-on to the Macintosh, especially if they could convince Apple Computer to modify the Mac so as to allow internal installation. Several months of development work brought General Computer to the point of a demo unit that they brought to Steve Jobs, Apple's CEO, only to be turned down, presumably because he did not like the minor noise produced by the fan that was needed to cool the internal unit. Left with no alternative, General Computer

repackaged its product as an external Macintosh peripheral and introduced Hyperdrive to the market, an instant success.

Despite many further unexpected problems in its continuing evolution,
General Computer, now renamed as GCC Technologies, quickly became a
dedicated developer and producer of a growing line of Macintosh peripheral
hardware products. Its current Ultradrive line of hard-disk memories is
complemented by a wide array of impact and laser printers, with other
peripherals being added regularly. Now doing over $30 million in sales,
the three founders still occupy the roles of chairman (Macrae), president
(Curran), and vice president (Tylko), and still own most of the company
they started as students. The unusual evolution from a preincorporation
service operation to a hardware then software then again hardware manufacturer is the only one of its kind that I have seen. But the attention early
on to customer desires is noteworthy throughout GCC's history. Not all
high-technology new enterprises show this degree of market orientation.

Getting Underway

People, technology, and/or an idea for a product or service, and money
enable a technical enterprise to get formally initiated. But what do the
founders do in getting underway? And how do their companies change
during the first several years of existence? While research on technology-
based enterprises has increased significantly, few research projects have
focused on aspects of change during the early life of these new technology-
based firms. Yet, it is safe to assume that evolution of both the founders
and their firms is necessary for companies to benefit from experiential
learning and to adapt to environmental changes in technologies and markets. This chapter adds to the evidences of change during the early years of
technology-based companies. How much occurs and in what ways? Can
prospective entrepreneurs learn from their predecessors any clues for more
effective launch and development of their companies? The companies will
continue to change and grow past this point of assessment, and later chapters distinguish those influences that affect corporate success and failure.
Here I focus upon the new firm's transition from technological orientation
to market orientation, an evolution I believe is essential to achieve eventual
corporate success.

Names and Places

Let me start with a couple of incidental observations on the companies'
first actions. First, more than half of the new companies in the research
samples chose names beginning with the letters A through D! I have no
idea how this distribution relates to the rest of the industrial world, but I
suspect that Exxon Corporation knew it was distinguishing itself from the
pack when it gave its high-technology venture operations names such as

Qume, Qwip, Qyx, and Zilog. Larry Liebson demonstrates the same affinity in naming his three start-ups Xylogics, Xylogic Systems, and Xyvision!

Second, almost every company sets up its initial location physically near the primary source organization that had been its incubator. This is true not just at the macro level, the Greater Boston area in general, but even at the micro level, holding heavily to their prior residential and commuting patterns. For example, MIT Lincoln Laboratory spin-offs concentrate in the towns around Lexington, Massachusetts, the home of Lincoln Lab, while Instrumentation Lab spin-offs focused on Cambridge as their home, the location of their "parent". Unique reasons tend to explain the approximately 15 percent company locations away from their technology origins, like a sole founder returning to his family's origins, or even more rarely (though thoughtfully) a company moving to be near an anticipated prime customer's location, such as a southward move into the NASA "Space Crescent". Feeser and Willard (1989) support this finding from their analysis of computer companies, 86 percent of their entrepreneurs not moving. This observation is relevant for the many regional development agencies that try to attract young firms with tax incentives or other inducements. The data suggest that these approaches may work only much later in the growth of a technical firm, when the company might be considering the location of a next engineering site or a manufacturing plant. Early-stage firms locate essentially where the founders are working at the time of company formation.

The Relevant Literature

Researchers have long been interested in how and why organizations change over their life cycles of birth, growth, maturation, and death. And yet, in their synopsis volume, *The Organizational Life Cycle*, Kimberly and Miles (1980) decry the absence of "the dynamic quality of organizational life ... from most research and writing in the area" (p. 3). Quinn and Cameron (1983) hypothesize four distinct phases, beginning with "an entrepreneurial stage characterized by innovation, creativity, and the marshalling of resources sufficient to survive." In an excellent identification of eighteen alternative "stage models" to depict the evolution of complex phenomena, especially in organization and management theory, Kazanjian (1984) provides further specifications that should facilitate concrete empirical research. Two studies that postulate organizational life cycle theories of the firm have attempted to relate psychological characteristics of entrepreneurs to the adaptation needed during a firm's transition, but neither study examined technological companies (Churchill, 1983; Smith and Miner, 1983). Only the few works cited have been found that applied these models toward data collection on technical firms.

Van de Ven, Hudson, and Schroeder (1984) classify the research on the creation of organizations into entrepreneurial (focusing on personal characteristics of founders), ecological (focusing on shifts in organization populations and their reasons), and organizational (focusing on managerial processes involved in initiation and early development). My research falls mainly into the first and third categories. Van de Ven et al.'s study of 12 educational software companies, treating all three research dimensions, identifies five stages in the development of these technical firms, of which two stages, gestation and planning, occur before actual company operation. On a number of dimensions Van de Ven documents major differences (few unfortunately of statistical significance due to small sample size) between the firms classified as in their early post-start-up stages and those in later stages. A critical organizational finding is that mean percentile time allocation of principals to working on products decreases from 26.6 percent in early stage companies to 21.1 percent in later stage firms, while time spent on customer contacts increases from 34.1 to 48.6 percent (p. 98).

Robinson and Pearce (1986) investigated the shifts in company strategy as a function of the life-cycle stage of its products for 77 small North Carolina manufacturing firms; they found little significant differences among the *relative* importance of ten factors affecting strategic management as the stage of the product life cycle changes. Rather consistent with Van de Ven's findings, they find a strong decline, as products evolve from development toward maturity, in CEO concerns for changes in process design; changes in product design; risk of producing the product; and emphasis on creativity. All these shifts contribute to diminished focus on technical issues as products evolve. However, Robinson and Pearce do not have complementary information available in regard to changes in concerns for market-related factors. Tushman and Romanelli (1985) trace the more dramatic changes that occur over the longer-term growth of companies from their emergence to maturity. Neither study looks at the early perhaps more modest changes that occur immediately postformation, as the primarily technically trained entrepreneurs begin to operate the companies they have founded. In contrast Teach, Tarpley, Schwartz, and Brawley (1987) examine the changes in key actors during the early years of evolution of software companies. "While only 12.5 percent of the original founders came from the marketing area, almost double that, 23.5 percent of the new principals came from a marketing position.... This shift was compensated in a large part by the almost disappearance of R&D as a source of new principals" (pp. 467–468).

In their recent paper Kazanjian and Drazin (1989) present and test a model of technology-based firms, "postulated to evolve through four discrete stages of growth—Conception and Development, Commercialization, Growth, and Stability" (p. 1489). Tracing 71 companies with tangible products over an 18-months period, they find that 28 ventures advanced and 14 regressed one or more stages during this relatively short observation period, providing partial support for their model. The dominant problems faced by management change significantly as high-technology firms proceed through the four stages.

From years of active involvement with the high-technology community of Cambridge, England, Matthew Bullock "has articulated a general model... a spectrum of risk, financial and technological, that faces a young high technology company. The company can get established at the low-risk end (a 'soft' start-up) and move along a development path that enables it gradually to take on bigger risks (i.e., to 'harden'); or it can start up anywhere along this path even, in exceptional cases, at the fully 'hard' extreme" (Segal Quince Wickstead, 1985, p. 66). Bullock sees that spectrum as including four typical stages of prototype businesses: (1) consulting; (2) reasonably standard analytical or design service, available on a custom-specific contract basis; (3) design and production of a particular product, probably on a customized basis; and (4) increasing standardization of the product (pp. 66–67). Cambridge, U.S.A., seems very similar to Cambridge, England, in these perspectives.

A Model of Organization Evolution

Inferences from the systematic research and observation cited earlier as well as that presented in the earlier chapters of this book, supported by intuition, personal experience, and anecdotal testimony from the likes of *INC.* and *Venture* magazines, provide bases for hypothesizing how a techno-logical enterprise changes during its early years. As indicated previously, technical companies are founded primarily on the basis of some techno-logical advance, rather than on the founders' presumption of competitive advantage in regard to marketing, sales, or distribution. The founders are for the most part engineers, with some marketing/sales and business expe-rience present in the multifounder teams. Not all technical entrepreneurs have unique ideas or a high need for achievement; some of them are initially merely fleeing dissatisfaction with a prior job or are pursuing inde-pendence, without specific ideas for product or market.

Consequently, reasonable expectations in regard to the early evolution of technological firms might include three notions:

1. Technology-based firms are initially divided between the intended sale of manufactured products (and/or repetitive services) and the intended performance of technical consulting and contractual de-velopment work, with evolution of the firms occurring over time toward more product focus,
2. High-technology entrepreneurs are initially oriented toward engi-neering and technology, not sales and marketing, with evolution toward marketing occurring over time, if the firm survives,
3. Multifounder firms show greater tendency toward both product and sales/marketing orientation initially, and also evolve more rapidly along both dimensions, than single-founder companies.

This chapter presents evidence from the research studies as to these three aspects of company change and development. In addition to the research samples used thus far in the book, a special subsample was generated to gather more detailed information about founder time allocation and company change over the first two years of new enterprise operations.[1]

Macroscopic Change: The Firm's Focus

The research provides insights into changes in the general focus and direction of the companies, in terms of what businesses they were in and their primary activities in achieving business objectives.

Initial Type of Business

The structured interviews my assistants and I conducted demonstrate that many entrepreneurial founders are unclear when they start their companies as to who will become the initial customers for their products or services; some are not even sure as to exactly what their companies will be doing. A number of the entrepreneurs, especially among the sole founders, might be regarded as rather flexible. In many cases the founders hope to do some consulting or contract research and development and, while living on the income from this work, plan to develop a product or find a product that they can exploit. It should not be surprising then to find a great number of the firms engaging in consulting and R&D work at the beginning, as the "Beginning Totals" row of Table 6–1 indicates. Forty-one of the 109 companies listed (or 38 percent) were solely engaged in consulting or R&D contracted to government agencies or to larger industrial firms. It is difficult to define the difference between these two categories of business, as various "consultants" describe their work as including analytical work, designing or developing a special system for a customer, or solving a particular technical problem for a customer, usually for an industrial client. Other entrepreneurs categorize the same work as being "contract R&D". The remaining 68 of the 109 firms (62 percent) were producing or refining for production software and hardware products, including 24 that were producing while concurrently carrying out contractual R&D work.

About one-third of the producers began as "job shops", making limited numbers of products to special orders; the others had proprietary products from the outset or under development. Occasionally this is a fine distinction, because some of the "standard" products of a technology-based firm are expensive machines or instruments with high unit cost (several thousand dollars each), that are adapted individually to each customer's requirements. Nevertheless the underlying product is the same in these cases.

As partial confirmation of the first dimension of change presented, note that only 64 percent of the technology-based firms in the sample began with some degree of product focus. In a separate study of 23 companies that had emerged from a large diversified technological firm, 61 percent started out with production focus, often also in conjunction with providing

Table 6–1

Business Orientation of Technology-Based Firms: Changes from Founding

Type of Business—Later	Type of Business — Beginning								Later Totals
	Production Hardware	Production Software	Production Hardware and Software	Production Hardware and R&D	Hardware Software and R&D	Production Hardware, Software, and R&D	Contract R&D	Consulting	
Production—Hardware	**28***	—	—	3	—	—	2	1	34
Production—Software	—	**8**	—	—	—	—	—	2	10
Production—Hardware and Software	—	1	**1**	—	—	—	1	1	4
Production—Hardware and Contract R&D	2	—	—	**12**	—	1	4	3	22
Production—Software and Contract R&D	—	1	—	—	—	—	2	1	4
Production—Hardware and Software and Contract R&D	3	—	—	2	1	**5**	5	2	18
Contract R&D	—	—	—	—	—	—	**7**	2	9
Consulting	—	—	—	—	—	—	—	**8**	8
Beginning Totals	33	10	1	17	1	6	21	20	**109**

*The bold numbers on the diagonal indicate how many firms remained unchanged in each business category, from the beginning until later.

R&D services. Far more severe are the results from research of Olofsson and co-workers on young technically-based companies originating from seven Swedish universities (1987). Half of the 90 "significant companies" in the Swedish data primarily carry out contract development work for customers; 30 percent do mostly consulting; and only 10 to 20 percent have their own products as their main line of operation. Presumably even fewer of the "less significant" Swedish spin-off firms have a product focus.

Initial fields of business of high-technology companies, of course, reflect in general their founders' technical backgrounds and the areas of emphasis of their source organizations. For example, MIT Lincoln Laboratory's major contributions have been in the digital computer area (both hardware and software) as well as in radar and communications. Not surprising, most of its spin-offs fall into related areas of business activity. They include firms in (1) computer software: technical programming, systems analysis, mathematical modeling, simulation, time sharing, computer services; (2) computer hardware and associated components: digital computers, digital logic modules, film readers, digigraphic equipment, display systems, power transistors, chemical compounds for semiconductors; and (3) radar and communications: components for troposcatter radio, troposcatter systems, radio detection devices, radio propagation, flicker free color display, radomes, crystal lattice filters, coding systems, airborne radar equipment, meteorological systems and instruments, microwave diodes, varacter diodes. Most of these products and services are based on direct or partial technology transfer from Lincoln. Other less expected spin-off firms were also founded by Lincoln alumni: a company that markets a device that automatically photographs the inside of sewer pipes to detect flaws or cracks; one that developed and manufactures a new type of shoe last; a manufacturer of fiberglass boats; a producer of cryogenic containers; a medical electronics company; a small glass-blowing shop; and two firms in different aspects of the personnel business. The latter group of "unusual" companies tend to have less or no technology transfer from Lincoln Laboratory.

Changes in Type of Business

The overall data enabled reexamination of the state of the companies at a time when they averaged five to seven years old. By this point in time these firms have certainly evolved from their initial condition and some have already demonstrated important growth and success. Other companies, while evolved and perhaps more stable, are less impressive in their achievements, except perhaps for their survival to date.

Table 6–1 also presents information on the later (average age, five to seven years) type of business activities for 109 companies. Underlying these numbers is a basic change in the nature of work done within the firms. For example, when operations began for these new firms only 57 of 109 (52 percent) of the enterprises possessed (or were developing) a hardware product. By the point in time several years later, documented here in the "Later Totals" column, 78 (or 72 percent) of the firms have at least one hardware product that has been marketed. In most cases the products were continuations and evolutions of their start-up activities, including for some bringing to fruition the product development efforts that were initially underway but not yet completed. For a few firms products resulted as an indirect (and sometimes unintended) consequence of contract R&D or consulting efforts carried out for governmental or industrial customers. One company became a major supplier of industrial electronic security systems as a result of bootlegged internal efforts by technical staff who had been working on somewhat related military contract developments.

Bernard Gordon, the MIT alumni founder of three successful companies in the analog/digital conversion field, started his first company, Epsco, in this pattern, an engineering company that would develop products for other companies. But Epsco quickly evolved into manufacturing products of its own development in areas of precision instrumentation and computing. Gordon Engineering, Bernie's second company, did technology and product development primarily in the same area, leading to his third and most successful firm, Analogic Corporation, a major player in sophisticated A/D conversion. Thus, Gordon's companies evidence the evolution from contract engineering to product both within each firm and across the set of three firms.

Occasionally even a consulting firm "gets lucky", and ends up being primarily a product-oriented firm. Such is the case with Management Decision Systems, Inc. (MDS is now merged into Information Resources, Inc.), a sophisticated market-research consulting firm started by my Sloan School colleagues, Professors John Little, Glen Urban, and Leonard Lodish (now at the Wharton School). Growing steadily as a state of the art academically based consulting firm under the leadership of our alumnus John Wurts, MDS exploded when, to avoid repetitive programming in their own projects, one of their programming staff developed the EXPRESS modeling and data-handling software system. EXPRESS took off as a software product, transforming the nature of the company and generating rapid growth of revenues and staff. My consulting firm, Pugh-Roberts Associates, experienced a similar phenomenon, although unfortunately not yet with the same extent of impact as MDS, when Kenneth Cooper developed PMMS (Project Management Modeling System) as a software product to service our growing number of customers who were needing help in large-scale project planning, cost analysis, and management.

In other instances the development of a product is the result of a conscious decision to change the character of a firm. This was especially true for several firms that had initially been engaged solely in contractual development work. Although this work often provided a stable source of entrepreneurial income, a number of the founders of these firms realized that far greater profit margins and better opportunities for corporate growth exist in the sale of products. These companies set about deliberately to develop products or to acquire ongoing firms with products that might complement their R&D service work. A few companies move in the opposite direction, supplementing their product sales with contractual R&D work, occasionally with the intent of using these funds and exposures to develop still further products.

The hierarchy of type of business preferred by the technical firms is suggested by the changes shown in Table 6–1, which presents a scattergram of the firms' beginning and later functions. Only eight of the 20 original consultants are still doing that kind of business; no firms in this particular sample shifted over time into consulting work. Firms originally doing just contract R&D (21 of the 109) remained intact, added production efforts to their R&D, or evolved into production alone. When shifts occurred for companies that were initially in both production and contract R&D, they were either to a more encompassing scope of those initial activities or into hardware production only. In the cases where initial producers evolved it was only to add contract R&D to their work mix.

Despite the changes in product status and manufacturing orientation described, the data in Table 6–1 present a relative lack of change for most of the firms in their overall types of business being pursued. The bold numbers along the principal diagonal, when compared with their column totals, reflect this constancy. Except for those 41 firms initially engaged only in consulting or in contract R&D, only 14 companies of 68, the off-diagonal firms (21 percent), deviated from the type of work they had initially undertaken. And in ten of those fourteen the change was to add other types of work to their original activities, five of them adding software production. Three of only four firms, which reduced their scope of work, went from hardware production plus contract R&D to hardware production only. The companies as a whole have clearly evolved toward more general business operations as firms engaged in their own product development, manufacture, and sales, moving from 62 percent initially engaged in production of some form to 84 percent.

In confirmation of the first area of expected evolution, note that the sample has indeed moved toward a more product-oriented focus. But also recognize that 16 percent of this sample, now on average five to seven years old, still do not qualify for inclusion in samples such as that of Kazanjian and Drazin (1989), companies with tangible products. A comprehensive "stages of growth" model of technology-based companies must take into account the large number of firms that never evolve into product manufacture and sales.

Supporting evidence is found in several other studies. Among nineteen computer-related firms, principal changes occurred over time in the two firms initially doing consulting and the two others initially performing only contract R&D. In that sample the number of firms involved with computer hardware and/or software production, alone or in combination with contract R&D work, grew from 79 percent at their foundings to 100 percent by the time of data collection. Comparable transitions occurred in the 26 biomedical companies studied. The two medical companies that started in sales and distribution only, plus the four that initially performed R&D and consulting work exclusively, moved into integrated operations covering the spectrum from R&D to manufacturing and sales, joining eleven other companies that started in that integrated mode. Nine of the biomedical firms maintained development and/or production focus throughout their lives, not undertaking any sales efforts, a pattern not unusual in the medical field where many smaller companies license or distribute their products entirely through other much larger corporations. And again in a cluster of 18 recently formed technological companies, R&D contractors went from four initially to only one six years later, while consultants went from three to one. In contrast, among the 23 spin-offs from the large diversified technological firm, several who were initially engaged at least in part in production dropped their hardware activities after encountering major problems in developing or selling products. For them a shift to consulting was undertaken to survive. These are among the only companies in all my entrepreneurship studies who "regressed" in the product-focused evolutionary pattern.

The many company changes in these several samples from original work in contract R&D or consulting indicate that much of this had been done merely "to get things going". For example, nine of the 15 firms that left the MIT Instrumentation Lab and started as consultants claimed that they intended this consulting activity to be transitional only. For a few firms, however, contracting and consulting still remain as the desired work. Several of the entrepreneurs have no inclination to expand beyond their consulting work, which provides for them substantial personal income even if their organizational size remains small. This is consistent with the finding in Chapter 3 that not all entrepreneurs have a high need for achievement. Unfortunately the data are inadequate to test whether the individual entrepreneur's motivational characteristics relate to the type of business he or she pursues. On the other hand, contractual R&D work is not necessarily done just "to get by" or to maintain independence. One firm in the sample, remaining entirely in the R&D contracting business, has already achieved significant sales and is growing rapidly. A number of founders of the seventeen companies that are still doing only R&D or consulting are genuinely disappointed. They had hoped initially to find or develop a product and transition into a manufacturing firm but they have been unable to make the change; some still assert they will shift their businesses in the future.

Microscopic Change: The Founders' Activities, Time Allocation, and Orientation

Turning to the second aspect of anticipated evolution, most technically-based companies start with technology uppermost in their minds and time commitments. But I presume that many entrepreneurs evolve toward more market-oriented perspectives and activities. A smaller but more detailed subsample (18 firms) was assembled to permit analysis of founder time allocation and activities that can evidence evolution of market-orientation, collecting data from the entrepreneurs regarding two time periods, the first six months of the company and eighteen months later, the end of year two. Except where specified, this smaller subsample is the source of data used in this section.

The First Six Months

Time Allocation. Aldrich and Auster (1986) note that young companies suffer "the consequences and strategic implications of two variables which affect metamorphosis—age and size." How technical companies begin to adapt to these variables is evidenced in Table 6–2, displaying for the four major operational areas the average percent efforts by company founders during the first six months of existence of the eighteen technical firms in the subsample. About 30 percent of their total working time is spent in engineering efforts, with a similar proportion in sales and marketing efforts. Twenty-five percent is devoted to manufacturing while the least percentage of time is used in financial and administrative activities. A discussion follows on the variations among the firms in these functional allocations.

The only companies spending more than 50 percent of their time in engineering both produce custom products. In contrast two of the three firms with no engineering activities supply services, while the third is selling a fully developed standard product. Three firms spent no time in manufacturing. One of these was still developing its product. Another had a product that was engineered for special uses. The third firm was solely involved in supplying programming services and classifies none of its efforts as manufacturing. Among the 18 firms providing detailed time allocation data manufacturing efforts are not statistically associated with other relevant variables, product manufacture being treated in the embryonic firm as largely an end result of all other activities.

Similar variation occurs in the sales and marketing time allocations. Only one firm did not spend any time on sales during its first six months. That company devoted its entire first year to developing a marketable product. Two other firms spent less than 10 percent of their founder time in sales/marketing; both made custom products.

The amount of initial capital correlates significantly with the early time

Table 6–2
Effort Allocation by Founders During
First Six Months ($n = 18$)

Operational Area	Percent of Total Work Time
Engineering	31
Sales/marketing	28
Manufacturing	25
Finance/administration	16

allocation to financial and administrative activities ($p = .02$). Perhaps more money requires more time to manage money, or technical entrepreneurs might just assume increasing financial responsibility as the funds dictate. It is not difficult to discover why several firms did not spend any time in this area. Four of the five firms with zero finance time had initial financing of only $1 thousand, with the fifth firm being unwilling to reveal its initial financing.

The monitoring of founder efforts during the first six months included an attempt to identify the extent to which they were aware of their competition, one aspect of market-orientation. This seems so obvious that some may be surprised to know that entrepreneurs frequently claim they have no competitors, their own products or services being so unique in their own opinion that no one else's outputs are relevant. These entrepreneurs sometimes discover, too late, that other firms are offering similar capabilities to the marketplace. Being aware of competitors should help to shape the course of entrepreneurial efforts, as indeed is shown simply in Table 6–3. Here the sample is split into two clusters based on the measure of competitor awareness and the mean percentage efforts are displayed for each group. Those who are aware of their competition reveal a more balanced effort allocation across the operational areas, even during the

Table 6–3
Early Effort Allocation Based on Awareness of Competition
(percent) ($n = 18$)

	Finance and Administration	Sales and Marketing	Engineering	Manufacturing
Aware	17.8	31.2	28.7	22.3
Unaware	5.0	17.6	40.7	36.7

first few months of the firm's existence, with about twice the orientation toward sales and marketing (at the expense of engineering and manufacturing), the primary prospective sources of still further insights to customers and competitors.

Sales and Marketing. The backup details on the early sales and marketing function are enlightening. This area of activity includes determining the existence and needs of a particular market, sales, and distribution within this market, customer service, advertising, and promotion. In a large company each of these activities is generally performed by a different individual or group who in turn reports to a director of marketing. These same functions must be accomplished somehow in the small firm, albeit with considerably less available manpower, and therefore, by allocation of scarce time, especially initially. The data here show that the percentage effort devoted to sales correlates negatively with the percent of customer contact made to take orders ($p = .13$), but strongly positively with the percent of customer contacts made to determine customer needs (.09) and to estimate market potential (.07). To some firms the sales/marketing function is narrowly proscribed, just the direct selling of the product. The more this order-taking perspective applies, the less is the time allocated overall to sales and marketing efforts. The more the firm views sales and marketing as a mechanism to observe and use market information, the greater the proportion of founder time devoted to this function. For example, some unusual founders believe that how customers might use their products is critical to eventual company success. They, therefore, spend lots of time visiting prospects to increase their own understanding. The presumed result is increased visibility of the total market by the company's leadership, with attendant impact on such issues as product targeting, design, selling approach, and accompanying service.

Fifty percent of the enterprises initially rely solely on direct contact by a founder to sell to its customers. While this is self-limiting it also seems beneficial in several ways. In the new technology-oriented firm the founding entrepreneur is very possibly in the best position to explain to customers the virtues of his company's product. In general he is not selling a standard product that the customer needs to run the operation, but rather one the entrepreneur strongly advocates for improving the customer's operation. Perhaps more important is that direct contact of a founder with a potential customer is an excellent means for synthesizing product ideas with market needs. One downside consequence is that the entrepreneur frequently responds to specific customer issues by redesigning the product on the spot, adding special features, and becoming a multi-product custom-oriented firm before he has given the intended standard line a chance to sell. This has been a common problem in several companies I served as a board member, including both Data Technology and Geophysical Survey Systems. The technically brilliant founders of both firms found solving unique problems of their customers both satisfying and somewhat easier to accomplish than selling their catalog items.

The numerous means of selling other than direct contact are almost nonexistent during this earliest period of company life. For instance, only three of the eighteen companies used a sales force and only two others used outside sales representatives. As we shall soon see, when the companies gradually begin ironing out their start-up difficulties their selling practices shift significantly.

The high incidence of direct contact selling by a founder (occurring in whole or in part initially in 96 of 109 companies in one sample) definitely reflects both the newness of the firms and their small size, the "liabilities" mentioned earlier by Aldrich and Auster (1986). But personal contact is not necessarily an ineffectual way to proceed. Indeed the personal representations of a founder to a potential customer are frequently perceived as the primary reason a project or product is sold. Of more than incidental note is the fact that the sales methods used do not vary significantly across the several different types of business initially undertaken, nor are there perceived differences in effectiveness of these approaches as a function of any other company characteristics.

It seems interesting to note these entrepreneurs' identification of the primary source of new product ideas during their first six months. One entrepreneur answered, "From the V.P.'s brain", communicating what I too often find to be the case in technical enterprises, the presumption that the uniqueness of a single person's ideas will suffice to generate product innovation and corporate growth. Research studies over many years (reviewed in Utterback, 1974) demonstrate far higher success for products that are responsive to "market pull" rather than "technology push". In many industries some customers go beyond "demanding" a new product and actually create products to meet their own needs. Indeed, as noted at the outset of this chapter, its own internal needs led General Computer into developing Hyperdrive for the Macintosh. Following this pathway can become the most assured means to achieving successful innovation for the responsive producer (von Hippel, 1988).

Numerous companies in the data samples, founded by technologically sophisticated entrepreneurs, "presume" market needs based on their own prejudices or "feel" rather than probe potential customers. Due to the high technical competence of most of these entrepreneurs, and the high regard in which they are often held in their particular technological field, such a feel is often a legitimate and sufficient basis for producing and selling a product or service. Too frequently, however, the engineer part of the entrepreneur overrules his as yet underdeveloped business sense, and he produces a product because it is technically appealing. Unfortunately no market may exist for this ingenious bit of engineering wizardry. The classical miscalculation of market "need" (and a stereotyped one) is the firm in the sample that decided to produce a circuit because it was a "cute" engineering design. When the company tried to sell it, it found dozens of competitors making similar circuits and few customers with a need for it. Another misperception was an MIT spin-off that made a small and inexpensive transistor tester to replace a popular unit costing several times as much. No sales resulted. "Everyone buys the expensive system", the entrepreneur decried. His market, the defense industry, is far more concerned with versatility than with price. This insensitivity to customer inputs does not characterize all strong technical companies. Interestingly, Bernie Gordon, successful founder of three firms in the analog/digital conversion field and many people's stereotype of "an engineer's engineer", in fact manifested almost a "pure marketing attitude" from early in his career. "You have to understand the customer's economic goals, his technical goals and then identify his engineering, economic and management point of view", Gordon observes. "At Analogic we go through this process and *then* we present the customer with a solution to his problems" (Loftus, 1978).

Table 6–4 shows the primary idea sources identified by the subsample companies. Those noting previous job requirements as a product idea source are indicating one form of sensing of market need. One example from the overall sample, Ken Olsen, the founder of DEC, started his company by producing high-speed transistorized circuit modules that he said he could have used in his MIT Lincoln Lab computer development projects. Table 6–4 suggests about a 50-50 split between external, that is market, "sourcing" of product ideas and internal, that is, founders, sources. In Chapter 9 the possible ties between product sources and new company success are assessed.

In partial confirmation of the second general expectation, note the variety of measures that indicate relative lack of market orientation by the typical technological entrepreneur at the time of company founding. The dominant time allocation to engineering and the lack of formal marketing and sales organizations reflect this condition. Yet, clearly some founders are more sensitive than their counterparts, aware of their competition,

Table 6–4
Primary Source of New Product Ideas
($n = 16$)

	Number of Firms
Previous job requirements	5
Customer	1
Customer-sponsored R&D	1
Founders, key employees	6
Product line evolution	1
Other	1
None	1

placing more investment in the marketing function, using market contacts from the outset of their companies to assess customer needs and develop new products based on customer inputs.

The Next Eighteen Months

Sales and Marketing. By the end of year two of existence fifteen of the eighteen companies studied in detail had a sales force (up from three during the first six months), used either alone or in combination with other methods, to sell to their customers. The number of potential customers contacted per week by these sales forces ranged from one to forty, with a surprisingly low median of three customer contacts per week. This small number might reflect a high concentration of customers. This contact number did increase as a function of the company's concentration upon the industrial market ($p = .077$), government or consumer orientations apparently tending to demand fewer direct selling contacts.

Sales representatives are also used increasingly as the new enterprises develop during the next eighteen months, nine firms employing them both to contact potential customers and to distribute the company's products and one company using reps just for distribution purposes. Five other methods of selling are used in varying degree by the firms, including mailing lists, new product "releases", attendance at technical shows, authoring of technical articles, and magazine advertisements. In addition those companies oriented primarily to government markets respond to formal government Requests for Proposals or Quotations and also submit unsolicited proposals to various agencies. All these methods tend to reflect the selling of a product's technical content or performance, an attribute usually associated with the industrial and governmental markets that are the dominant customers of the firms studied. The low use of magazine advertising is consistent with the lack of consumer products.

As the company develops, its understanding and use of various aspects of customer contact also change. Contacts can be divided into three categories: selling, servicing, and researching. Selling includes both direct sales efforts and taking orders. Servicing covers discussing technical or delivery problems and procuring product specifications. Researching involves evaluating competition, determining customer needs, finding other possible customers, and estimating market potential. Table 6–5 shows the average percentage contacts in each of these categories as computed from the data collected from the eighteen companies subsample.

Only by knowing the needs of the market can an enterprise develop highly saleable items. By knowing its potential market size the management can better decide if undertaking a development program is worthwhile. Equally important, the market is a haven for suggestions for new products. Yet technical entrepreneurs often do not particularly appreciate these perspectives. Those teams with prior experience in marketing operations best understand the need to research the market and not just sell to it. Years of

Table 6–5
Purpose of Customer Contacts ($n = 18$)

Contact Category	Percentage
Selling	*38.4*
Selling	31.9
Taking orders	6.5
Servicing	*28.7*
Technical/delivery problems	18.2
Procuring product specifications	10.5
Researching	*32.9*
Evaluating competition	3.6
Determining customer needs	19.3
Finding new customers	4.2
Estimating market potential	5.8

prior founder experience in sales correlate closely with the time allocated toward contacts made to estimate market potential ($p = .065$) as well as with contacts made to determine customer needs (.047), but negatively with contacts made solely for selling (.082). Interestingly this focus on getting insights from the customer, rather than just selling to the customer, is strongest for those firms that serve the defense market. Close statistical associations are found between the overall percentage contacts with the military market and the percent of customer contacts made to determine customer needs (.006) as well as the percent contacts made to estimate market potential (.082). Technical entrepreneurs trying to sell to the military seem more willing to accept that customer's demands as "givens" that need to be uncovered, rather than assuming that the market will beat a path to the door of "brilliant" entrepreneurial ideas.

Evolved Operations

Turning back now to the larger sample helps assess the extent of further evolution of company operations as the firms reach an average age of five to seven years. As a baseline a study of Michigan technical entrepreneurs provides some details that suggest possible implications of the shifts that occur in business type, as shown earlier in Table 6–1, on the founders' time allocations. Braden (1977, p. 45) divides 69 firms, that are on average eight years old, into standard products/ services, custom products/services, and R&D/consulting. The percentile distribution of founder time differs substantially among these three different business types: (1) engineering/R&D: 16.3, 23.2 and 47.3 percent, respectively; and (2) marketing: 20.1, 17.4, and 16.7 percent, respectively. To the extent that the Michigan cross-sectional data can be applied to changes over time among the

Greater Boston entrepreneurs, the founders in my sample might be expected similarly to shift their time allocation heavily away from engineering/R&D and to somewhat increase their marketing percentage as the mix of their business activities shift over time from R&D/consulting to custom products to standard products. These founders would also be expected to increase their attention to production and finance activities as their mix of business types changed.

Sales and Marketing. The sales methods employed by the technological firms at their later stage of evolution remain heavily dependent on the founders. Forty-two percent of the MIT spin-off firms still use only the founders for selling contacts, but 19 percent have added sales reps and 9 percent utilize a sales force as founder supplements. The other 30 percent of the companies split between sales representatives and the company sales force, alone or in combination. Sales representatives serve to bring together the products of a number of companies, usually small ones, for hopefully efficient and complementary presentation to prospective customers. This "shared sales force" is seldom seen by the entrepreneur as the most effective means of achieving a significant penetration of the market but rather is regarded as a "necessary evil". A broad product line and some level of achieved sales are usually perceived as necessary economically for a firm to afford its own direct sales force. The increase of direct sales forces is, therefore, an indication of both growth and transformation of the firms. Initially only nine companies in the sample had sales forces; now thirty-two have their own sales forces. The sales methods used do not vary greatly among the different types of businesses.

About 30 percent of the enterprises have still not used any means to advertise or promote their work. For those that do advertise the methods characterize the technical nature of the work being done, including direct mail, trade shows, trade advertising, product releases, or combinations of these four. None of the statistics on promotional approaches used relates significantly to any other major variable associated with these technological enterprises.

This lack of use of sophisticated marketing is not limited to the Greater Boston high-technology community. A recent study of R&D-intensive companies finds primarily technology-related aspects of the firm as their key "marketing tools" (Traynor and Traynor, 1989). In their cluster of firms that are most comparable to the entrepreneurial companies in my studies, Traynor and Traynor find that high-technology companies under $10 million in sales rate their most important marketing approaches to be product image (reputation), personal selling efforts, and having state of the art technology. Among promotional tools those companies see greatest importance in sales and sales management activities and trade shows.

Although existence of a formal marketing department is neither necessary nor sufficient for establishing a marketing-orientation in a technological firm, its use here as an index of organizational evolution seems reasonable. Only 46 percent of the 110 responding companies in this sample have created marketing departments, apparently in all types of businesses. Sixty percent of the 110 firms do sales forecasting, primarily those engaged in hardware production. Formal sales forecasts are generated by 85 percent of the firms doing solely hardware production, 60 percent of those doing hardware plus contract R&D, 50 percent of those producing both hardware and software products, and 80 percent of the companies engaged in contract R&D along with both hardware and software production. Only 20 percent of the companies without any hardware production do sales forecasting. Thirty-five percent of the firms carry out analyses of market potential, again done primarily by hardware producing firms.

The data do not indicate the exact timing of establishment of the specific marketing-related operations just described: the existence of a marketing department, the development of sales forecasts, and the performance of market analyses. Some of these conceivably may have begun at the very outset of the firm; others were no doubt established later as the firms evolved.

The data analyses at the three time periods—founding, year 2, and years 5 to 7—provide strong confirmation of the second hypothesis regarding company development and change. A small percentage of technological enterprises begin with an orientation toward their markets and toward serving their customers' needs. Many of the companies gradually evolve in this direction, manifesting their shift in both time allocation and formal market-related activities. Clearly not all technical entrepreneurs have made this transformation by the latest time studied in these samples, five to seven years postfounding. Some may change at a still later time in their company lives but skepticism on this issue seems reasonable. This evolution toward a market orientation "split" among technically trained entrepreneurs has held over several decades of new company formations: A similar 40 percent of my most recent sample of new computer-oriented firms has also developed marketing departments at some point in time after they were founded, moving away from initial 100 percent dependence on one or more of the founders. The cluster of spin-off companies that originated from the large electronic systems firm examined also includes 50 percent with marketing departments, the slightly higher ratio perhaps reflecting a benefit of their industrial experience.

Effects of Multiple Founders

The third expectation is that the larger-size founder groups reflect stronger product and market focus at the outset of their companies and evolve more rapidly in these dimensions. Both the overall sample and the more detailed time data from the subsample help test this hypothesis.

Smaller founding teams that, therefore, have available and work fewer total hours are constrained by the requirements of engineering. Being a necessary part of initiating a product-producing technological enterprise, engineering activities, therefore, consume a large proportion of a small group's time. If the founding group is larger, the entrepreneurs can fulfill the requisite engineering activities and still have some time remaining for filling the other "buckets", ending up devoting an increasing portion of their time to contacting the market. The actual hours spent in engineering are not diminished, but the proportion of their time spent on engineering is. Statistical findings from the detailed subsample are in accord with this explanation. The more total founder hours worked, the more the number of hours spent on engineering efforts ($p = .13$) but the less the percent of time on engineering (.08).

As the size of the founding team increases greater proportions of their time are spent in efforts to sell the company's products (.04). Another measure of the same phenomenon is that total founder hours worked per week correlates positively with percentage time spent on sales (.018). This presumably draws on the increased years of prior sales experience encountered within the larger founding teams (.004).

Data from the overall sample confirm and amplify the findings from the smaller but deeper analysis presented previously. In particular the advantage of multiple founders is evidenced in several characteristics of business operations, even during the early months. For example, none of 43 single-founder companies initially had a sales force. Ninety percent of them relied on the founder to do the selling, with the remainder depending on sales representatives alone or in conjunction with the founder. Nine of the multifounder companies set up sales forces immediately and 30 percent were able to use selling methods other than personal contact by a founder. The single founder has to sell along with performing all the other tasks. In most multiple-founder companies, the sales/marketing task is the responsibility of one (or on rare occasions several) of the founders who devotes his principal time to that job, reflecting a natural division of labor.

The figures in Table 6–6 from the overall sample do demonstrate that the carrying out of these market-oriented operations is strongly associated with the number of company co-founders. Thirty-five percent of the companies founded by one person versus 50 percent of the multifounder firms set up marketing departments; 41 percent of sole entrepreneur firms do sales forecasts in contrast with 70 percent of the multiple entrepreneur companies; and market analyses are performed by only 26 percent of the enterprises founded by one person as opposed to 41 percent of the other firms. Teams of founders obviously can undertake more tasks in parallel than a single founder, which indeed may well have motivated the establishment of many of the multiple founder groups. The detailed subsample data shows the shift of founder time allocation toward sales and marketing as the size of the founding group increases. Table 6–6 supports the perspective that when limited by available human resources the technical entrepreneur gives priority to the technical aspects of the business with which

Table 6–6
Marketing Operations as a Function of the Number of
Founders ($n = 110$)

Number of Founders	Marketing Department		Sales Forecasting		Analyze Potential Markets	
	Yes	No	Yes	No	Yes	No
1	12	22	14	20	9	25
2	8	16	15	9	8	16
3	13	9	14	8	9	12
4	4	2	4	2	4	3
5	3	3	6	0	3	3
6	3	1	4	0	1	3
7	1	0	1	0	1	0
8	0	0	0	0	0	0
9	0	1	1	0	1	0
Totals	44	54	59	39	36	62
*	$p =. 027$		$p = .000$		$p = .038$	

* Mann-Whitney U significance levels, indicating the strong association of each of the three characteristics of marketing operations with multiple founders.

he is most familiar, and the development of a marketing organization and its related activities is delayed by default. The evidence strongly confirms the third hypothesis on the positive impacts of multiple founders.

Operating Problems During Evoluation

Regardless of how successful they may be, few of the entrepreneurial firms have evolved their businesses from start-up into successful ongoing growth without encountering some major difficulties. Perhaps this explains why 45 percent of the companies have hired management consultants to assist them at one time or another. The major problem area identified by most of the firms, consistent with the theme of this chapter, has been sales, with 55 percent of the companies claiming this as their first or second most important area of concern. Interestingly only one company stated that marketing is the activity that caused the greatest need for capital funds. This contrast speaks to the continuing relative marketing naivete of many of the technical founders. To many marketing is still viewed solely as the getting of orders or contracts. As shown in Table 6–6 less than 40 percent do market analyses, even after five to seven years of evolution of their firms, and many still do not seem to understand the concept of market feedback. The founder as seller is still dominant and is seen as the marketing functionary; with that perspective, of course, marketing requires little capital.

Two thirds of the entrepreneurs feel that the marketing of their products is not accomplished in a satisfactory manner. However, when asked what they might do differently to improve sales, most indicate they would hire more salesmen or do more advertising. Almost none indicate an intent to do more market research. They often admit to ineptness in commercial marketing techniques, and tend to develop products in line with their own lofty ideals as to what they feel their customers really need, rather than what the customers themselves might state they need. Further support for this skeptical assessment of the entrepreneur is found surprisingly in the answers to our question, "What has given you the most satisfaction as a principal in a new enterprise?" Almost half of the respondents indicate that the production of high quality goods or services is most important, a reasonable response for a technical director perhaps. An entrepreneur concerned about integration of all the elements of a business—technical quality, finance, marketing and general management—might be expected to derive satisfaction more from the effective integration of all these into a well-administered going concern.

Table 6–7 presents the frequencies of the major problems perceived by the entrepreneurs. Beyond sales issues, personnel acquisition, motivation, and management are identified as key issues, particularly for technical professionals. My most recent study of computer-oriented entrepreneurs finds the same ordering of key identified problems: sales, personnel, and capital support. Throughout the research studies founders have expressed major problems in locating and retaining qualified engineers, technicians, and technically oriented salespeople. The Greater Boston area has a strong supply of technical personnel but accompanied by such strong demand when economic conditions are good that skilled workers are often unavailable except by pirating from other firms.

Personality conflicts, especially among the multiple founders, have also loomed large as problems. Frequently, founding teams are established under implicit (sometimes even explicit) assumptions of equality. Recall the example in Chapter 1 of equal ownership established initially in Transducer Devices. When it becomes necessary for one founder to assume the position as principal decision maker, as inevitably occurs, resentment by other founders often follows. Internal quarrels often lead to open hostility with both indirect and direct impact on the business. Departure of one or more co-founders following such disputes is common, such as the withdrawal of Harlan Anderson from Digital Equipment Corporation, the firm he co-founded with Ken Olsen, or Matt Lorber's movement from Analog Devices after his co-founder Ray Stata took over as CEO. Partner tiffs, followed often by partner split-ups, are rather commonplace among these technology-

Table 6–7
Major Business Problems ($n = 93$)

Area of Problem	Primary Problem	Secondary Problem
Sales	34	17
Personnel	17	14
Personalities	7	10
Initial financing	2	1
Additional capital support	14	9
Production reliability	2	5
Research and development	2	1
Other	15	7
Number of companies	93	64

based firms, frequently becoming the most negative aspect of their entre-preneurial experience recounted by founders in later years. Differences among our General Partners in personal priorities and styles led, for ex-ample, to Arthur Obermeyer departing from Zero Stage Capital and being replaced by Jerry Goldstein, who also added desired experience in the biomedical arena. In several companies, I have advised the co-founders to provide an easy mechanism for reviewing and possibly shifting ownership percentages among the founders after one, two, or three years in operation. This provides a formal means that encourages and enables "blowing off steam" among the co-founders, leading to adjustments that frequently save the partnership.

Another form of personal conflict that commonly arises as the com-pany develops is not exactly a business problem but often was discussed during the interviews. An apparently high rate of marital problems occurs, often leading to divorce of the entrepreneur. This happens even though in most cases the entrepreneur's wife was very supportive of the company formation. Frequently the wife helped out at the business during its early days, doing secretarial, even janitorial, work as needed. Some professionally trained wives did the bookkeeping or the legal work. One wife handled not only all key negotiations but took charge of raising the finances through-out the company's life. Several entrepreneurs exclaimed, "The business became my mistress", remarking on the priority demands on time and attention generated by the firm. Many of these entrepreneurs indeed gave their greatest love to their businesses, feeling badly later that they had not spent more time with their wives and children. Despite marital breakups, most of the affected founders essentially admit they would not have changed their actions, had they a chance to do it over again. On more than one occasion the divorce was accelerated by long hours spent in the office with a co-worker of the opposite sex, leading to entanglements. When this was discussed openly, the entrepreneur often dismissed the importance of the extramarital relationship per se and chalked up the divorce as an unfortunate cost of doing business.

Despite the generally small amounts of initial capital, as shown in the previous chapter, only three of the founders see initial capital as being a major problem, while twenty-three see additional capital support as a critical area. Company efforts to raise secondary and tertiary funding following the founding of the firm are discussed in depth in the next chapter.

Summary and Implications

Primary changes occur in technology-based companies during their first several years of existence. Those documented in this chapter relate principally to the firms' orientation to product-based businesses and to marketing and selling activities within them. The sample that was analyzed here of 114 MIT spin-off companies went from 62 percent initially engaged in development and sales of their own products to 84 percent over the course of the first few years, the bulk of change coming from a number of companies departing from or supplementing initial focus on consulting or contract R&D work. At least those changing firms have shifted from engineering and technology as their almost total initial involvement toward some mix of product and market orientation.

These early years also witness increased formal commitment by the entrepreneurial founders to marketing and sales activities. Evidences include a reduction in the number of firms that are solely dependent on their founders for direct customer contact, paralleled by a dramatic increase in direct sales forces, and even larger growth in the use of sales representatives. Awareness of competition is a strong influence on company orientation, with those sensitive to competitive environments placing far more effort in the direction of marketing activities. Evolution of these technological firms, however, leads to still less than half with their own marketing departments, even after five to seven years of company growth and development.

Companies founded by more than one entrepreneur devote a larger proportion of their efforts to marketing and sales even from the outset, with less effort going into engineering. This tendency relates in part to the increased presence of prior sales and marketing experience in the backgrounds of multifounder teams. The multifounder firms more quickly employ sales forces than single founder firms, and also are more likely to develop marketing departments, carry out sales forecasts, and perform analyses of potential markets.

Most firms that started without product either are slow to evolve or never get to production status. Half the companies that originate without a product at least in development stage never make the conversion into manufacturing. Similarly, lack of market orientation at the outset is frequently not corrected merely by the passing of time, with formal market-related activities still missing from the majority of technology-based enterprises even after several years of company existence.

Both entrepreneurs and investors need to recognize the more balanced and accelerated approach to company development that is typically undertaken by the multifounder firm, which initially targets an explicit product market. Single founder firms are especially slow in developing formal sales and marketing approaches that go beyond their own personal skills and effort, thereby retarding the firm's evolution. Including sales and marketing skills in the initial founding group seems especially appropriate. Confirmation of the possible relations of these variables to eventual success and failure of the technology-based firm awaits later chapters. But in the meantime prospective entrepreneurs should strongly consider taking the rather low "risk" of adopting a product/market-oriented team approach toward company formation and development.

Notes

1. To assure reliable data recall on the first years of existence an original sample group of 96 MIT laboratory spin-off companies was screened to exclude those companies over five years old at the time of data collection. Limiting the sample further to be within the Greater Boston area for ease of data gathering, and omitting the high outliers in regard to size of initial capitalization to assure more comparability, twenty prospects were identified. Eighteen of the twenty companies (90 percent) agreed to cooperate with a more probing set of structured interviews using a detailed questionnaire that covered the first two years of company activities. The resulting small group constitutes 16 percent of the initial larger sample. The research design thus provides a form of longitudinal study for this chapter, with data gathered for three time periods in the lives of the participating companies: founding, two years, and five to seven years.

References

H. Aldrich & E. R. Auster. "Even Dwarfs Started Small: Liabilities of Age and Size and their Strategic Implications", in *Research in Organizational Behavior, 8* (Greenwich, CT: JAI Press, Inc., 1986), 165–198.

P. L. Braden. *Technological Entrepreneurship: The Allocation of Time and Money in Technology-Based Firms* (Ann Arbor, MI: Division of Research, Graduate School of Business Administration, 1977).

N. C. Churchill. "Entrepreneurs and their Enterprises: A Stage Model", in J. A. Hornaday, J. A. Timmons, & K. H. Vesper (editors), *Frontiers of Entrepreneurship Research, 1983* (Wellesley, MA: Babson College, 1983), 1–22.

H. R. Feeser & G. E. Willard. "Incubators and Performance: A Comparison of High- and

Low-Growth High-Tech Firms", *Journal of Business Venturing*, 4, 6 (1989), 429–442.

R. K. Kazanjian. "Operationalizing Stage of Growth: An Empirical Assessment of Dominant Problems", in J. A. Hornaday et al. (editors), *Frontiers of Entrepreneurship Research, 1984* (Wellesley, MA: Babson College, 1984), 144–158.

R. K. Kazanjian & R. Drazin. "An Empirical Test of a Stage of Growth Progression Model", *Management Science*, 35, 12 (December 1989), 1489–1503.

J. R. Kimberly, R. H. Miles, & Associates. *The Organizational Life Cycle* (San Francisco: Jossey-Bass, Inc., 1980).

C. Loftus. "The Driving Force Behind Analogic: Bernie Gordon", *Electronic Business*, May 1978, 20–24.

C. Olofsson, G. Reitberger, P. Tovman, & C. Wahlbin. "Technology–Based New Ventures from Swedish Universities: A Survey", in N. C. Churchill et al. (editors), *Frontiers of Entrepreneurship Research, 1987* (Wellesley, MA: Babson College, 1987), 605–616.

R. E. Quinn & K. Cameron. "Organizational Life Cycles and Shifting Criteria of Effectiveness: Some Preliminary Evidence", *Management Science*, 29 (1983), 33–51.

R. B. Robinson, Jr. & J. A. Pearce, II. "Product Life-Cycle Considerations and the Nature of Strategic Activities in Entrepreneurial Firms", *Journal of Business Venturing*, 1, 2 (1986), 207–224.

Segal Quince Wickstead. *The Cambridge Phenomenon*, second printing (Cambridge, England: Segal Quince Wickstead, November 1985).

N. R. Smith & J. B. Miner. "Type of Entrepreneur, Type of Firm, and Managerial Innovation: Implications for Organizational Life Cycle Theory", in J. A. Hornaday et al. (editors), *Frontiers of Entrepreneurship Research, 1983* (Wellesley, MA: Babson College, 1983), 51–71.

R. D. Teach, F. A. Tarpley, Jr., R. G. Schwartz, & D. E. Brawley. "Maturation in the Microcomputer Software Industry: Venture Teams and their Firms", in N. C. Churchill et al. (editors), *Frontiers in Entrepreneurship Research, 1987* (Wellesley, MA: Babson College, 1987), 464–473.

K. Traynor & S. C. Traynor. "Marketing Approaches Used by High Tech Firms", *Industrial Marketing Management*, 18 (1989), 281–287.

M. Tushman & E. Romanelli. "Organizational Evolution: A Metamorphosis Model of Convergence and Reorientation", in L. Cummings & B. Staw (editors), *Research in Organizational Behavior*, 7 (Greenwich, CT: JAI Press, Inc., 1985), 171–222.

J. M. Utterback. "Innovation in Industry and Diffusion of Technology", *Science* 183, 4125 (February 1974), 620–626.

A. H. Van de Ven, R. Hudson, & D. M. Schroeder. "Designing New Business Startups: Entrepreneurial, Organizational, and Ecological Considerations", *Journal of Management*, 10, 1 (1984), 87–107.

E. Von Hippel. *The Sources of Innovation* (New York: Oxford University Press, 1988).

CHAPTER 7

Finding Additional Financing

While beginning to evolve and grow, the postformation technological firm develops increased needs for funding. This chapter examines the subsequent financing of technical companies, including the sources and amounts raised during second and third rounds. It covers the entrepreneurs' search for additional capital and those pathways that are most productive. Additional financing increasingly depends on outside investors and, consequently, entrepreneurs frequently prepare business plans, whose characteristic deficiencies are assessed here. Finally, due to their greater importance for later stage financing, venture capitalists' decision making in funding high-technology firms are evaluated. The decision to go public, the process, and the consequences, are treated in Chapter 8.

Secondary and Tertiary Financing

Chapter 5 provides a picture of the evolution of technological firms through multiple stages of growth and development, with the attendant changes in capital requirements. It also reviews the various sources of company financing, from personal funds all the way to public stock offerings, and puts into perspective the likely differences in use of these sources during a company's life. As shown, most companies find that their initial funds are insufficient to support their operations and/or their growth during the early years. Typically 60 percent of the companies in the samples raise capital a second time (secondary financing) and about 30 percent obtain funds a third time (tertiary financing). This is in a sense fortunate, since most investors strongly prefer those later stage investment opportunities (Ruhnka and Young, 1987). Not only does 75 percent of the venture capital money go into second round and later investments (*Venture*, 1989), but even informal investors place over 50 percent of their funds in post-start-up situations in both New England and California (Wetzel, 1983, p. 26; Tynes and Krasner, 1983, p. 351) and slightly less in the Great Lakes region (Aram, 1989, p. 338).

While the first funding of a company tends to be memorable to its founders (see Chapter 5), details on subsequent funding often are less clear in retrospect. In part this is because the financing received after initial capital, especially when debt (often the principal source of subsequent funds) is considered in addition to equity, often occurs in small, frequent, and apparently "forgettable" amounts. Of course, this is not true for a major round of venture capital financing or a company's initial public offering. Such events are also remembered in detail. Unfortunately, at least from a researcher's perspective, these memorable occasions are not the norm for technology-based companies. An indication of this tendency is that the entrepreneur usually personally provided answers immediately during interviews on most aspects of the detailed questionnaires. For financing information, he frequently referred the interviewer to the accountant, controller, or vice president of finance or agreed to send those details when he could pull the information together. Consequently, some of the specific numbers in this chapter are probably less reliable than the quantitative indicators used in previous chapters.

The numbers, rough as they may be, do indicate that a technological enterprise may initially need less than $50,000 to get started with funds for working capital requirements for payrolls, accounts receivable and, most important, product development. Production-oriented companies and some R&D contractors require additional funds at the outset for fixed assets, such as facilities and test equipment. Having met the initial crisis and developed initial products or services, many young technology-based enterprises find their growth capital requirements to be much larger, typically several hundred thousand dollars or more to permit expansion. The timing of second-phase financing means that most companies in question have by then developed some history of sales and performance that, rightly or wrongly, is usually required for favorable decisions by banks, venture capitalists, and similar institutions.

For perspective note that fully 40 percent of technology-based companies never obtain financing beyond their initial capitalization. Some do not need it, their continuing support and growth requirements being met from self-generated cash flow. This is especially true of consulting firms, contract development organizations, and software product companies. Other start-ups have needs for more capital but are unsuccessful in raising additional funds. Overall 20 percent of the firms in my primary sample that did succeed in raising additional capital received less than $50,000; more than 40 percent generated between $50,000 and $500,000; the other 40 percent received more than a half million dollars, including 23 percent that got over $1 million in follow-on funding. The median firm that generated subsequent financing added between one-quarter and one-half million dollars. This varies greatly by type of business, both in terms of the percentage of firms obtaining follow-on capital and the amounts they raise. Few of the consulting firms (25 percent) raised additional capital, and in all cases obtained less than $100,000. More of the software developers and producers (but still only 40 percent) received subsequent financing, $500,000 or more in each situation in the data base. For example, Meditech (see Chapter 1), which obtained $500,000 initially from EG&G for 30 percent of the company, raised a second $500,000 one year later for about 10 percent more of the firm. In fact Meditech never needed to dip into that second half-million dollars, financing the rest of its growth requirements from self-generated profits. The majority of the hardware manufacturers (59 percent) and even more (67 percent) of the mixed business firms (e.g., some combination of hardware, software, and possibly contract development) did get growth financing, more than one fourth of these companies raising over $1 million. A sample of young energy-related manufacturers also includes 57 percent that added subsequent financing.

A side note on impressions about capital funding comes from analysis of one group, 20 technical firms that had once been seriously evaluated for possible investment by one large Boston venture capital fund. By the time my assistant and I interviewed those 20 firms, one to three years after they had been seen by the venture capital fund providing the original data, the companies had added an average of $1,400,000 per company to their initial funding, the range being from 0 to $9,600,000, including both debt and equity. The average bank financing of $730,000 to 15 of those 20 companies ranked second in magnitude only to the average funds received by the few who had publicly issued stock or convertible debentures. Truth, as beauty, is often in the eyes of the beholder, and a major venture capitalist, personally experiencing financing amounts of the extent just cited, might well form the opinion that funding requirements of *typical* technological firms are large. This data set is seriously misleading. Only relatively few technical companies, usually biased toward the biggest ones by far or at least those with the biggest potential, ever get to be carefully assessed by most large venture capital investors. The larger data samples presented previously, consisting of close to all companies emerging from particular university and industry laboratories, are far more representative of the total population of technology-based start-ups, and show far smaller size financings as typical.

Table 7–1 displays the sources of second round financing for 71 companies of one sample in which 110 MIT spin-off firms had received initial financing, as well as the sources of tertiary financing for 31 of those 71 firms. The distributions for follow-on money are considerably different from that observed for the initial capitalization of similar companies (see Table 5–1). Only 7 percent of the companies that obtained secondary financing raised it from the personal savings of the founders, and only 5 percent received additional money from their family and friends. Some of these particular companies still lack knowledge of how to approach other more professional financial sources; others maintain a desire to retain full ownership of their companies; still others had tried but failed to raise capital from the more independent sources and had to contribute their own funds or risk company stagnation and/or death.

Table 7–1
Principal Sources of Additional Capital

Source	Secondary Financing (71 companies)	Tertiary Financing (31 companies)
Personal savings	5	4
Family and friends	4	—
Private individual investors	24	9
Venture capital funds	9	2
Commercial banks	11	3
Public stock issues	7	8
Nonfinancial corporations	11	5

As indicated, several companies that had initially received funds from "outside" sources were supplied additional funds by these same sources, sometimes owing to prior commitments, occasionally from attempts by the financiers to protect their vested interests. For most of the other technical enterprises, however, the primary sources of additional financing are new.

Only 21 percent of the companies had been initially financed by the more sophisticated categories of investors. Now these professional financiers account for approximately 90 percent of the subsequent financial sources. The increased proportion of funds received from the more professional sources reveals the greater willingness of such sources to invest in on-going firms, as opposed to raw start-ups. In several instances the financial source approached the new enterprise and not the other way around. The technical firms apparently now have something to offer these investors—a share in a prospectively profitable and growing firm. To get this share the financial source is obviously willing to supply money. Occasionally this happens because the entrepreneur, becoming more sophisticated about the world of finance, has made his needs known "in the right places". More often, though, the firm approaches the financial sources directly. Having established operations the owners are in a more advantageous position to receive money than when their companies were merely ideas.

Banks supply debt funding, using as collateral the firm's equipment and more often its accounts receivable. Many of the Boston-area banks have had long and good relationships with high-technology firms and they are "understanding" of the specific needs that arise in these companies. Yet, it was not always this way. As EG&G grew its previous concentration on development work changed to include production, and EG&G attempted to get commercial and industrial customers rather than just contracts from government agencies. Although additional financing was not necessary the thought that the government might withdraw some advanced funding led the EG&G founders to discuss with their bank (a well established Boston bank) the possibility of borrowing money. "We don't lend money on brains: What's your collateral?", was the response that precipitated an immediate change in bank. It is important to note that this attitude toward not investing in "brains" no longer exists at this bank or at most other banks in the Boston area. Yet the old cliché that "banks are most happy to provide funds once it has been established that such funds are not needed" is not refuted by these data.

Private individuals and venture capital funds invest in the future of the company, waiting for the time the company will go public or be sold to a larger corporation. In these subsequent stages the private investors are of a somewhat more varied nature than the "angels" described in the Chapter 5 discussion of initial financing. Later stage private financing more frequently takes the form of a privately placed stock issue, with a larger number of investors subscribing for stock. Most of these new individual investors expect only financial relationships with the young technical firms. Private investors account for about 34 percent of the secondary sources and about 29 percent of the third-round financings, making them by far the most used source of subsequent funding. Freear and Wetzel (1989) find nationally that private outside individual investors maintain a prominent role in these later stages of financings, especially for secondary funding. Nonfinancial corporations, having helped to create some of these new firms, continue their investments or take new investment positions in young companies that have become their suppliers or that have already begun to look like attractive future acquisition candidates. For seven of these evolving technological companies in Table 7–1 the source of secondary capital is the issuance of stock to the public marketplace, joined by another eight companies at the tertiary financing stage. Chapter 8 focuses on technological firms "going public".

The companies that had emerged from the large electronics systems firm in the comparative samples show similar results. About half subsequently added more capital, none from founder personal savings, and about one-third of these later raised tertiary funding. As with the MIT spin-offs described earlier, private individuals, venture capital funds and commercial banks are the principal subsequent funding sources.

The evidence clearly indicates a tendency for firms initially financed by outside investors to return to the same sources for additional funds, highlighting the importance noted in Chapter 5 of the initial investor's ability to provide those additional funds. Sixty-three percent of those new enterprises initially financed by nonfinancial corporations return to them for growth financing; 50 percent of the firms financed by private individual investors go back to them for more; and venture capital firms provide "next round" funding to 37 percent of the companies in the sample that had been in their portfolios from start-up. Dean and Giglierano (1989) find that 76 percent of venture capital firms in Silicon Valley tend to finance multiple rounds for their earlier recipients, but the variance in their practices is very high.

Table 7–2 indicates the amount supplied by each source for 68 additional capital transactions in the sample of 110 companies. As was true of initial funding of these firms, personal savings and funds from family and friends are again concentrated in the lower amount ranges (less than $50,000). The other sources provide more varied amounts. Private stock issues to individual investors (30 percent), public issues (26 percent), and venture capitalists (22 percent) supply most of the amounts exceeding $500,000. In addition, seven of the companies that had been financed by nonfinancial corporations were eventually fully acquired by that "parent". Only five other firms in this sample were acquired during the timeframe of study (i.e., average age of the companies about five to seven years).

Some subjective information was also gathered about the entrepreneurs' attitudes toward their financings. For example, despite the typically small initial capitalization of these technical firms over 63 percent think that their initial capital was sufficient. The majority do not feel that they had been hindered by too little capital. Some even suggest that too much capital has been a problem source, tempting them into expensive and/or unfocused activities that might better have been avoided. Most of the entrepreneurs feel that their financial backers have been reasonably understanding and responsive to the companies' needs. Some hold strong negative views, one claiming that his investors "don't have a clue" as to the unique characteristics and requirements of a highly technical organization. All of the companies feel that financiers who are competent to do so should provide the firms with financial and business advice, but how much is disputed. Being primarily technically oriented the entrepreneurs generally

Table 7–2
Source of Growth Financing by Amount
(n = 68 transactions)

Amount (thousands of dollars)	Personal Savings	Family and Friends	Private Investors	Venture Capital Funds	Commercial Banks	Public Stock	Non-financial Corporations	Total
< 50	4	1	3	1	0	3	1	13
50–250	1	2	8	2	1	2	1	17
250–500	1	0	4	0	1	3	2	11
500–1000	0	0	5	3	3	1	2	14
> 1000	0	0	3	3	0	6	1	13
Totals	6	3	23	9	5	15	7	68

feel that their investors' experience in the industrial and financial community is an asset to the company.

For one subsample I quantified these attitudes and tested for significance among these "soft" relationships. Those entrepreneurs who came from more research-oriented work backgrounds (i.e., higher "Bullpup" ratings) feel most strongly that financiers should provide financial and business advice to their firms ($p = .03$). No doubt this relates to the fact that these individuals have the least prior business experience among the entrepreneurs (.016). Entrepreneurs who feel they have understanding and responsive financiers also rank highest the amount of financial and business assistance their backers should provide (.02). Especially significant is that the investors that entrepreneurs rate as most helpful are perceived as tending to exert less pressure on the entrepreneurs than the less helpful investors (.0008). Investors often provide advice along with pressure for its adoption, whether desired by the entrepreneur or not; but only those entrepreneurs who want it in the first place view the guidance as helpful. This point has not escaped venture capitalists, who often maintain as a criterion for investment an assessment of whether the entrepreneurial team appears willing to accept advice. It may be that most entrepreneurs just want to do their own thing, regardless of advice. One study of the use of outside consultants by prospective entrepreneurs indicates that "the recommendations of the consultants did not have a significant influence on the propensity of the clients to go into business" (Chrisman, 1989, p. 411).

The Search for Capital

If the entrepreneur is going to go beyond the personal funds of himself, his co-founders, and their families and friends, whether for initial or subsequent financing, how does he locate prospective investors? Venture capitalists are usually listed in directories, which inevitably indicate at least some criteria being applied, such as stage of company, size of investments, geographic preferences, and industry specialization. But the names of wealthy individual investors do not appear in such directories, and these individuals have proven to be critical especially for initial financing as well as for additional funding. Larger nonfinancial companies are still another prospective funding source not so readily identified. And how does the entrepreneur approach these sources to maximize his likelihood of gaining support?

All of my studies of technical entrepreneurship, including two that focus explicitly on the search for capital, fail to identify any optimal approach to capital search. Most first-time entrepreneurs seem to be stumbling in the dark, even when they have drawn up detailed and careful plans for acquiring venture financing. Entrepreneurs who are funding their second (or later) company have the benefit of their prior track record and the inevitable relationships with investors or at least banks from their previous companies.

The most common approach taken by the new entrepreneur is "shopping around", contacting all the founders' supposedly knowledgeable friends and business acquaintances to seek their suggestions and referrals. Friends and relatives, lawyers and accountants, are all sought out and frequently provide less than useless advice and delay. Some of the major law firms and accounting firms are actively involved with venture capital investing and professionals at these organizations have been very helpful to company founders. This level of knowledgeable involvement is often less true for the smaller service organizations or for the individual lawyer or accountant "friend of the entrepreneur's brother-in-law" who seems often to get into the act, especially with naive entrepreneurs. This "shopping" process sometimes leads to introduction to a "finder", an individual who specializes in enlisting financial support for small companies, who typically wants as a fee 5 to 10 percent of all funds raised. A number of the entrepreneurs interviewed unfortunately testify that some finders they worked with were unethical, most were ineffectual, at least in part because the financial community often finds them distasteful to deal with. When the shopping around includes other, preferably already successful, entrepreneurs, helpful experiences are shared and useful contacts within the investment community often result.

When the entrepreneur fails to turn up any potential investors from the circle of his own acquaintances, all is not lost. Typically, his next step is to make informal approaches to individuals in the financial community, often starting with his own local bank. Loan officers may be aware of individual investors among their bank's clientele and be willing to provide introductions. More likely, at least in the Greater Boston area, the local bank may be a branch of a larger commercial bank and the entrepreneur gets referred "downtown" for screening at the home office, where senior lending officers have close contacts at several venture capital firms. Several of the Boston banks operate their own Small Business Investment Corporations (SBICs), federally chartered venture capital funds mentioned in Chapter 5, which occasionally participate in start-up financing or which certainly have collaborative links to other early-stage investors. By this stage the entrepreneur is now talking to the potential investor community and needs to be concerned about his overall approach.

This description indicates that the use of referrals is the means by which most entrepreneurs enter into discussions with eventual investors. Alternatively, the entrepreneur might directly approach a venture capital firm listed in a directory or introduce himself to a venture capitalist at one of the numerous "entrepreneurship" meetings taking place in the Boston area. These "networking organizations" (Kanai, 1989) certainly include the monthly sessions of the MIT Enterprise Forum or the 128 Venture Forum, the two groups mentioned as part of the Chapter 3 study of entrepreneurial motivations. Table 7–3 shows the frequency with which various "approach" mechanisms were used for each source that supplied funds (of both initial and additional capital) in 54 transactions involving 20 technical firms in one sample. Direct approaches and references from personal acquaintances tend to be the more common means for establishing what became successful contacts with banks. Business acquaintances and personal friends were most helpful in successful entry to private individuals and private investment groups. A variety of sources, including finders, led to investments from venture capital funds. The realistic appraisal of this small sample is that the wide variety of both sources used and referral mechanisms that produced successful entry leads to no conclusions possible as to an optimal means for getting in the door of prospective investors.

Two additional studies shown in Table 7–4, both examining decision processes of venture capital firms, reaffirm this variety of effective entries.

Table 7–3
Referral Mechanisms Leading to 54 Investments in 20 Firms

| Source of Referral | Type of Capital Sources which Invested Funds | | | | Total Uses |
	Private Investors	Venture Capital	Com- mercial Bank	Invest- ment Banker	
Direct approach	2	2	5	0	9
Personal acquaintances:					
Previous business or university acquaintance	4	1	3	1	9
Personal friend	4	0	2	1	7
Knew person at investor	0	2	4	0	6
Other referral sources:					
Same type as investor	0	3	2	0	5
Different type as investor	0	1	0	1	2
Bank	1	3	0	0	4
Finder	2	3	1	0	6
Lawyer	0	1	0	3	4
Approached by investor	0	0	2	0	2
Total use of each source	13	16	19	6	54

Table 7–4
Referral Mechanisms Leading to 23 Decisions to Invest
by Two Venture Capital Firms

Source of Referral	Venture Capital Firm #1	Venture Capital Firm #2
Direct approach	3	2
Private individual	1	2
Other venture capitalist	1	4
Investment banker	5	0
Commercial bank	3	2
Total	13	10

Many paths can lead to success in securing a venture capital investment. The inverse of this table, the referral sources leading to rejections by these two venture capital organizations, shows similar variety and also no clear pattern.

Another study traces the experiences of 19 firms that sought help in acquiring venture capital by subscribing to the Investor Advisory Service of Venture Economics Inc., a consulting firm in the Boston area that provides various data-based services relating to the venture capital industry. The companies in this survey may well represent an extreme of desperation or alternatively may reflect an extreme of sophistication in their choice of this "dating game" approach to the search process. Of course, they do not limit themselves to use of the Venture Economics assistance; they try all the other conventional approaches as well. I did not attempt to codify the companies on a "norm" relative to other early-stage firms. Seven of the 19 (37 percent) did obtain funding prior to completion of the analysis, seven had already given up on their search, and four (21 percent) were still trying. Even if this group of firms turns out not to be representative of typical entrepreneurial situations, their search process experiences provide some interesting perspectives.

The length of time spent looking for venture capital financing averages 14.4 months for all 19 companies in this study. Even the typical "quitter" had spent 13.6 months before giving up, while the successful searcher kept going for an average of 15.1 months before gaining funds, but these numbers are not statistically significantly different, given the small sample size. During their search period firms allocate around half their managerial time to the search, clearly communicating what they see as Job #1! Searchers typically devote between twelve and eighteen man-months to looking for money, contacting large numbers of potential investors, often 25 or even more. My own experiences in searching for funds for several different ground zero start-ups, including Carousel Software, an educational software firm initiated by my wife and several colleagues that never received external funding, are not dramatically different from all these numbers, albeit we were perhaps a bit faster in general. One study of California technological companies (Bruno and Tyebjee, 1982,1985) finds the median time spent in search of a round of financing as only 4.5 months, but the first round took nearly 68 percent longer than other rounds (1985, p. 66) and complete denial by the venture capital industry adds approximately four more months (1983, p. 299). Moreover the authors find "surprisingly the search took nearly as long on the average for entrepreneurs seeking $250,000 or less as it did for those seeking over a million dollars" (p. 293). My special group of 19 companies demonstrates significantly that high-technology firms are more likely to be successful in their searches than low-technology or nontech companies; successful searchers also have larger-sized founding teams, as might be anticipated by the findings in Chapter 5. Overall these data suggest that entrepreneurs who work harder at finding money are the ones who succeed at it. But again, as shown in Table 7–3, no optimal routing to funding, beyond hard work and patience, is identified in this analysis.

Business Plans and the Search
for Capital

Formal business plans have taken on such a large role in the financing of start-up and early-stage firms that many "how to" books have been written on the subject. Most university "entrepreneurship" courses are dominated by their preparation and evaluation. Yet, little objective analysis has been done on their contents or impact on investor decision making. I report here some findings from examination of twenty business plans submitted by start-up high-technology product-based companies to several Boston venture capital firms. Ten of the firms had no prior operations and only one had significant product sales, but most had received at least prior founder investment. One firm already had three-quarters of a million dollars in earlier funding. The plans are not statistically representative of any population of plans but rather were volunteered by venture capitalists in response to a request for written materials submitted by early-stage technology-oriented product businesses. Venture capitalists do lots of informal screening to avoid even receiving questionable deals and/or ill-prepared plans and represent a high hurdle for entrepreneurs. Consequently, the plans examined are likely to be more sophisticated and comprehensive than would be prepared initially by a "typical" new technical enterprise. Assessment of these plans was carried out by a careful and systematic diagnosis of the plan documents themselves, performing the equivalent in each case of a plan "audit".

Plan Perspectives

The first observation, gratefully, is that all twenty plans have some kind of stated overall objective, presenting broadly and understandably what the companies are trying to do. Only 14 of the 20, however, have a specific strategy, formulated and clearly explained, that appears rational and achievable at least at first glance. The other six firms' documents only contain an implied strategy, not clearly indicating the overall approach they intend to take toward achieving their overall objectives. Table 7–5 shows the classification of the central thrust or orientation of the plans into three different emphases, allocating those that had mixed priorities on a weighted basis. In 47 percent, the central thrust was the product itself. While product emphasis can be a good feature for a plan, this sometimes comes across as "technology push". The entrepreneurs propose that they can do *it* better than anyone else, but may forget to demonstrate that anyone wants *it*. Occasionally, a product-oriented plan focuses so strongly on how wonderful the product is that the plan omits critical elements of attractiveness from a market perspective as well as aspects of the founding group's capabilities to carry out company development and implementation. A large fraction of the proposals we see at Zero Stage Capital seems to have this same technology or technology-based product thrust, with similar questions immediately arising as to whether there really is a market as well as an ability of the founders to build a successful company.

Twenty-nine percent of the 20 plans reviewed do indeed focus on the market. Many of these, however, primarily address how attractive and rapidly growing it is. The plans frequently do not persuade the reader that the start-up group has a clear competitive advantage in that marketplace, or why the new company is going to be able to develop a particular share of that market.

Twenty-four percent of the plans concentrate on the people who are

Table 7–5
Central Thrust or Orientation of Plan

Type of Firm	Percent
Product	47
Market	29
People	24
Total	100

the entrepreneurial team. Despite all the talk by venture capitalists and successful entrepreneurs about people being the most important ingredient for company success, as I appraise later in this chapter, when people are the key thrust of a business plan serious questions are still raised. Some plans say, "We have assembled a superb team", leaving by implication the perhaps unintended side comment, "Don't worry about our product and market; we have the skills and flexibility to shift as needed." This situation does not create confidence for the potential investor. The team sometimes does not appear to have yet committed itself to the specifics of targeted activities and participation in the marketplace. What seems needed in a new business plan is a central thrust, whatever it may be, but with clear elaboration of that thrust in both of the other identified dimensions.

Deficiencies in Plans

Analysis of the plans for possible deficiencies uncovers major gaps in the details provided to support the strategy. Forty-five percent lack emphasis on economic performance. They just do not talk about profitability and growth in a significant enough fashion. Profits often appear to be either implicit or just happen to be included in the numbers. Profits seldom are the focal point of the founders' discussion of the company they are trying to build. A venture capitalist enjoys at least seeing an attempt at making and supporting promises that the company will make a profit. A business plan enthusiastically devoted to the generation of profits would be a refreshing change from the usual.

In over half the cases, plans lack adequate discussion of the economic environment—the business climate and the extent of competition in the market. And yet only 30 percent are rated as inadequate in regard to assessment of the technical environment. In general the plans contain strong coverage of technical dimensions, and I might comment overly strong, relative to anything else. The plans reflect what the entrepreneurs understand best, but by omission they also reflect what the founders do not understand. Plans frequently describe the general technology, where it is going and who is doing what within the technology, with much less attention to nontechnical aspects of the business environs.

In terms of the specifics of competition, in 75 percent of the business plans it is almost an impossibility to identify anyone who competes with the proposed company. These business plans communicate that the company has come forth in full fruition out of the head of Zeus, in a brand new marketplace that no one has ever entered. This is likely to be wishful thinking, and is usually incorrect and dangerous planning. Especially absent is the recognition that a new technology-based product or service is an intended substitution for doing something by an old approach, manual for example, rather than a proposed computer-based method. This "substitution competition", when it exists, is almost always omitted from plan discussion.

A final broad deficiency is that almost half of the company plans suggest that they are trying to do too much broadly or too many different things at once. Plans often enumerate multiple product lines at the outset. They list large numbers of markets that the company will somehow instantly attempt to enter and conquer, without really communicating a rational allocation of resources or priorities so as to demonstrate focus. Chapter 10 assesses in depth the need for focus in launching and building technology-based companies.

Functional Planning Deficiencies

Four clusters of functionally oriented plan components—the marketing plan, the management team, the technology plan, and the financial plan—were evaluated and rated for moderate or significant deficiencies. As shown in Table 7–6 this analysis identified serious inadequacies in all areas, with marketing plans being seen as deficient in 70 percent of the cases.

Marketing. Let us look at marketing first. In only 40 percent of the cases has the company formulated a specific marketing strategy, stated or implied. In half the cases the plan communicates that the company is targeting a very specific market segment or specialized niche. At least it provides an understanding of how to assess marketing, in terms of its specific orientation. The plan reader (often a venture capitalist) can then worry about whether or not that target constitutes a real and attractive market opportunity.

In only 30 percent of the cases have the companies developed a detailed sales (not marketing) plan, stating details on salespeople—why, where, when—means of compensation and the like. In 20 percent of the situations a broad sales plan is implied but not specifically detailed. 50 percent of the cases do not contain any sales plan. By implication, sales are to be realized from a good product aimed at a great market, but with no information provided as to how the firm plans to get customers to place orders. In general, this type of deficiency reflects a lack of appreciation of the selling function by people who form companies based on beliefs in the techno-

Table 7–6
Inadequate Functional Plan in Critical Area ($n = 20$ plans)

Type of Plan	Percent of Plans Reviewed		
	Moderate Deficiencies	Significant Deficiencies	Total Plans with Inadequacies
Marketing plan	25	45	70
Management team	35	30	65
Technology plan	40	5	45
Financial plan	25	35	60

logical advantages they can provide to a market. One company was founded by five MIT graduates who, for several years, rotated among themselves the position of vice president of sales. None of the founders wanted that function. The firm got off the ground and advanced to an early low threshold, staying there for about eight years. Then, the firm totally reorganized, brought in an experienced sales executive, and grew substantially.

In terms of market research (Table 7–7), 70 percent of the plans evidence only broad brush market data. In 25 percent of the cases at least one team member is shown to have had personal sales experience in the particular market and can relate to the market personally. That adds some real credibility to the plan's claims. A number of the plans show more details on the market and potential customer requirements.

I must admit that you do not always have to demonstrate market research to have a business plan that gets funded or even to develop a successful business. Several years ago, as part of these overall entrepreneurship studies I interviewed the founder of a company that had started in the business of high-speed transistorized electronic modules for use in the assembly of digital systems. I asked him how he knew his product would sell. The entrepreneur responded, "I was a project engineer at MIT working on digital systems. I knew I would have used those circuit modules if they were available." "Did you talk to people elsewhere?", I persisted. "I did not need to", he replied. "I knew there were hundreds of guys just like me who would have been delighted to have had such a product." I am not about to disparage that founder, Ken Olsen, or the accomplishments of his firm, Digital Equipment Corporation, despite the absence of clear market research at the initiation of DEC. But prospective investors do gain confidence from a plan that indicates the entrepreneurs have attempted to gain specific insights about the customer and market at which they are aiming.

Management. The next area of analysis is the skills composition of the management team. This is more a potential deficiency of the entrepreneurial group, not just of their planning, but plans can acknowledge skills issues and include intentions about curing any deficits. In only three of the twenty business plans reviewed is there an obvious gap on the research and

Table 7–7
Market Research Issues

Research Issues	Percent
"Broad brush" approach to establishing existence of the market	70
Personal sales experience of team member	25
Detailed customer survey	25
Plan indicates detailed knowledge of potential customers	45

product development side of the group, again emphasizing the clear technological origins of these firms. On average two of the four founding team members have R&D or design backgrounds. In 40 percent of the firms, however, marketing and sales are absent from the team. What frequently annoys potential investors in these cases is the oft-found comment in the business plan that a marketing/sales person will be recruited as soon as the product development has been completed, clearly indicating the lack of understanding of the proper up-front role of marketing input. Thirty-five percent of the companies have a marked deficiency in operations in cases where operational skills seem to matter. I am not referring to a software organization where the initial "factory" consists of desks and people sitting behind them. I mean cases where the company is proposing an activity that needs a manufacturing capability yet has no manufacturing skills represented among the founders. This deficiency is amplified by the fact that 60 percent of the plans do not even discuss production plans. Any of these omitted key roles can doom the business plan as well as the business itself.

Technology. In 45 percent of the business plans, the initial technology appears to be unique (Table 7–8). Chapter 4 indicates that many firms of this type are heavily dependent on the advanced technology base of university or industry laboratories at which the founders had recently worked. The new firm might by this means be bringing to the marketplace a technical edge on its competitors. Companies whose plans reflect less or even no technological uniqueness have a burden of proof to demonstrate that they have some other unique basis for overtaking larger already entrenched competitors. Yet, in only 40 percent of the plans is the basic product already developed, ready for production or already in the market. Seventy percent of the plans do indicate an intent to continue R&D work beyond the first product and to introduce new products in the future.

One final aspect of technology is the extent to which it is protected. Only 45 percent of the companies even discuss the protection of technology, 25 percent have taken steps to patent their technology, and in 10 percent of the cases the technology cannot be patented or the patents have expired. In most high-technology firms (other than biotechnology companies)

Table 7–8
Degree of Uniqueness of Technology

Degree of Uniqueness	Percent
Unique technology	45
Some characteristics of uniqueness	40
Not unique	15
Total	100

patents per se are not vital. Yet, discussion of the technology's protectability is important. The key questions are whether the unique qualities of the firm can be maintained and the firm generate continuing profits based on that uniqueness, not whether the product can be patented.

Finance. Let me turn finally to the financial aspects of the plan (Table 7–9). In 10 percent of the business plans no financial statements are included. This is inexcusable. Entrepreneurs should not expect potential investors to deal with them financially and invest money in their proposed businesses if the plans do not treat expected financial outcomes. Five percent of the plans include bare outlines of financial statements but the specifics are not included, claiming that the data are not available. Several of the plans promise that more financial information will be provided in a later submission. In 10 percent of the cases a one- to three-year income statement is available, despite the fact that many technical entrepreneurs do not know what an income statement is. Mitchell Kapor, the founder of Lotus Development Corporation, says that accounting is one of the most important things he learned while a student at the MIT Sloan School of Management. "Try running a business without knowing accounting", he exclaimed. In one of four situations both multiyear income statements and balance sheets are included in the business plans. Cash flow statements are included in only a few plans, despite the fact that investors are usually more immediately concerned with cash flow than with any other financial parameter. Paul Kelley, the true entrepreneur of Zero Stage Capital and its managing general partner, frequently semi-jokes, "Cash flow is more important than your mother!"

Only 10 percent of the proposals provide financial plans based on multiple sets of assumptions, and these only reflect alternative sales projections of "pessimistic", "most likely", and "optimistic", along with their corresponding staff build-up rates. In these days of Lotus 1-2-3 or similar spreadsheet availability, an entrepreneurial team can, with little effort, roll out as many variations on a financial theme as anyone might request. Indeed one almost has to hold back, since thick appendices of computer spreadsheet outputs are sure to turn off most prospective investors. But

Table 7–9
Projected Financial Statements Included (%)

None	10
Data not available	5
One- to three-year income statement only	10
Four- or five-year income statement only	40
One- to three-year income statement and balance sheet	15
Four- or five-year income statement and balance sheet	10
Total	100

these tools can be used to alter market penetration assumptions, start-up costs, or competitive responses and generate alternative financials based on whatever the team wants to put forward in its plan. When we were trying to raise initial funds for Medical Information Technology, Inc., back in 1969, long before computer spreadsheet programs, I built a simple simulation model to generate alternative financial projections based on key issues that had arisen in our preliminary discussions with prospective investors. That approach to embodying financials as a direct reflection of the entrepreneurs' business thinking still does not appear in many business plans.

Only 20 percent of the cases, in my opinion, include adequate supporting detail for the financial plans. The other 80 percent lack appropriate back-up information. In this regard one more Ken Olsen story bears repeating. When he and Harlan Anderson approached American Research and Development (AR&D) with their proposal to get start-up funding for Digital Equipment, they were told that their project was interesting but needed more financial details in the plans. As two engineers from MIT's Lincoln Laboratory the entrepreneurs did not know much about how to create financial plans. Then Olsen remembered that he had taken introductory economics at MIT and that he still had his copy of the 1951 edition of Paul Samuelson's classic text. In his section on accounting Samuelson tells how to put together an income statement and a balance sheet, using the so-called Pepto Glitter Company as example. Olsen told me that he and Anderson were not really sure what AR&D wanted, but they copied out all the entries for Pepto Glitter, changed the name to Digital, and plugged in a few different numbers. AR&D executives later told Olsen that his plan was one of the most sophisticated they had ever seen. The $70,000 that AR&D so wisely invested for over 70 percent of DEC turned out to be worth billions.

Business Plans and Venture Capital Funding

At the start of the firm sophisticated business plans say less about whether the company will be successful eventually and more about whether it will be successful initially. Initial success consists of getting the money sought from those venture capitalists or other investors. In Chapter 9 I comment on the relationship between initial business plans and later corporate evolution. Table 7–10 indicates the relationship between the assessed adequacy of twenty business plans and whether or not the companies receive venture capital financing. Of course, a proposed financing would seldom be decided solely on the basis of the plan itself. Despite this, as shown, when the overall plan was evaluated (by my assistant and me) as having only minor deficiencies, four of the five were financed by venture capital firms. One of the plans was not initially financed by a venture capitalist, although such financing might have occurred considerably later, following the study. With moderate deficiencies in the business plans, four of seven received venture

Table 7–10
Relationship between Deficiencies in Business Plans and
Success in Obtaining Financing

	Number of Plans and Percentage of Plans Reviewed			
Overall Assessment of Business Plan	Enterprise Financed by a Venture Capital Firm		Not Financed by a Venture Capital Firm or Data Unavailable	
Minor Deficiencies	4	(20%)	1	(5%)
Moderate Deficiencies	4	(20%)	3	(15%)
Significant Deficiencies	2	(10%)	6	(30%)
Total	10	(50%)	10	(50%)

capital funding. In the cases where our detailed evaluation of the plan generated a significant deficiency rating, then only two of those eight firms were funded. I do not know what later happened to the ten firms in this group that were not venture capital funded. In all likelihood some went out of business lack of financing; others no doubt survived despite the initial turndowns by the venture capital community, possibly with additional personal financing, perhaps even to prosper later. Bruno and Tyebjee (1985) find that more than two-thirds of the companies denied funding by a venture capitalist are still in business a few years after their initial rejection, 60 percent eventually raising outside capital elsewhere, albeit frequently receiving much less funding than they had originally sought (p. 70).

Venture Capital Decision Making

As indicated earlier in this chapter venture capital firms play a prominent, though not dominant role, in the later financing of technology-based companies. This section presents the results of two in-depth studies of Boston-area venture capital firms, focusing on their decision making criteria. In the next chapter I describe the overall process of public stock market financing of technical companies, including the entrepreneurs' decision to go public, the decision making of investment banking firms to take them public, and the outcomes of their actions.

Research on venture capital investment decision making, combined with years of personal experiences as a participant in these decisions, suggests that no stereotypes are in order. Venture capital firms are as different from each other as are individuals. Some venture capitalists prefer technology-oriented companies for their investments; others do not. Some specialize by industry; most do not. Some prefer later round financings; fewer prefer early round investment and very few prefer seed round investments. Some will not make investments smaller than \$500,000; others will not make investments larger than \$500,000. Some tell you they focus their evaluations on the people; others say they concentrate on the markets or products and technology. The research literature on venture capital decisions reflects the variety of criteria that are perceived as important influences. Tyebjee and Bruno (1984) conclude that "venture capitalists evaluate potential deals in terms of five basic characteristics: Market Attractiveness, ... Product Differentiation, ... Managerial Capability, ... Environmental Threat Resistance, ... [and] Cash-Out Potential" (p. 1059). Assessed lack of managerial capabilities significantly increases the perceived risk ($p = .05$), whereas attractive market conditions have the strongest effect on expected rate of return ($p = .01$) (p. 1060). MacMillan, Siegel and Subba Narasimha (1985) also find many different dimensions considered important by venture capitalists, but stress that "five of the ten criteria most commonly rated as essential have to do with the entrepreneurs themselves" (p. 123). Goslin and Barge (1986) find the management team to be preeminent in significance, followed by consideration of the product and then other factors. MacMillan and Subba Narasimha (1986) warn of the impact of optimistic forecasts on credibility, and Bruno and Tyebjee (1983) identify "deficiencies in the venture's management" as explaining one-third of the rejections by venture capital companies (p. 290). Dean and Giglierano (1989) differentiate venture capitalists' criteria between first round and subsequent financing. They agree with the gist of prior studies that venture capitalists focus on management team, market need, and technology/product characteristics, in that order, for first-round investments. For subsequent financing they find that venture capital firms emphasize the young company's "performance-to-plan-to-date" over all other criteria. Rea (1989) clusters decision criteria into five groups also, although again somewhat different from factors identified in other studies, and finds that "business factors are more important than product characteristics for successful negotiations" (p. 149).

At one point in my research on technological entrepreneurship I carried out a four-hour interview with the famed head of a pioneering Boston venture capital firm. During the course of our "let's get-acquainted" meeting he went through his entire investment portfolio, giving rather subjective reasons in my opinion for each of the investments. He was particularly emphatic on the characteristics of the people. As one example he exclaimed, "These boys were from the farm. Got up early in the morning. Used to hard work. Bright eager eyes, enthusiastic about what they were doing. These are the kind of boys we invest in." When I pointed out rather cynically that the description might also fit millions of other "young boys" in whom he might not choose to invest, the guru responded, "Well, of course, there is also the idea. You certainly wouldn't want to invest in someone proposing foolish things, like going into competition with General Motors." This meeting obviously occurred before the Japanese proved it was possible to compete rather successfully with GM. But it reflects the great difficulty of discovering whether venture capitalists apply consistent and learnable criteria to their investment decisions.

This section seeks to provide some further enlightenment on the variety of venture capitalist criteria and decision processes by describing results of two in-depth studies, carried out sixteen years apart, of individual venture capital firms, called VC1 and VC2. Both VCs were reasonably large Boston-area firms that had been actively engaged for a number of years in investments that include but are not limited to technological companies. Both use a multistage process of evaluation, beginning with cursory assessment of investment opportunities presented, followed by a meeting with the entrepreneurial team for a personal presentation and discussion, followed by far more thorough detailed analysis of those situations that "pass" these initial reviews. In both studies rather complete samples were developed of those recent cases that went through the full review process, are technologically oriented, still had the principal venture capital staff reviewer accessible for interview, and producing for research purposes a more-or-less balanced set of approval and rejection decisions.

Venture Capital Firm 1

VC1's management claims that it prefers investing in the very early stages of a company's development. This is obviously a matter of definition since review of its past investments indicates that the typical company age at the time of VC1 investment is two to five years. VC1 has a stated investment philosophy to seek investments in companies with unusual growth potential and technical leadership, in which it can acquire a significant equity interest either immediately or in the future, and maintain close and effective working relationships. VC1 says that its careful investment analysis places particular emphasis on management capabilities of the prospective investee company. Interviews with its managers indicate VC1 no longer invests in companies with merely ideas or even prototypes, but now requires a working product as a minimum. The product must be technically better than anything presently available in an established market, not restricted by industry. The company should be founded and operated by a group, not an individual, and VC1 prefers that each of the two or three co-founders have both technical and business experience. It looks for realism in the business plans, attributing overly optimistic sales and profit forecasts as indicative of poor management. At the time of study of its decision making VC1 preferred second- or third-stage financing amounts in the $250,000 to $1 million range and would not generally invest less than $250,000.

In reviewing VC1 my assistant and I found that over 2,000 firms had previously been brought to its attention, 150 had passed first-phase review, and 45 decisions to invest had been made. We carefully examined 24 situations that VC1 analyzed in-depth, carrying out a comprehensive structured interview with the principal investment officer relevant to each case.

Thirteen of the 24 companies had been chosen for investment and 11 were rejected. (Two of the "accepted" turned down VC1's offer and were funded by other venture capital firms.) Technical fields of these companies include computing systems and support products, power supplies, high temperature materials, crystal technology, electronic test equipment, vacuum process equipment, and plastics.

As indicated earlier in this chapter the general type of referral source that brings the opportunity to VC1 is not important. However, that source's prior experience with VC1 turns out to be a good predictor of likelihood of investment. The frequency with which a source has previously brought projects to VC1's attention is significant ($p = .10$) in determining if the project is accepted or rejected. Similarly, and even more strongly, the sources of the accepted companies among this sample of 24 cases had significantly (0.01) higher percentage acceptance rates in their previous referrals than did the sources of newly rejected situations. As shown in Table 7–11 a positive feedback loop of growing compatibility seems well established between VC1 and its effective sources.

Looking into the characteristics of the companies themselves, VC1 does invest in young firms, the average age being 2.1 years, not significantly different from the age of the companies it rejects. It appears to prefer slightly larger founder teams, consistent with its management claims, those accepted for investment having a mean size of 3.2 persons (median, 3)

Table 7–11
Frequency of Sources and Source Acceptance by VC1
($n = 22$)

Number of Projects Brought by Sources of Accepted Firms	Percent of Prior Projects Invested In	Number of Projects Brought by Sources of Rejected Firms	Percent of Prior Projects Invested In
6	0	3	0
12	0	7	15
1	100	6	0
1	100	1	0
5	20	10	15
30	10	5	15
4	25	3	0
1	100	2	0
4	50	3	0
5	20	1	0
2	50		
1	100		
Mean 6.0	48	4.1	3

against the 2.1 persons mean size (median, 2) of the rejected firms. Its favorable decisions also reflect statistically significant bias ($p = .01$) toward slightly older entrepreneurs (mean age of the group 39, in contrast to 36.9 for the rejected entrepreneurial teams), with a tendency for at least one person in the accepted teams to be much older and more experienced. The average years of individual commercial experience of each entrepreneur in both the accepted and rejected companies is about the same, at 13 years, comparable to the experience level of the MIT-based entrepreneurs described in Chapter 3.

A detailed evaluation was performed of those factors that VC1 staff perceives to be important in potential customer purchase decisions for each of the sampled companies. Clear investment preference is shown for those firms operating in markets perceived to place high importance on new technology, special purpose products, special customer specifications, strong service requirements, and company–customer interaction that would maximize impact of high quality personnel. VC1 rejected firms for which price is seen to be a key customer purchasing criterion, seeing such markets as unattractive and perhaps better served by larger companies.

The intended use of the funds varies substantially across the entire sample, with half of both approved and rejected proposals including additions to working capital for at least part of the money. The only significant difference ($p = .10$) is that six accepted firms plan to use funds to support product development while only one rejected company has similar plans. This use is consistent with VC1's announced bias toward technical leadership for the firms in which it invests.

The VC1 investigators rated the overall quality of each plan that has been submitted. Six of twelve accepted plans are rated at 5 (out of 7) or better, while only one of six rejected plans receives a comparably high rating. Finally, rejected companies tend to be rated higher in risk than accepted companies (nine rejections rated at 6 or 7 on the risk scale, while only seven accepted companies are similarly rated), although this might be seen as after-the-fact rationalization. In explaining the perceived risk, risk in product, market, and personnel are independently evaluated. For rejected companies the product as a cause of perceived risk appears significantly more often (0.15) than for accepted companies. Combined with the findings of Table 7–12, plus inability to find significant differences in the people dimensions of VC1's investment decisions, VC1's investments appear to be more influenced by product issues than people issues, despite the VC1 management's general assertions. Conceivably the entrepreneurs' personal characteristics are more important in the initial broad screening done by VC1, but I have no data available to evaluate that possibility.

Venture Capital Firm 2

VC2's publicly stated investment objective is to provide "equity financing for smaller companies in the early stages of development, where there is

Table 7–12
VC1 Perceptions of Key Factors in Customer
Purchase Decisions

Factor	Accepted Companies	Rejected Companies	Significance Level (one-tailed)
New technology	+		0.10
Price		+	0.05
Special specifications	+		0.10
Service	+		0.10
Quality of personnel	+		0.01

significant potential for long-term growth and capital appreciation", with emphasis on high technology companies. Despite these written words in the five years prior to study VC2 has made 26 percent of its investments in high-technology firms, 35 percent in low-technology companies, and 39 percent in nontechnical firms. It, too, says it prefers groups or teams over sole entrepreneurs, and looks for a strong leader who is energetic and also understands the value of money. Next to VC2's impression of the people its officers claim they seek a well-defined company focus and manageable approach to the market. Generally it invests in companies in the northeastern United States for reasons of control and communications and avoids investing in government contractors as "not sufficiently profit oriented". VC2 says it wants companies to be in the stage of first sales of its product, so as to facilitate evaluation while still offering high upside potential to the investor. At the time of study VC2 management claimed that it prefers to make first (not seed) round investments in the $100,000 to $500,000 range, in combination with other outside investors. VC2 does not usually undertake the role or responsibilities of prime investor in a deal. It is willing to consider investing more than $500,000 in more mature companies, brought to it by a co-investing venture capital firm.

In reviewing VC2 my assistant and I found initial contacts from phone, mail, or walk-in from approximately 1,000 aspiring entrepreneurial teams annually, with about 50 percent of those being turned away immediately without review of business plan. Only 75 to 100 of the 500 plans that are reviewed generate an invitation for a personal conference and presentation, leading to about twenty in-depth analyses per year, and eight to 12 decisions to invest. These percentages at each stage are different from the VC1 experiences described earlier but in both cases represent high degrees of selectivity by the venture capital firms. We carefully examined 20 technologically oriented situations that VC2 had analyzed in-depth, split evenly between positive and negative investment decisions, carrying out a comprehensive structured interview with the principal investment officer relevant to each case. Technical fields of these companies include microcomputers, laser inspection technology, programmable controllers, fiber optics applications, genetics, and infrared imaging technology.

All types of referral sources lead to both acceptances and rejections by VC2, with some positive bias apparent toward those prospects referred by other venture capitalists. This bias is especially reasonable for VC2 in that it specifically seeks to be part of other organizations' deals, in contrast to VC1's greater willingness to be the principal investor. Carrying out similar analyses to those performed at VC1 again produces a statistically significant difference ($p = .036$) between accepted deals being brought to VC2's attention primarily by sources who had brought many previous deals, while the rejects are referred more often by sources with little or no prior contact. The accepted firms' sources have also more frequently accounted for prior actual investments by VC2. This is a natural process of collaboration development over time, and I am not surprised to be able to evidence this phenomenon in the venture capital community.

Turning to the companies themselves, VC2's ten accepted companies are older on average, 5.7 years, than the ones they rejected, 3.2 years, but both groups are on average well beyond the "first product sales" or "first-round financing" VC2 says it desires. VC2 invested in only one of the four start-ups that were carefully reviewed in the study. In contrast with VC1, no size differences show up between accepted and rejected teams, with 18 of the 20 cases coming from multifounder groups. While statistically not significant VC2 does display obvious preference for teams balanced between technical and business orientation, accepting eight that it says had "good balance" while rejecting three with comparable team composition. Not unrelated is VC2's tendency to reject higher educated teams (0.05), including rejection of four of five teams that include a Ph.D., with occasional side comments recorded during the interviews about lack of open-mindedness and/or flexibility.

As part of the detailed study VC2 staff members rated the twenty entrepreneurial teams in the sample on a variety of personal characteristics. Significant differences arose between the approved and rejected entrepreneurs: the accepted ones appear to be more open-minded (0.001); they are rated as more aware of their personal limitations (0.05); and they are more concerned about long-run direction than short-term goals (0.01). These impressions by prospective investors may rest on shaky grounds, but may influence the investment decisions. VC1 does not evidence comparable results.

In evaluating VC2's impressions of the key factors affecting the firms' potential customers, the only striking finding is that all ten of the rejected companies are perceived as having customers for whom price is an important purchasing criterion, statistically more significant ($p = .01$) to the rejected firms than to those firms approved for investment. VC2 agrees with VC1 in seeing this kind of price-sensitive market as generally inappropriate for the young small firm.

The VC2 investigators rated several aspects of the business plans. Five of those rejected have marginal to poorly done plans, in contrast to two of those that were eventually accepted. Claims are perceived as not well supported by facts in four of the rejected firms against two of the accepted companies. Statistically significant ($p = .002$) is the perception that all but two of the rejects have overly optimistic forecasts, while the accepted plans tend to present reasonable forecasts. Finally, rejected companies are seen as far riskier than those accepted (.05), with the greatest source of perceived risk in these VC2 decisions coming from uncertain markets, in contrast with VC1's greater concern about product/technology risk.

A Tale of Two VCs

Comparison of the two venture capital firms' decision making shows both responsive to their previous sources of deal referral, investing in particular in companies brought to them by the same sources as their prior investments. Both VCs practice high degrees of selectivity, following multiple stages of increasingly intensive review, leading up to their final investment decisions. Despite both claiming interest in "early stage" companies their practices define their focal points to be quite different, with VC1 investing in two-year-old companies on average and VC2 typically approving investments in firms five-years-old and more. Indeed the age distributions of approvals of the two venture capitalists are almost nonoverlapping, to cite one striking difference. The age of a technologically oriented company is a surrogate for many other facets not explicitly discussed in our studies, such as whether the firm has tested its ideas, its market, or its managerial organization. VC1 is clearly investing in far more uncertain situations than VC2.

Both venture capitalists show strong preference for founder teams, rather than solo entrepreneurs, agreeing with the data on initial capitalization displayed in Chapter 5. In Zero Stage Capital we see many cases of single founder seed stage situations, often strongly encouraging that founder to strengthen his organization promptly by finding one or more partners who can occupy critical roles. My partner Paul Kelley often says, "We frequently invest in two-legged deals". This actually has a double meaning: the two legs refer on the one hand to single entrepreneur situations, more seen at the zero stage than later; but Paul is also referring to a two-legged rather than a three-legged stool, testifying to the rather obvious requirement for supplementing the key players among the several other aspects of needed development.

While both investment firms studied claim a primary focus on certain personal characteristics of the entrepreneurs, the studies discover consistent actions only in the VC2 data. Both investors evidence strong prejudice against companies operating in markets that seem to be price sensitive.

And VC1 shows consistent investment patterns that favor markets in which advanced technology and technology-based customer servicing appear to be the most critical influences affecting purchase decisions. VC1 backs up this preference with investments that are earmarked for product development support.

Rather self-fulfilling is the finding that both firms rate the quality of business plans higher for the companies in which they have invested, and also rate them less risky than their rejected proposals. Perhaps related are their assessments that the rejected companies have been overly optimistic in their sales and profits forecasts, although both VC1 and VC2 complain about poor support for key elements of the rejected plans.

Summary and Implications

Once up and operating nearly two-thirds of the technology-based companies develop, discover, or admit to larger and growing needs for funds beyond their initial financing. Their prior dominant sources, their own savings and family and friends, are no longer able to support most of their needs, usually amounting to several hundred thousand dollars up to several million dollars. Private investors, venture capital firms, the public stock market, and larger nonfinancial corporations account for the highest frequency as well as the largest amounts of follow-on funding, whereas only seldom had they been the sources of the initial financing.

Entrepreneurs search for capital using a wide variety of approaches, with unsystematic "shopping around" being the most common. Unfortunately, careful assessment of the pathways that actually lead to funding do not illuminate the search process. No optimal pattern seems to exist, except for the clear research finding that venture capitalists (and probably other investors too) tend strongly to invest in deals referred to them by people and institutions who have been sources of their prior investments. The long time and large effort spent in fund raising are impressive, leading to eventual dropout from financing efforts for many entrepreneurs, causing company termination in some cases and reappraisal and reshaping of plans and directions for others. No measures exist to assess whether the "best" or the "not so best" opportunities are funded, with the large number of alternative sources of venture funding in the United States producing at least a working marketplace, though one that is probably far from "efficient" in economic terms.

As part of raising funds, especially follow-on monies from venture capitalists, entrepreneurs often put extensive efforts into the preparation of business plans. Despite all the "how to" books and seminars on preparation of these plans, thorough diagnoses of twenty plans uncover significant deficiencies in most of them. Lack of clear communication of company strategy and several needed functional substrategies is evident, with both marketing and financial plans being especially weak. Data support of many claims contained in the plans is inadequate in 80 percent of the plans, this feature showing up as critical in venture capital decision making. The research reveals a strong relation between overall objective assessment of the business plans and whether or not those firms receive early venture capital backing. This finding is also supported by the significant difference in appraisals of approved versus rejected plans by two venture capital organizations studied in depth.

Detailed examination of 44 technology-related investment decisions by two venture capital organizations reveals striking differences in their tastes and consequent investments, beyond both VCs preferring entrepreneurial groups to solo entrepreneurs. One VC firm puts its money into primarily product development efforts by two-year old firms, the other company participates in the further growth financing of companies nearly six years old on average. Although both investment companies proclaim as critical the personal characteristics of the people in whom they invest, only one VC displays investment behavior consistent with its claims, the other evidently principally affected by characteristics of the markets being served by the candidate firms. Both investors shun market situations in which customers appear importantly influenced by price, preferring cases in which the asserted technological advantages of the firms in which they invest might become dominant factors. With each venture capitalist the "funnel effect" is most evident, transforming 1,000 to 2,000 initial inquiries for funds into 100 to 200 that receive careful screening, and eventually into ten to twenty or so actual investments being made. In this selective environment technological entrepreneurs had better carefully map out their strategies for growth, the subject of the next part of this book.

References

J. D. Aram. "Attitudes and Behaviors of Informal Investors Toward Early-Stage Investments, Technology-Based Ventures, and Coinvestors", *Journal of Business Venturing,* 4, 5 (1989), 333–347.

A. V. Bruno & T. T. Tyebjee. "The One that Got Away: A Study of Ventures Rejected by Venture Capitalists", in J.A. Hornaday et al. (editors), *Frontiers of Entrepreneurship Research, 1982* (Wellesley, MA.: Babson College, 1982), 289–306.

A. V. Bruno & T. T. Tyebjee. "The Entrepreneur's Search for Capital", *Journal of Business Venturing,* 1 (1985), 61–74.

J. J. Chrisman. "Strategic, Administrative, and Operating Assistance: The Value of Outside Consulting to Pre-Venture Entrepreneurs", *Journal of Business Venturing,* 4, 6 (1989), 401–418.

B. Dean & J. J. Giglierano. "Patterns in Multi-Stage Financing in Silicon Valley", in *Proceedings of Vancouver Conference* (Vancouver, BC: College on Innovation Management and Entrepreneurship, The Institute of Management Science, May 1989).

J. Freear & W. E. Wetzel, Jr. "Equity Capital for Entrepreneurs", in *Proceedings of Vancouver*

Conference (Vancouver, BC: College on Innovation Management and Entrepreneurship, The Institute of Management Science, May 1989).

L. N. Goslin & B. Barge. "Entrepreneurial Qualities Considered in Venture Capital Support", in R. Ronstadt et al. (editors), *Frontiers of Entrepreneurship Research, 1986* (Wellesley, MA.: Babson College, 1986), 366–379.

T. Kanai. *Entrepreneurial Networking: A Comparative Analysis of Networking Organizations and Their Participants in an Entrepreneurial Community.* Unpublished Ph.D. dissertation. (Cambridge, MA: MIT Sloan School of Management, 1989).

I. C. MacMillan, R. Siegel, & P. N. Subba Narasimha. "Criteria Used by Venture Capitalists to Evaluate New Venture Proposals", *Journal of Business Venturing*, 1 (1985), 119–128.

I. C. MacMillan & P. N. Subba Narasimha. "Characteristics Distinguishing Funded from Unfunded Business Plans Evaluated by Venture Capitalists", in R. Ronstadt et al. (editors), *Frontiers of Entrepreneurship Research, 1986* (Wellesley, MA: Babson College, 1986), 404–413.

R. H. Rea. "Factors Affecting Success and Failure of Seed Capital/Start–Up Negotiations", *Journal of Business Venturing*, 4, 2 (1989), 149–158.

J. C. Ruhnka & J. E. Young. "A Venture Capital Model of the Development Process for New Ventures", *Journal of Business Venturing*, 2 (1987), 167–184.

T. T. Tyebjee & A. V. Bruno. "A Model of Venture Capitalist Investment Activity", *Management Science*, 30, 9 (September 1984), 1051–1066.

E. R. Tynes & O. J. Krasner. "Informal Risk Capital in California", in J. A. Hornaday et al. (editors), *Frontiers of Entrepreneurship Research, 1983* (Wellesley, MA: Babson College, 1983), 347–368.

Venture. "Looking Back", May 1989, 54–56.

W. E. Wetzel, Jr. "Angels and Informal Risk Capital", *Sloan Management Review*, Summer (1983), 23–34.

CHAPTER 8

Going Public

On the day Tyco Laboratories (see Chapter 1) went public, Arthur Rosenberg was ecstatic. Now he had a company to run. In his unusual case, going public provided the funds to buy his research laboratory from its parent corporation. On the day SofTech went public, Doug Ross was also ecstatic, but for the more usual set of reasons. Now his growing firm had the funds to help finance its efforts to build its microcomputer software product line. But in addition Doug had also realized about $600,000 in cash from the offering and had remaining shares valued at the public market offering price at $4 million.

For many technological firms going public is a logical step in their continuing growth. The capital made available from the public offering helps fund accelerated product development programs, enables the broadening of their distribution channels, and generates financial strengthening through debt retirement. Yet success is a many-colored fabric. To some technological entrepreneurs "going public" *is* success, not just part of further "growing up". As one entrepreneur philosophized, "I built this company up from scratch, made it profitable and growing, and brought it public. Now it's time for me to step aside and let these other fellows run the company." Especially from the perspective of personal fulfillment, bringing the firm from being privately held into public ownership, with the company's stock traded and reported daily (more-or-less), engenders strong feelings of pride of accomplishment. For those entrepreneurs with high n-Ach, going public creates new tangible measures of attainment.

Whether another part of corporate growth or a first measure of company success, personal financial success of the entrepreneur is also usually solidified, and even somewhat enhanced, by going public. Some of the entrepreneurs sell a portion of their ownership as part of the initial public offering and transform paper wealth into cash. These and other entrepreneurs, as well as the early investors in the company and the usually many stockholding employees, soon begin to sell portions of their stock into the public market, as allowed by the Securities and Exchange Commission (SEC) regulations. The realized liquidity of their previously illiquid assets generates for many of them thousands and even millions of dollars. For all entrepreneurs and their stockholding associates, going public makes the paper assets they still hold much more real, valued tangibly by an existing outside market into which they at least conceptually can sell their holdings. And almost always, the pricing of their shares of stock by the public market is considerably higher, even at the time of initial public offering, than their prior in-house prices, thereby increasing their perceived wealth.

As indicated going public produces for almost all entrepreneurs capital needed by their firms, enabling them to continue to grow toward fulfilling their company goals. Corporate success may thus follow going public as a consequence in part of the strengthened financial capacity of the firm. This occurs directly by the increased working capital produced by the public offering. It may also occur indirectly through enhanced access to capital markets, both debt and equity, for the future needs of the firm. The publicly traded stock certainly makes it easier for the company to attract and/or hold on to key employees through the stock-based incentives that become available. That public stock may in addition facilitate acquisitions of other companies or product lines, if these are part of the firm's strategy. Many entrepreneurs report that the enhanced image and reputation derived from being a public corporation even improve their ability to sell products and services.

Such a stream of benefits does not come without costs. The most obvious is that going public is itself a very costly process, consuming not only significant time of the key managers of the firm but also a substantial part of the proceeds of the public stock sale for commissions to the under-writers and brokers and for the sizable expenses of lawyers, accountants and printers. Less visible initially, but for some entrepreneurs a greater real cost after the fact, are the continuing requirements for the changed conduct of a public corporation, with quarterly reports, annual meetings, continuous public visibility and scrutiny, demanding time and patience of officers, as well as increased overhead costs. "Living in a fishbowl" is a new and often unwelcome consequence for the entrepreneurs who may previously have run their firms in their own private ways. To many entrepreneurs the most painful aspect of this public existence is the perceived pressure they sense for short-term performance. This is an anathema to those who believe that their company's destiny and competitive advantage are only achievable through long-term technology and product development. For some entrepreneurs the benefits with customers of their improved public relations seem countered by the increase of company information now made readily available to competitors. And to others the fear of loss of control of their "baby" to outside stockholders looms large, whether realis-tic or not.

Many books have been written for many years on the whys and wherefores of going public. The technology-based firm is not without specific guidance in this regard. Among the most recent and most thorough of these guides is the Peat Marwick publication, *Going Public: What the High Technology CEO Needs to Know* (1987). This chapter does not present a how-to perspective. Rather, consistent with the book, it reports the results of research studies of technological enterprises that have gone public. The chapter covers decisions by entrepreneurs to seek public market funding, their search process for investment bankers/underwriters, the negotiations with the underwriters including the underwriters' decision criteria, and the outcomes of the public offerings, both in terms of stock performance and impacts upon the companies. Surprisingly, despite the obvious significance of going public to the entrepreneurial firm, little research on the subject appears in six recent annual volumes of *Frontiers of Entrepreneurship Research* and only one paper appears in the issues of *Journal of Business Venturing.* The experiences of Japanese firms that go public are described in one recent report (Systems Science Institute, 1989).

My data sources for this chapter are two focused studies carried out fourteen years apart, each covering approximately a three year period of "going public" activities. For each study comprehensive lists were prepared of all the New England area technologically oriented companies that had their initial public offering (IPO) during the time period that ended about one year prior (so as to obtain some record of postissue stock market results and impacts upon company management), generating 30 companies on each list. Development of these lists was difficult as no single source of information, including the regional office of the SEC, could reliably identify the companies. Indeed the two primary sources on IPO activities nationally, the *Investment Dealers Digest* and the *IPO Reporter,* disagree in their numbers over the years by as much as 40 percent, due in large part to different definitions of IPOs. The lists used here should be reasonably reliable, given all of the cross-checking done to assure completeness. Half of each list was included in the actual interviews, providing detailed data on 16 and 15 firms, respectively. The Boston and New York underwriters involved in these public issues were also studied, producing information from nine different investment banking firms in each of the two analyses, occasionally from more than one office of these firms.

What seems remarkable is the essential sameness of the findings from the two studies, despite their decade plus separation. For example, even the spread of company interviewees turns out to be the same, although originally the CEO of each company was approached for cooperation: seven presidents, seven treasurers, and two outside board members are the sources in the first analysis; eight presidents and seven treasurers provide the data in the second study. While the results are not necessarily representative of what might be found today, or in other parts of the United States, they reflect the types of issues, the process, and the outcomes that at least Greater Boston technical entrepreneurs have experienced over the past twenty five years. A review of recent IPOs with high market valuations (Davis et al., 1989) indicates that 13 out of 100 came from Massachusetts, second only to California, which originated 39 of the IPOs. Those data suggest that a comparison of these samples with California high-technology public offerings might be the most beneficial route toward generalization.

However, this chapter provides no insights on technological firms that wanted to go public but failed, a group that must be very large given the information from the underwriters in both studies who claim that as few as 1 to 5 percent of the companies they review eventually go public. This percentage screened out of the public market is comparable to the turndown rate for funding by venture capitalists that was demonstrated in Chapter 7.

Who Goes Public?

The previous discussions of financing in Chapters 5 and 7 provide some incidental indications of initial public offerings among young Greater Boston-area technology-based companies. For example, of one sample of 110 MIT-based companies only two had gone public for their initial financing and, up to the time of the data gathering, an additional seven received public funds as their means of secondary financing along with eight that went public as their tertiary-stage funding. While others from this cluster may well have gone public at some later date, these data suggest that not much more than 15 percent of the total population of technological companies went public, probably generally indicative of the extent of public market financing of technical firms. For perspective, notice from the data in prior chapters that venture capital firms eventually get involved as the principal financiers of a comparable percentage of technical companies, some of whom of course later go public. And nonfinancial corporations are key funding sources of about 20 percent of the firms, some of which later go public but more of which are eventually purchased outright by the nonfinancial corporations.

An old adage in going public is "sell the sizzle or sell the steak". Some potential stock purchasers are perceived as likely to be attracted by the glamour of very early-stage firms, which have great promise but are too young to be assessed in terms of actual performance. Other potential buyers are assumed to want to be able to evaluate actual company performance, that is, products, revenues, and profits, before they buy stock. As shown in Table 8–1 the two going public studies, supported by the MIT spin-off information cited earlier, indicate that technical firms tend to split 50-50 between trying to sell "sizzle" or "steak". In contrast only two of 79 Japanese firms studied went public with less than ten years of existence (Systems Science Institute, 1989).

Clearly those firms going public at the time of their founding have only promise and no performance. The entrepreneurs personally may have established a prior track record at some other firm that constitutes a "substitute" measure of performance, but their new companies are as yet untested. The postformation firms selling public stock at an early stage are similarly lacking in tangible measures of performance and are relying primarily upon "sizzle". They have no sales or very little sales, a maximum of

Table 8–1
Stage of Development for Technical Companies Going
Public

Stage of Development	MIT Spin-offs		Early IPOs		Recent IPOs	
	n	%	n	%	n	%
Initial stage financing	2	12	1	6	1	7
Early stage, but						
secondary financing	7	41	7	44	6	40
Later stage, tertiary or						
later financing	8	47	8	50	8	53
Totals	17	100	16	100	15	100

$453,000 in the prior year for the largest "early issuer" in the more recent study. The eight early-stage issuers in the first study all have fewer than 25 employees and average three years old.

In contrast, the eight firms in the recent IPO study that went public at a later stage have revenues from over $1 million to over $50 million in the latest fiscal year prior to their public offering. Typically the later-stage issuers have several hundred employees and an average age of eight years. This latter group is still rapidly growing, however, showing sales growth rates of up to several hundred percent over their previous year. Throughout this chapter the company's stage of development and size are shown to be the critical parameters that affect many aspects of the going public process. Small/large, sizzle/steak are the differentiating elements for technical firms going public.

Why Go Public?

All the companies in the two IPO studies have at least one factor in common: They had decided to issue new equity to the public rather than choose other available sources of capital such as bank loans, private placements, venture capital, or even selling the company. To the extent that the firms are at various stages of development they have differing financing opportunities available to them. As contrasts, a couple of the firms went public at inception while others had long stable records of sales and earnings. The former have the choice between going public and raising funds from the more common private investors and venture capital companies; the latter are already much too large to seek further venture capital. For smaller, earlier-stage firms the principal issue is the likely higher cost, in terms of proportion of the company to be given up, to raise the needed capital privately. Alternatively, a number of these younger companies worry about whether they are far enough along in their own development to cope with the consequences of being a public corporation. Several firms have already exhausted other sources. For example, one company was started with founder seed capital, privately placed $300,000 of equity one year later, entered into a number of contract development programs with customers to expand its resources further, and had already privately placed additional securities twice before making the decision to go public. Another early stage company sought advice from an investment banker for sources to expand beyond the financial capacity of the original founders. The investment banker presented the alternatives (his firm also had a venture capital arm) and the decision was made to go public. Advice on alternatives seems readily available, especially to the larger companies in which venture capital funds tend to have prior investments ($p = .02$). For example, venture capital companies have previously invested in six of the seven largest firms in the early IPO study.

A majority of the larger companies have seriously considered making a primary issue previously but have not gone through with it for various reasons. Abandonment due to poor stock market conditions is the principal explanation but some firms had been advised that they were too small to go public. Two of the larger companies in the more recent IPO study had previous unsuccessful attempts at nonunderwritten "Reg A" offerings. (Reg As are less formal and smaller public issues carried out under the SEC's Regulation A provisions, and can be underwritten or sold by the company itself without using an underwriter.) Those two firms proceeded this time through to an immediate search for an investment banker "who would do it right".

When asked to rank the advantages they had perceived in advance of going public, the entrepreneurs in the first study responded:
1. Fill an immediate need for cash and working capital
2. Create a public market to facilitate acquisitions
3. Create a public market to permit sales of their own holdings
4. Improve their firm's debt–equity ratio.

Seven of the eight smaller firms regard their capital needs as urgent; 11 of 15 firms rate the need for working capital as No. 1, agreeing with 14 of 15 companies in the second study. The urgency issue caused several CEOs and CFOs to laugh when asked this question. One responded, "Our bank refused to extend our line of credit, even at 2 1/2 percent above prime, and we forecast that available working capital would be completely gone by the end of the year. A firm in a similar business had just gone to the market at some crazy multiple. We would have been nuts *not* to go public." Another CEO remarked, "Going public was not the best alternative, it was the only way left to us to raise real high risk capital." Others put this in perspective by saying that the public market provided the lowest cost source of needed working capital.

Special capital-related reasons show up in both IPO samples. In one case the public offering permitted the sale of a large fraction of stock of the major shareholders who are relatives of the original but no longer dominant founder. This enabled the present management to end an uncomfortable situation of stock control by a small group of foreign owners. A second case permitted a negotiated settlement to be paid by a spin-off from another high-technology company that was in bankruptcy proceedings.

The larger technical companies perceive the creation of a public market as a more important advantage than do the smaller companies ($p = .001$). This is true both for the purpose of facilitating acquisitions as well as permitting the sale of founders' stock in the after-market. (During any three-month period SEC Rule 144 allows the sale by insiders of the greater of 1 percent of the class of stock outstanding or the average of the four most recent weeks' trading volume.) The larger firms also express less need for working capital than the smaller companies (.04). The desire to improve debt–equity ratio through the significant increase in their equity base is common among firms of all sizes, the managements recognizing that their ability to borrow funds will be enhanced. Half of the entrepreneurs specifically mention advantages accruing from their customers' interests in becoming stockholders.

A comparable number of disadvantages is also perceived by the entrepreneurs before deciding to go public. A major issue with some is the direct cost of going public, which is discussed in depth in "The Deal" section of this chapter. Some entrepreneurs in the more recent study were concerned in advance about potential loss of control, although other entrepreneurs specifically point out that private investors can exert a much higher degree of control than the public. Several entrepreneurs remarked that the public financing would effectively strengthen management's control by widely distributing stock to more passive investors. The principal difference between the earlier and later IPO studies is the more recent group's worries about undue public pressures on short term results (10 of the 15 entrepreneurs) and the somewhat related cost of managing ongoing shareholder relations, especially in terms of key persons' time (8 of 15). The earlier group senses essentially no overall disadvantages of going public, expressing minimal concerns in all these areas, with no one worried about loss of control.

Recent technological entrepreneurs who are going public may be more sophisticated than their earlier counterparts, or there may indeed have been a real change in the stock analyst, regulatory, and other public pressures upon the publicly owned corporation. Not all entrepreneurs knuckle under to these pressures. One proclaimed, "We make it very clear in all our reports that our goal is *long*-term growth. In the prospectus we said that no dividends would be paid and that we did not expect to earn a profit in the next three years. As long as we make this clear, then if our investors are so concerned about quarterly results they can find another company!"

Another remarked, perhaps more thoughtfully, "Growing at our rate means we're going to have to issue more debt or equity periodically to satisfy our needs. If the investment community is unhappy with our short term results, even if we think this is nonsense, it will affect our ability and our costs to acquire funds in the future. We have to take this into account."

The timing of an issue is jointly determined by the companies' internal needs for funds and the conditions of the securities markets. Overall these are ranked comparably by the entrepreneurs in both studies, 13 or 14 in each sample regarding each as paramount in importance. The large companies rate market considerations as a more important factor than the small companies ($p = .002$), for which capital had generally become a critical requirement. At least 20 percent of the companies would have gone under in the absence of an immediate public offering, since no other source of funding seemed available to them. The small companies that had gone public during a "hot issues" market admit after the fact that the timing of their issue was luck, but large companies had tended to wait for favorable market conditions. One biotechnology company admits it was awakened by the hot biotech market into quickly developing an ambitious business plan requiring a large capital input to take advantage of the market opportunity. Often the three to six months delay from the beginning of preparation for going public to the final effective date of the offering causes loss of an attractive hot market. Fortunately for it, the biotech firm in the recent study completed its public offering while the market window was still open, enabling the company to launch an extensive program of product development. But the company did have to suffer with lots of stockholder discontent in the eroded stock market that soon followed.

Finding an Underwriter

Occasionally a company decides to go public after being approached by an underwriter who persuades the firm to issue public stock. In this case the underwriter choice and the company's decision are concurrent events. In most situations a technical firm needs to find an investment banking firm that will agree to underwrite a public offering. This involves decision making by two parties—the technological firm and the investment banking firm—both of which apply their own search, evaluation, and decision criteria. This section first examines the firm's efforts and then turns attention to the underwriter.

In the first study three of the five smallest companies, all of which were still developing their first products, did not seriously attempt to find an underwriter. They considered it a hopeless task since they were so small and risky and decided to prepare and sell their own small "Reg A" issues of stock. Two others in this sample failed to find a satisfactory underwriting deal and also went the Reg A route on their own. All the rest of the first

group and all but one of the second sample of companies were fully underwritten.

The Technological Firm's Search and Decision

The legal, financial, and organizational intricacies of going public are complex and not well understood by other than those professionals who are
directly engaged in these activities. The need for help becomes apparent
quickly to anyone contemplating a public offering. Entrepreneurs of the
smaller companies rely on a diverse group of professional people for guidance, including accountants, lawyers, bankers and personal friends. Two of
these companies were advised by members of their Boards who eventually
managed the offerings that were made directly to the public. The larger
companies rely heavily on the outsiders on their boards for advice, especially
in the many cases in which venture capital investors are on the boards.
After extensive consultation with its advisors, one entrepreneur felt so
helpless that he reported, "I was willing to let anyone underwrite the issue."
The other entrepreneurs communicate more definite criteria to evaluate
possible underwriters.

In the first study three factors emerge prominently. Most important is
a preference for an established investment banking firm that had built a
good reputation. The larger the technical firm the more important is this
reputation criterion ($p = .04$). Second is the importance of national distribution, this capability also being weighed significantly more heavily by the
larger companies (.05). Finally, in each study several of the entrepreneurs
prefer underwriters who specialized in or had extensive experience with
technology-based issues. Interestingly, the smaller firms tend to see this
criterion as more important. In describing the ideal investment banker the
more recent sample of entrepreneurs agree on the desirability of the large
national firm or the old established firm with good reputation. These
entrepreneurs support the concern for gaining a wide distribution, but
more strongly emphasize the desire for smooth working relationships, as
well as for an investment banker with an ability to maintain an after-market
for the company's stock.

While the larger companies carefully seek out underwriting houses
that meet their concerns, the smaller companies are far less sophisticated,
approaching investment banking firms more or less at random. Among the
eight smaller companies in the early IPOs two did not bother to approach
any underwriter while at the opposite extreme two presented formal proposals to more than five firms. The larger companies are often more selective, in part due to the presence of a venture capitalist or banker on their
boards or due to dealings with an investment banker in regard to earlier
private placements. Two in the early study and eight in the later group
approached only one investment banker each and secured its services. A
number of the companies report that they were engaged in negotiations
with two underwriters simultaneously, and several entrepreneurs made

contacts with three or more underwriters. One large company discussed its proposed offering with eight special and major bracket underwriters who all seemed eager to perform underwriting services. The company was in the pleasant position of not needing to go public and believed that competition among the underwriters would encourage each to outperform the other, both on the IPO and on any future transactions.

Three of the smaller companies in the first study obtained an underwriter after a long and discouraging search. They were continually refused due to their small size and lack of earnings. Each of them ultimately found an underwriter after enlisting the services of a "finder". The larger companies in that study and all of those in the later study were more successful in attracting proposals from underwriters. Most of the enterprises that talked to more than one underwriter received more than one proposal, leaving final choices to be made that turned out to be quite subjective. Ultimately, the interviews suggest that each company's decision is closely tied to its objectives in making a public issue. As in many gray areas of business the entrepreneurs have trouble justifying their decisions on rational grounds.

The Underwriter's Perspective

Detailed structured interviews with the 18 investment banking firms in the two studies generate perspectives on the process followed in deciding to underwrite. These firms vary considerably by size (regional versus national), reputation (old established versus new), specialization (full line services versus technology specialist), and attitudes toward IPOs as a business in general. These differences cause few conclusions of general validity. One clear generalization is that especially in times of "new issues" markets underwriters are besieged by opportunities. They then need to be very selective in matching their own capacities to the potential deals they see. Even in less hectic periods only a small fraction of companies wanting to go public are given underwriting. The principal sources of referrals to these investment banking firms are, in descending order, friends, current clients, and venture capitalists. Direct approaches by entrepreneurs do result in some underwriting but are seen by the underwriting firms as far less important overall because of the lack of prior screening.

Although hundreds of underwriters exist in the United States (and many more servicing new opportunities not included in this study for going public in overseas markets), the market is clearly dominated by a few specialist firms for which the IPO is a major source of income. At least one firm in the sample has IPOs accounting for over 75 percent of its total revenue. For another full service investment banker in the sample, IPOs might produce only 5 to 10 percent of its revenues. All the underwriters interviewed find IPOs profitable. Naturally those specializing in initial public offerings are generally more active in directly marketing themselves to potential IPO candidates, but in recent years most of the national firms have set up technology groups specifically aimed at attracting and servicing these smaller companies. Each underwriter has its own general and specific criteria for agreeing to underwrite a company. But their testimony in the interviews, evidenced by their documented practice as well, indicates that these criteria may be ignored if the company looks exceptionally "good".

In choosing to underwrite an issue the firm must balance its own perceptions of the key factors determining the strength of the company and the characteristics that they feel investors view as most important. Ultimately the underwriter is an intermediary between the investor and the issuer. Therefore, the underwriter must be concerned with the marketability of a given security. If the underwriter feels that the market is currently responsive to service industry issues (e.g., restaurant chains), it will select these. If, on the other hand, technology issues seem "hot" and are expected to continue to be, the underwriter may select companies within this industry cluster. With industry choice determined by the market the underwriter then looks for individual companies within the favored industries that satisfy its taste and judgment.

Table 8–2 tabulates the number of underwriters who view particular company characteristics as most important to stock buying customers, based on those participating in the more recent study. Sound company management and future growth prospects are the factors most frequently cited as very important. The earlier study finds the same ordered priorities toward the capabilities and depth of management, the future growth prospects of the firm and its industry, and the past record of the company. This agreement on the importance of company management is not surprising, yet investors rarely have the opportunity to meet management before buying the securities. As few investors read beyond the summary and first pages of a prospectus, the investors ultimately depend on the underwriters' evaluation of management, implicit presumably in the decision to underwrite. Despite the claimed relative importance of historical earnings, many of the companies studied have little or no previous earnings (including seven in each of

Table 8–2
Underwriters' Views of Factors Most Important to
Customers Buying Shares in an IPO ($n = 10$ underwriters)

Criterion	Not Important	Moderately Important	Most Important
Historical sales growth rate	1	7	2
Historical earnings per share	0	6	4
Sound company management	0	2	8
Future growth prospects	0	0	10
Size of company	5	4	1
Technological glamour	0	7	3
Price–earnings ratio	0	8	2

the two samples), several expecting none for at least a few years. This is an example of a criterion that many underwriters are obviously willing to forgive for the hot prospect, even though they feel it is very important.

Whether only a rationalization of the opinions expressed in Table 8–2 or not, it is comforting to note in Table 8–3 that assessment of weak management, in addition to weak market conditions, are the key reasons cited by underwriters for rejecting a company. Beyond these general selection criteria most underwriters also have specific rules of thumb used in their evaluations. Minimum size of the company, and minimum (for the smaller underwriters, also maximum) size of the stock offering are among the scoping criteria applied by underwriters to screen initial interest in the potential underwriting.

Eleven of the 16 technological companies in the sample of early IPOs and all but one of the later sample ended up with full commitment underwritings. The smaller firms often seemed lucky to get one underwriter, the larger companies usually got to choose. The larger companies all obtained the services of reputable investment bankers including a few of the most prestigious firms in the industry. Investment bankers for the entire sample range from the lesser lights of the industry up to Paine Webber and Alex Brown. All of the investment bankers violate their own stated guidelines regarding the minimum size of companies, some of the situations not even coming close. Despite this the size of the technical company and the size of the investment banker are highly correlated ($p = .001$), indicating primarily the preferences of the larger investment bankers.

The Deal

While the initial proposals of both the companies and the underwriters outline the general terms of an issue, the two parties must negotiate the

Table 8–3
Primary Reasons for Not Underwriting an IPO Candidate
($n = 10$)

Criterion	Not Important	Moderately Important	Most Important
Market conditions (timing)	1	2	7
Weak management	1	2	7
Weak earnings	1	4	5
Premature	1	5	4
No agreement on terms	6	3	1
Company found other sources of financing	9	1	0

exact terms of the deal as it will be offered to the public. Key areas of negotiation include the market value of the technical firm, the percentage of the company to be offered, the price per share, and the underwriter's compensation. Other less prominent issues also need to be resolved, such as secondary shares, warrants, board seats, and rights of first refusal on subsequent offerings. Overall both the entrepreneurs and the underwriters report that these negotiations are relatively harmonious based on trust, respect, and the general feeling that neither side tries to gain the upper hand. A few of the entrepreneurs report extensive hard bargaining in which they felt themselves at a disadvantage due to their inability to find another investment banker. Yet only one entrepreneur believes that the underwriter and deteriorating market conditions placed him in a position of doing a deal on terms with which he was less than comfortable. For nonunderwritten issues the company's officers and advisors unilaterally set the terms.

After futile searches for an underwriter five of the companies in the earlier study decided to offer their issues directly to the public, without an underwriter. All of these were small Reg A issues, offered by companies that were new, small, unprofitable, and undercapitalized. The size of the offerings were set to meet the projected capital needs of these companies. One firm nominally sought twice the amount it needed because its management assumed they would only be able to sell about half the issue. Prices were set on a subjective basis with three of the companies choosing $10.00 per share because they thought that price would convey a good image to the public. Despite their companies' real condition as seed-stage firms, the entrepreneurs did not want to create the impression that they were "penny stocks".

Company Valuation

In the underwritten cases joint decision making, rather than negotiation, is a better way to describe the process for establishing company valuation. Stated differences of opinion, with few exceptions, range only up to 20 percent. For instance, eight entrepreneurs say they entered the discussions with a firm idea of the market value of their own company, based largely on their analyses of market valuation of similar companies. In only one case was the company eventually valued lower than management's initial assessment.

The companies with no earnings generally left the valuation up to the underwriter. Textbook solutions call for projections of a company's future earnings and payouts, accounting for all associated taxes, discounting the yield back to the present based on a market interest rate that reflects the riskiness of the investment. Needless to say no underwriter indicates it adopts the textbook approach to valuation. One development stage company studied, with no earnings but with sophisticated management, utilized the underwriter search process as a mechanism for evaluating itself. It garnered five offers from financial sources that helped establish an envelope for company valuation in its public offering. The management and underwriter of another development stage company for which there is no public market industry equivalent patterned its public offering, in terms of dollars raised for a given percentage of the company, on a basis comparable to what a private venture capital transaction would cost. Thus, a valuation was simply transferred between financial market segments.

For companies with a record of earnings, especially the more established firms, the underwriters tend to apply loosely price–earnings (P/E) ratios of "similar" firms to reflect current market conditions. Table 8–4 lists the offering P/E ratios for the companies that have earnings, assuming full dilution based on the stock being sold. The more recently issued stocks have lower P/E ratios overall, but a wide variation of a five to one range is reflected in both samples. Statistical tests find little correlation between these ratios and the size of the companies as measured by sales, total assets, or net worth. Nor are they correlated with sales growth rates. The most cogent explanation is that the spread in the P/E ratios shows the effects of industry fads, special circumstances of the companies, and different timing relative to hot markets. The issue of glamour of a particular industry adds considerable volatility to the pricing of IPOs. Security analysts told the manager of one biotechnology company that it was too bad his company had earnings, because everyone would "attach" a P/E ratio to his income stream. The analysts presumed that due to market hype the company might have received a higher valuation without any earnings. As a further statement of the quandary surrounding the pricing of new issues the price–revenues ratios calculated for these firms show a range of 1.2 to 4.8, not correlated with the P/E ratios or with any other other performance measure. Examining a large number of IPOs issued during 1978–1985, McBain and Krause

Table 8–4
Fully Diluted Price–Earnings Ratios
for Underwritten Offerings

Early IPOs	Recent IPOs
18	12
21	13
23	14
30	19
30	37
40	43
42	67
90	

(1989) find that the P/E ratio is directly related to the percentage of the firm's equity retained by the insiders.

Price and Proceeds

The offering prices themselves for the underwritten companies range from $0.50 per share in two cases up to $22.00 per share, with a median price of $9.50. The higher the offering price the higher the total proceeds of the primary issue ($R^2 = .70$, $p = .01$). Price also correlates with several aspects of size of the company, including net worth, total assets, and revenues, with revenues the strongest explaining factor ($R^2 = .63$, $p = .005$). Since the larger investment banking firms underwrite the larger issues, the size of the investment banking firm is positively rank correlated with the offering price ($p = .002$).

As expected the total proceeds from the public offering highly correlate with the size of the company (.005), as well as the size of the investment banker (.02). Total proceeds equals price per share times number of shares. Since both the proceeds and the price correlate with company size, it is not too surprising to discover that the number of shares issued also positively relates with the total assets of the company ($R^2 = .79$, $p = .005$) and is rank correlated with the size of the underwriter (.002).

In eight of the 31 cases companies had combined offerings comprised of new shares and the registration and sale of some insider shares, called secondary shares. Underwriters monitor this process closely to avoid the appearance that owner–managers are bailing out. Half of these cases involve primarily nonmanagement shareholders desiring to achieve some liquidity. In the other four cases the company's own cash needs are modest and the investment banker felt that sale of additional shares was necessary to make the size of the issue large enough. In these latter circumstances the majority of the preissue stockholders are somewhat reluctant to offer their stock as part of the primary issue since they anticipate being able to sell at a much higher price in the after-market. In three cases insiders finally agreed to sell an equal "taxed" proportion of their holdings, feeling mollified by the fact that all shareholders were being treated uniformly.

Underwriter Compensation

Many entrepreneurs find the underwriter's discount and commissions and the total accounting, legal, and underwriting expenses quite high relative to their prior expectations. However, these issuing costs are not heavily negotiated at the outset. The underwriter's discount or spread (the difference between the offering price and the net proceeds to the company) is the underwriter's main source of compensation for all of the deals and the sole source of compensation for the nine largest issues (no warrants on them). The spread, expressed as a percentage of the offering price, varies from a low of 6.0 percent to a high of 18.4 percent. Table 8–5 shows, not surprisingly, that in the fourteen years between the two studies the underwriter's spread as a percent of total proceeds grew about 10 percent (as measured by the median), while the total expenses increased about 40 percent. The spread is negatively correlated with the price of the stock ($R^2 = .72$, $p = .005$), indicating a higher spread for the lower priced offerings. Since the price of the stock supposedly reflects the quality and riskiness of the issue, the underwriter is expected by financial theorists to demand higher compensation for assuming the risk inherent in a low priced offering. By widening its spread on the more risky issues the underwriter provides greater margin for himself to be able to sell off the inventory of company stock if the security meets with a poor reception on the offering date. The same outcome of higher spread for the smaller issues also results if one believes that the underwriter is merely allotting its fixed costs of service over a small base. As another measure of the same phenomenon, when the two samples of IPOs are each split at their medians by asset size (as listed in Table 8–5), the smaller companies are found to have incurred a significantly higher mean spread of 10.6 percent (10.9 percent in the more recent

Table 8–5
Total Costs of Underwritten Issues (excluding warrants)
(Percent of total proceeds)
(Companies rank ordered by increasing assets)

	Early IPOs		Recent IPOs	
	Underwriter's Discount (%)	Total Costs (%)	Underwriter's Discount (%)	Total Costs (%)
	12.5*	31.4*	10.0	14.0
	10.0*	26.7*	10.0	14.8
	9.0	18.7	18.4	25.5
	12.5	19.9	10.0	18.8
	9.0	13.2	10.0	14.8
	7.7	10.8	7.5	15.9
	6.7	10.4	10.0	18.3
	7.3	9.9	9.0	15.9
	8.1	9.7	14.0	21.2
	7.0	8.6	8.5	12.1
	6.4	8.0	7.2	8.6
			6.1	15.4
			6.0	8.2
			7.2	10.1
			7.3	8.7
Medians	8.1 (7.7†)	10.8 (10.4†)	9.0	14.8
Means	8.7 (8.2†)	15.2 (12.1†)	9.0	14.8

* Indicates the two underwritten Reg A issues.

† Indicates the medians and means omitting the two Reg As.

group) versus only 7.2 percent in the larger companies (9.9 percent for the recent IPOs).

Multivariate analysis reveals only one other factor that helps explain the variance of the spread. Spread tends to be less during periods of hot new issues markets than for a similarly priced stock issued at some other time. Practically, the underwriter's own cost of selling no doubt decreases when investor demand is high. This is also theoretically reasonable since the underwriter's risk presumably declines when the investors' demand for primary stocks increases.

In addition to the underwriter's spread, the company must pay the other direct costs associated with the offering, including fees for lawyers and accountants, printing costs, and other direct costs incurred by the underwriter in connection with the issue. The rapid increase of these costs in recent years, partially to pay for new selling costs such as "road shows" and color inserts in the prospectus, partially to pay for significantly increased legal scrutiny, makes the other costs a large add-on to the underwriter's base commission. These "other costs" are especially significant in the two underwritten Reg. A issues in the earlier sample (* in Table 8–5), where they represent 18.9 and 16.7 percent of the total proceeds. Real direct costs should not have been so high for these small issues. It is likely that these two companies, in urgent need of funds, were charged unreasonably for the services rendered by others in connection with the issue. The median and mean costs shown in Table 8–5 are recalculated (†), omitting those two special cases, to provide a possibly more representative cost picture.

The resulting total costs of the issue by company are also presented in Table 8–5, indicating that the underwriter's spread is the major component of the total cost for all but a few cases. The smaller companies, or those with low issue price, experience much higher direct costs. Splitting the two samples again at their medians by asset size confirms the effect of company size on the total cost of going public. Smaller companies, including the two Reg As, incur a significantly higher mean total cost of 22.0 percent (17.4 percent in the more recent IPOs), while larger firms average only 9.6 percent (12.6 percent in the recent sample). The same statistically significant result is found when the firms are divided into clusters in accord with their previous year's sales, that is, the smaller technological companies pay considerably more proportionately than the large firms to generate their public funds.

Matching total costs against underwriters shows that the most prestigious underwriters engage in the larger transactions with lower issuing costs, while the less prestigious and "best efforts" financiers with smaller transactions have comparatively higher expenses. These may alternatively reveal the prestigious underwriters as being more cost competitive or, more likely, the smaller technical firms as having less bargaining power. Then again, some entrepreneurs do not regard these total costs as costs at all, because they are not reflected in their own company's income statements.

A final important compensation are the "warrants" that are sold to the underwriter at a nominal charge, giving the underwriter rights to purchase shares of stock from the company at a given price some time into the future. These warrants are required by underwriters of the smaller issues, requested by some other underwriters but not by those doing the largest underwritings. The six largest underwritings in the early IPO sample and the three largest in the recent sample do not include warrants for the underwriter; all but one of the remaining underwritten issues contain warrants. In most cases the warrants are for 10 percent of the number of total shares offered in the primary issue, although the sample does include one firm with only 2 percent warrant coverage and several with up to 20 percent. The prices at which the warrants can be exercised in the future range from the offering price itself in most situations, to 120 percent of the offering price in one case, to another deal that provides for a 10 percent per year escalation over the offering price during the next five years. In general the warrants are not exercisable for a year but their life lasts as long as five years. While it is very difficult to attach a monetary cost to the warrants, it is sufficient to note that the largest technical firms do not incur this cost as part of their going public episode.

Board representation by the underwriters, rights of first refusal on subsequent offerings, and even consulting services are the remaining terms and conditions negotiated. About one-third of the sampled companies elected a new board member from the underwriter, contingent on the offering, while several had preexisting board members from those firms, sometimes owing to prior venture capital investments. The smaller underwriters sometimes negotiate a right of first refusal and consulting fees. The weaker companies in the sample, whose horizons are dark without the public monies, are hardly in a position to negotiate vigorously these minor issues or indeed may believe that the ongoing outsider presence is beneficial. Larger investment banking firms not only do not require board membership, but often decline invitations to serve except in unusually attractive companies. Similarly, the larger underwriters rarely argue for a clause in their contracts to guarantee future underwritings, presuming that their performance and position will generally suffice for assuring future business.

Outcomes

Preissue Activities

Concluding a deal is only the beginning of "the rest of the story", as Paul Harvey says. Much work is now needed to "clean up" the company, prepare the prospectus, and gain SEC approval, and market the stock to prospective investors, before the new issue can become effective. In the firms in both samples employment agreements with key employees are often revised to provide assurances demanded by the underwriter. Recapitalization of the company's stock frequently occurs, with stock splits or consolidations to generate the right number of shares desired for pricing considerations. Liquidations of product lines, rewriting agreements among stockholders, and persuading debtors to alter their terms are common preissue requirements.

More subtle changes also take place in preparation for the public offering. Many companies restructure their boards of directors to include outsiders and individuals with expertise missing from top management. A substantial number change law firms and/or accountants as part of image projection, while also bringing in more expertise for the public offering. Some spin-off firms need to project better arms-length relationships with their original parent, some enter licensing agreements to ward off fears of possible law suits, and others tie down formal relationships with outside customers or key suppliers. A number of the larger firms, which have been waiting for years for the right opportunity to go public, already had formally audited financials, prestigious outside directors, and disclosures prepared even when not required. Several other companies have earlier entered into joint development contracts specifically to bring a large and respected company into a financial relationship before going public. Another firm entered into a marketing agreement with a large overseas partner to "clean up our balance sheet". These types of changes by the early stage companies are obviously more limited due to time constraints and lack of prior planning to go public. Pre-issue cleanup is widely practiced by Japanese firms as well, beginning deliberately several years before the planned public offering (Systems Science Institute, 1989).

Preparation of the prospectus, in close working relationships with the underwriters and lawyers, consumes much management time. The SEC has specific disclosure guidelines about securities, and rules controlling undue promotional activities by management around the offering date. The prospectus tells part of the company's story, but is usually highly stylized and replete with caveats insisted upon by the SEC and by the company's and the underwriter's counsel. The prospectus cannot contain any forecasts but only a bland and boilerplate section entitled "use of proceeds", which may mention working capital requirements, the intent to retire debt, or funding for a new product generation, hardly enough information to project a vision of the future. Historical earnings are not always helpful because as shown earlier a large fraction of the technological firms going public have little or no past earnings. There is seldom opportunity to "sell" within the prospectus except in very subtle ways. In the more recent sample one third of the prospectus have color pictures of company products and even fold out presentations to improve communication and image.

The conservatism embodied in these prospectus is reflected in the fact that none of the companies encountered any really serious problems with the SEC during the registration process. The entrepreneurs attribute their success to their lawyers and accounting firms as well as to the expertise of the investment bankers in preparing the registration statement and prospectus. Ironically, although the Reg A filing is intended to simplify going public for a small issue of stock, those few smaller firms, which used the Reg A, were also more likely to lack expert legal advice. Consequently two of the nonunderwritten companies required more than six months to gain SEC approval after submitting their initial statements, in contrast to the typical delay of two to three months encountered for the full preparation and SEC approval of the more complex filings. Only one firm got into trouble with the more local regulatory groups that also control public offerings, the so-called blue sky commissions of the various states in which the offerings were registered. In this one case the commission ultimately limited the price of the issue by constraining, to 25:1, the price–earnings ratio of the offering. Since a major stockholder of the company resided in that state it was necessary to obtain the commission's approval. Much to the chagrin of the company and its underwriter, the issue was forced to be priced substantially below their intended level. It later became one of the hot stocks of the year.

The "road show" is often the mechanism for selling. Not all underwriters organize this effort, but several entrepreneurs made presentations about their companies around the country, occasionally also overseas in Europe and Japan. The entrepreneur's own salesmanship talents are put to use in promoting the company's securities at meetings in major cities of local brokers, their clients, and institutional investors. Management is usually rehearsed by the underwriters and legal counsel to assure that statements and responses to questions are within SEC guidelines. One underwriter, summing up the importance of the road show in convincing prospective buyers, exclaimed "No story, no deal." One entrepreneur describes with glee his zealous underwriter, who videotaped management's discussion and demonstration of company products and then flew a private jet around the country stirring up investor interest. These methods do provide some potential investors, or at least their stock brokers, a chance to see and hear the company president in advance of the offering, perhaps satisfying their investment concerns for sound management that are highlighted in Table 8–2.

In organizing to move the stock the underwriter creates a distribution network of retail and institutional brokers. The underwriting syndicates range from a single firm, carrying out a "best efforts" underwriting, to a full commitment co-managed transaction that includes 83 underwriters. The median number of underwriters in the recent IPOs is 33. As expected the larger dollar volume transactions have the larger number of underwriters. Before setting the final price the syndicate may accept "indications of interest" from potential purchasers, in theory to help the underwriter determine how he should set the final price. Actually the underwriter usually tells prospects that the price has been set at close to a certain level and asks how many shares the prospect would like if it is possible to get that number. If demand exceeds supply, the underwriters allocate their shares to their customers as they choose. Should supply and demand be out of balance by several orders of magnitude, the offering price may be changed, but rarely is there an iterative process in an attempt to find a market equilibrium. In only one case in the samples was there a last minute significant decrease in both the offering price and the number of shares offered by the underwriter, upsetting the entrepreneur who felt trapped into a more costly and less beneficial underwriting.

Sales of the Issues

The underwritten issues were sold in their entirety on the effective date of the offering. The five nonunderwritten Reg A issues among the earlier IPOs were slow to sell, Figure 8–1 showing the sales records of the four that only sold part of their intended offerings. Only one of those firms placed its entire issue, and then only after a seven-month selling effort. Three nonunderwritten companies suspended their issues within one year of the effective date without completing the offering. One of these stopped after receiving 55 percent of the proposed total proceeds, the same firm mentioned earlier that had set its stated proceeds for the issue at twice its estimated requirements. A second firm's stockholders voted to suspend stock sales when 75 percent of its initial target was reached. Since amendments to the filing notice and offering circular have to be filed with the SEC after one year, a third firm suspended its offering at that point, with 75 percent of its issue sold. The fourth company that did not sell its entire issue did file the required amendments and continued to sell stock for a

Figure 8–1
Partial Sales of Nonunderwritten Issues

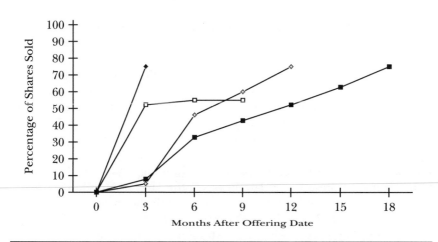

total of eighteen months, when it too finally gave up at the 75 percent completion level.

Stock distribution varies enormously across the issues. The large underwriters tend to distribute the stock far more widely, as evidenced by the fewer average shares held by an individual ($p = .04$). Again the nonunderwritten cases stand out as perhaps unfortunate exceptions. In three situations the entrepreneurs managed the offerings personally, selling stock to friends, relatives, business acquaintances, and professional contacts. Initial sales often generated a chain reaction: Individuals who bought a stock recommended it to their friends who contacted the company and bought shares. One issue, managed by an influential member of the company's board, was placed almost entirely with doctors and other professionals in Alabama.

After-Market Performance

An after-market for trading cannot develop while the company is still offering the stock at the issue price. Consequently no trading occurs in any of the nonunderwritten issues until well after the companies suspend their selling efforts. And even then a rather inactive trading market has developed for only two of those five companies, prior to later refinancing activities of some of the firms.

In contrast markets and trading started immediately for all the underwritten issues in both studies, the prices at first day's closing ranging from 11 percent below the offering price to 181 percent above it, as shown in Table 8–6. The gains on offering day for the earlier set of underwritten IPOs are significantly greater than those that went public more recently, due primarily to the generally hotter market conditions affecting the early IPOs. An unanticipated positive side effect on employee morale and esprit de corps often results. At one company employees were dancing and singing in the halls as the stock price more than doubled on the effective date of the issue.

Analysis of the data shows that larger companies escalate more on opening day than smaller firms ($R^2 = .33$, $p = 0.10$). Also issues made during a hot new issues market appreciate more on opening day than those made outside this time period ($R^2 = .27$, $p = .10$). Neither the size of the underwriting firm, the sales growth rate of the company, nor the price–earnings ratio at the offering price relates significantly with the first-day price change.

Analyses of the stock price data after ten trading days, after one month, after 40 trading days, and after three months show an increasing spread of the results over time. Some of the companies continue to decline from their initial prices, the biggest drop after 40 trading days being 32 percent. Other firms continue to escalate in price, the largest after 40 days being 313 percent above the initial offering price, with the mean change in the more recent set of IPOs being a 53 percent increase. In contrast, the range of the NASDAQ Index during the 40 trading days following each of these

Table 8–6
Percent Change from the Offering
Price at First Day's Closing

	Early IPOs	Recent IPOs
	8	(11)
	8	(5)
	26	(4)
	33	0
	33	0
	35	4
	50	6
	51	10
	127	12
	139	12
	181	36
		100
		100
Mean	63	20
Median	35	6

initial offerings ranges only from –6 percent to +16 percent, with a mean range of +5 percent, showing the far greater volatility of the new stocks. No easy way exists to evaluate the risk of these IPO securities although risk is certainly greater for them than for the market composite. But to outperform the market average by a factor of ten seems remarkable. Sales growth rate of the company before going public shows up as statistically related to price growth of the stock in these slightly longer period market studies. A perhaps negative short-term effect on company efficiency is an unexpected by-product of the stock price volatility. As one entrepreneur comments, "I couldn't get an outside telephone line for about a month after the issue. Everyone was calling his broker to get the latest price!"

Perhaps the most important observation about the stock price is that the general increase over the offering price is sustained over these longer but still early periods of stock trading (i.e., up to three months post-issue), clearly indicating that these technology-based company IPOs are not overpriced. When questioned in this regard, many of the underwriters indicate they believe in "leaving something on the table" for the market investors. Somewhat surprisingly the entrepreneurs do not often object to this strong suggestion that their companies might have raised as much as 20 to 50 percent more funds by a higher offering price. All the companies have experienced some periods of selloff in their stock prices since their offering and entrepreneurs often seem relieved that they went public at what both before and after the fact seem like reasonable valuations. The principal

exception is the entrepreneur whose offering price had been limited by
the state blue sky commission. The others feel that they received a good
price for their stock and do not begrudge the public its profits. Of course,
the entrepreneurs' own considerable stockholdings have also gone up in
value. Clearly related is the observation that the price–earnings ratio at
offering does not show up as statistically significant in any of the after-
market regression analyses.

Effects of Public Ownership

Despite the concerns raised by many of the entrepreneurs before going
public, few of them think that public ownership now has an important
effect on the way they run their businesses. Only three of them remarked
that it has affected their long range growth goals, one commenting that his
firm has to grow faster now to justify the "inflated" stock price. The other
two, in contrast, say they have to be more conservative and cautious in
making strategic decisions. The majority of the entrepreneurs claim that
they are not doing anything now that they would not have done had they
remained a privately held corporation. In terms of operations several CEOs
affirm that they do feel short term pressures to accommodate investors, but
one entrepreneur calls this process a forced focus rather than a planning
and operating constraint. Several observe that managements had better
pay attention to their business rather than to their stock prices, one speaking for this group in proclaiming, "My goal is to make money for the
company, and that is what the stockholders' goal should be too."

The investment banking firms are now represented on the majority of
boards of directors of the companies, with all of the entrepreneurs reporting that this relationship is beneficial and that the underwriter's advice is
welcome. The investment banker does have the opportunity to influence
the long range goals of the company, but the entrepreneurs feel that an
investment banker has little real knowledge of or influence on the internal
operations of their companies.

All the entrepreneurs agree that being public has added significantly
to their companies' accounting, legal, and public relations expenses, especially due to the quarterly SEC filing requirements. Many of the companies
have added a public relations or investor relations director as a result,
while some have incorporated these new responsibilities into the expanded
job of the treasurer. The CEOs report many more phone calls from investors,
investment analysts, and the news media due to their public status, but
most treat the annoyance as only at the "noise" level and not a significant
interference. The positive side of this is increased name recognition with
prospective suppliers and customers, salesmen in several firms being well
received by some who previously would not talk to them. Indeed none of
the entrepreneurs regrets the decision to go public, usually appraising the
realized advantages as the same as the entrepreneur had perceived before
going public. Officers of the three smallest underwritten companies do

feel that they had gone public prematurely, leading to high costs relative to the funds raised.

In the small group of nonunderwritten entrepreneurs, two now think they should have made a private placement instead of a public issue, believing they should have waited until their firms were bigger and with a better track record before going public. Two others think they should have worked harder to find an underwriter for their issue. Surprisingly, despite their lack of success in selling their issues, as documented in Figure 8–1, these entrepreneurs also do not regret having gone public.

Both large and small company entrepreneurs now think the most important advantage of being a public company is increased access to additional capital. They anticipate, and some have already realized, that future public issues and bank borrowings are much easier once they are publicly held.

The second key advantage is enhanced ability to make acquisitions, with a small number of the firms having already made noncash acquisitions since their primary public stock issues. While these acquisitions would not have been impossible before going public, the entrepreneurs attest that being a publicly held concern facilitates negotiations. The larger companies rank the advantage of a public market for help in their acquisitions higher than do the small companies ($p = .001$), no doubt because the smaller firms are still more internally oriented in their product and business development strategies.

Two other after-market advantages are considered important by many entrepreneurs, again ranked higher by the larger technical enterprises (.07). The creation of a public market both enables them to sell small parts of their equity holdings and also increases the value of their employees' stock options.

In this regard it should be noted that the vast majority of the technical entrepreneurs who have gone public are multimillionaires today, at least on paper. Included in the samples are two different entrepreneurs each with market worth of several hundred million dollars. While it might be difficult for many entrepreneurs to realize the current paper value of their holdings if they try to liquidate their entire positions through sale of stock in the market, the market value is gradually being turned into "cashed in" wealth by most of the entrepreneurs. Furthermore, the market value sets a public price for beginning negotiations toward being acquired by still larger firms, the outcome that eventually occurs to many of these technology-based companies.

Many of the short-term and long-term pluses and minuses of going public are illustrated by the case of EG&G. Ken Germeshausen said they went public for three main reasons: (1) concern about the liquidity of the founding partners' estates; (2) realization that sooner or later the company would need more money; and (3) a desire to establish a position in the marketplace. Germes felt strongly that public ownership brought pressures from stockholders both inside and outside the company. He regarded it as a burden to have investment analysts constantly prying into the business and to have to report frequently to the SEC. Pressure to grow earnings kept the company from embarking on major R&D programs and forced it instead into an acquisition program. Initial acquisitions, concentrated in the two areas of ocean-related research and technology and electronic instrumentation, had only mixed results. But over time it was EG&G's acquisition program, perhaps induced by public market pressures, that eventually made it a multibillion dollar firm.

Summary and Implications: Sizzle or Steak

The initial observation in this chapter that technology-based firms can sell the sizzle or the steak is borne out in the evidences introduced throughout the analyses of the two samples. Whether recently or fourteen years earlier, technical firms that go public have different motives and different consequences, depending significantly on their stage of development at the time of public offering. The data show that technical firms that go public split about 50-50 between early stage and later, the half selling sizzle only going public at founding or typically within their first three years of existence, with fewer than 25 employees and at best a few hundred thousand dollars in sales. Those that are "selling the steak" are much older, averaging eight years, have several hundred employees, and sales revenues from $1 to 50 million.

Table 8–7 lists the statistically significant benefits accruing more to the larger firms (compared with their smaller counterparts) in the two studies, as enumerated in the chapter, clearly one-sided in supporting the gains from waiting for the further growth to be achieved. The larger companies are better prepared for and in greater control of their decisions to go public. They more carefully search for and find higher quality underwriters. The larger companies' deals cost them less in direct and total costs, as well as in warrant dilution of their stock. Their stock sales go smoothly and they even gain surprisingly in immediate after-market price appreciation, although this differential benefit is not sustained over the smaller companies. After the fact the larger firms feel that they had also benefited more in regard to acquisitions, personal entrepreneurial liquidity, and employee perks.

But if the principal purpose of going public is to raise needed capital, then both large and small firms meet their requirements. And neither group feels it has incurred meaningful disadvantages in the process. What is not measured in the formal data collection presented in this chapter is the impact of going public on survivability of the firms. Here, in contrast, the clear advantage is gained by the smaller companies, many of which in their own founders' judgments would have gone under had the public offering not succeeded. For the smaller companies going public by itself is not equivalent to success, but rather a crucial step enabling the process of building a technological enterprise to continue.

Table 8–7
Differences in Advantages to the Larger Technological Firms That Went Public

Why Go Public?

Advice on alternatives readily available; prior investments in larger firms by venture funds

Perceive public market as a greater advantage, for both acquisitions and sales of founder stock

Rate market considerations as more important factor in timing the offering

Have far less working capital urgency for going public

Finding the Underwriter

All seek and find underwriters; some smaller firms have to do direct nonunderwritten Reg A deals

Prefer underwriters with good reputations, as well as national distribution capabilities

The Deal

Larger companies have higher offering prices

Larger number of shares issued and higher total proceeds gained

Much lower underwriter spread for larger companies, as well as for the higher priced stocks usually issued by them

Timing effect on spread, with hot issues market leading to lower spread, of greater benefit to larger firms that have better timing control of their public issues

Total direct costs much lower for larger sales and larger assets companies

No warrants as part of compensation for larger company deals

Outcomes

Larger dollar volume transactions have larger syndicates of underwriters

Sale of underwritten stock completed expeditiously; selling problems for nonunderwritten issues of a few smaller firms

Large underwriters distribute stock more widely

Higher first day price appreciation for larger companies, but difference not sustained over longer time periods; timing of hot issues market also affects first day appreciation

No differences in perceived effects of public ownership, although some smaller firms feel they had acted prematurely

Larger companies rate after-market advantages regarding help in acquisitions as more important than small companies; larger companies feel more positive about benefits from sale of founder stock and enhanced attractiveness of employee stock options

References

E. Davis et al. "The IPO Fast-Track", *Venture*, April 1989, 25–39.

M. L. McBain & D. S. Krause. "Going Public: The Impact of Insiders' Holdings on the Price of Initial Public Offerings", *Journal of Business Venturing*, 4(6) (1989), 419–428.

Peat Marwick. *Going Public: What the High Technology CEO Needs to Know* (Chicago: Peat Marwick Main, 1987).

Systems Science Institute. *Growth Strategies of Up-And-Coming Enterprises* (Waseda University Business School, 1989).

CHAPTER 9

Survival Versus Success

A Framework

The eventual goal of an entrepreneur, whether technology based or not, is success for his or her enterprise, at least in terms of the entrepreneur's own standards of success. Thus, the foremost consideration in my research program on technological entrepreneurship has been to determine the causes of the high-technology company's success and failure. The problem is that the performance of an enterprise is the culmination of intricate interaction of a large number of factors. Although each alone has its effect on company performance, the proper mix of those factors is no doubt of critical importance and perhaps cannot be discovered. Several different groups of factors, combined in different ways, may produce equally successful companies, each one successful for fundamentally different reasons.

The complexity inherent in the process leading to company performance precludes the possibility of an all-explaining or even all-encompassing "model" of successful technological entrepreneurship. Several prior empirical studies of success and failure tend to emphasize specific characteristics of the entrepreneur or the enterprise, such as motivation (Smith and Miner, 1984), the "incubating" organization (Feeser and Willard, 1989), prefunding factors (Roure and Maidique, 1986), capitalization (Bruno and Tyebjee, 1985), competitive tactics (Slevin and Covin, 1987), strategy and industrial structure (Tushman, Virany and Romanelli, 1985; Sandberg and Hofer, 1987; Dubini, 1989; Fombrun and Wally, 1989; McDougall, 1989; Feeser and Willard, 1990), with a few studies examining many different variables (Cooper and Bruno, 1977; Van de Ven, Hudson and Schroeder, 1984; Bruno and Leidecker, 1988). References to these and other works are made as appropriate throughout this chapter. In the preceding chapters a number of the key factors that might be related to entrepreneurial performance have been analyzed in depth. Now their potential influences on the eventual success of the technological enterprise need to be evaluated.

A general model of enterprise performance that provides the framework for analysis in this chapter follows the flow of the earlier chapters of this book. One of the most important foundations of a new enterprise that determines whether it succeeds or not is the entrepreneur himself. Chapter 3 explores in depth what produces a technological entrepreneur. Now the question is, "What chain of factors is likely to produce a *successful* entrepreneur?" Clearly, family background factors such as religion and father's occupational status, are important in the development of an individual's orientation. Furthermore, they influence where a prospective entrepreneur goes to college, what educational level he attains, what courses he takes, and for what career he prepares. As demonstrated earlier these individual variables influence first whether or not the person will start his own venture. They may also affect what the individual seeks to achieve and how the entrepreneur organizes and conducts his company.

Some of the personal characteristics of the founders, including their work experience, their combination into teams, their initial company directions, may also affect the initial technological base and the first financing of the company, as shown in Chapters 4 and 5. Once established, these nonpersonal founding resources no doubt also have independent influences upon the later achievements of the company.

The early evolution of the company, obviously affected by the entrepreneurial founders as well as the initial technical and financial resources, also contribute to the further growth and corporate development. Chapters 6, 7, and 8 suggest the need to look to additional financing, market orientation, and overall management as keys to later company success and failure.

All these potential, in fact likely, influences upon company success and failure are pictured in Figure 9–1, which forms the outline of this chapter. The comparative data gathered in the overall research program on technological entrepreneurship permit a statistical determination of the impact of these factors on the firm.

Measuring Success

Much of my multiphase research program is an attempt to relate various aspects of the entrepreneur or enterprise to the success of the firm. As such, a difficult but necessary task has been to define a usable measure of success, despite some reluctance to disagree with any individual's definition of success. To the extent that personal values and motives differ, the yardsticks by which success is measured differ accordingly. The variety of aspects recounted by the entrepreneurs when asked to assess their own success indicates the problem. Some find success from their products becoming widely used. Others report success in terms of money made for themselves, and for friends and relatives who had made investments. One entrepreneur feels pride (i.e., success??) in having supplied jobs to many people in the community. Some of the most technically oriented entrepreneurs regard themselves as having successfully explored new areas. Despite this multitude of views my assistants and I decided to develop measures that correspond somewhat to how the public might judge the success of a young high-technology company.

For MIT and other source organization spin-off companies, the primary information sources for this chapter, detailed data are usually fully available.

Figure 9–1
Influences Upon Success of Technology-Based Companies

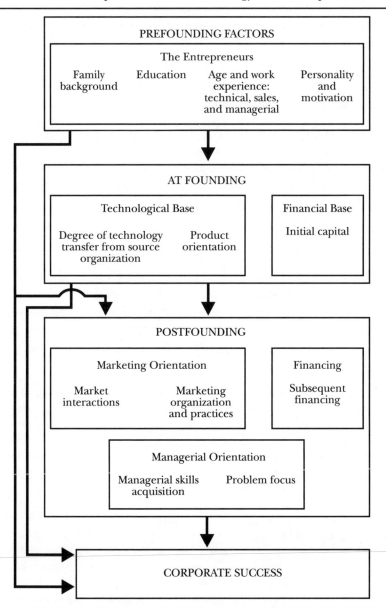

The overall performance measure used for them is determined by integrating three considerations: (1) the average sales growth over the life of the enterprise; (2) modified to account for the number of years that the company has been in business; and (3) incorporating the fact of profitability or nonprofitability of the company.[1] This "weighted overall performance rating" is used throughout the analyses reported in this chapter.

The overall research program tested a wide variety of other success measures in attempts to indicate consistent trends and relationships among measures that are related but not completely independent. Strong statistical correlation exists among five different company performance measures. These are: the weighted overall performance rating scheme defined previously; the average sales of the company over its life; the annual trend in sales growth in terms of dollars increase per year; an adjusted version of the present sales rate; and sales projections using a ten-year period.[2] The strong intercorrelations of the several measures indicate that any of these performance indices might be used reasonably to represent enterprise success. Similarly, in the electronic systems company spin-offs that are analyzed separately, sales growth correlates very strongly with the overall performance measure ($p = .0001$). (Reminder, these numbers in parentheses are the probabilities that the relationships found in the research might have resulted from chance alone; the lower the probability, the higher the confidence that the reported finding is real.)

It would be desirable to be able to include the present value of capital invested as some part of the company's overall performance measure. However, that information is not available reliably in a sufficient number of cases for broad use. Return on capital measures are used in the special study of high-growth companies that is reported in Chapter 11. Net market value of the companies would be desirable and usable measures if only more companies in the studies had gone public. Asset data are used throughout the analyses reported in Chapter 8, and sales revenues are shown there to be highly correlated to the assets.

As another validity evaluation for the composite performance measure developed and used in this chapter, that index has been compared with Dun and Bradstreet ratings for one sample of companies. Two commonly used and accepted measures, the Dun and Bradstreet Estimate of Financial Strength and the Dun and Bradstreet Composite Rating, correlate positively and significantly with the overall performance measure described here ($p = .006$ and $p = .02$), respectively. This adds some dignity and credence to the composite measure developed earlier.

Any measure of company performance is likely to be incompatible with the exact measure the entrepreneur has in his own mind. More subjective than not, the entrepreneur's own performance measure rarely lets him rate his company, and more importantly himself, as a failure. Asked to rate their performance on a 1 to 7 scale, with 1 being complete failure and 7 being complete success, 95 of 107 technological entrepreneurs who were questioned in one sample responded. Only 16 of the 95 (or 17 percent) rate their companies at less than the mid-point of the range, 4. Twenty-nine entrepreneurs are somewhat noncommittal, rating their firms as 4s. Twenty-nine give their companies ratings of 5, seventeen a rating of 6, and four a rating of 7, the highest degree of success. Similarly, the 58 founders of new enterprises from the large diversified corporation rate their firms an average of 5.4, with relatively little differentiation between the more versus the less successful firms.

I find very interesting that in all but one study no significant correlation exists between the overall company performance measure as determined previously and the degree of success (either at present or potential) attributed to the company by the entrepreneur himself. (The unexplainable exception are the electronic systems company spin-offs, whose self-assessed success correlates significantly ($p = .04$) with the overall company performance. Perhaps those entrepreneurs' industrial background causes them to adopt implicitly the same sales- and profit-oriented values as embedded in the overall success measure.) In general the entrepreneur's self-assessment of success does not correlate significantly with any of the several reasonable and objective performance measures tried. This may mean that these performance measures are not good. More probably, however, this simply confirms my previous contention that people view and measure success by different yardsticks, especially when self-assessment is the task. I vividly recall one interview with the head of a firm then doing several hundred million dollars in sales, with continuing growth and profitability. When I questioned his self-assessed company rating of 5, rather than my assumption of 7, he replied, "We could have done so much better." Other entrepreneurs, who had been forced out of their companies by outside investors, are somewhat negative about the degree of success of the firms they had founded, despite objective indications of significant corporate achievements.

For a subset of 142 companies in the research sample, comprising mostly spin-off companies from MIT and from the electronic systems firm, this performance measure generates the distribution of ratings shown in Table 9–1. In contrast with the 17 percent that are subjectively self-assessed by their founders as being below the midpoint of the performance range from low to high, 69 percent are evaluated by the more objective overall performance measure as being below the range midpoint. These companies might be labeled as "survivors", rather than "successes". However, relatively few firms deserve the lowest rating of 15, denoting absolute company failure (closure or bankruptcy) or at least declining sales. Only 15 percent of the companies in this sample merit that status, a rather remarkably low rate of failure for companies that average about five years from their dates of founding. This low rate of absolute failure is evidenced throughout the studies of technology-based companies. For example, only five (or 25 percent) of the twenty companies that were studied in regard to their search for venture capital (Chapter 7) had terminated their businesses by the

Table 9–1
Tabulation of Technology-Based Company
Performance Ratings ($n = 142$ companies)

Performance Rating		Number of Companies	Percent of Companies with Rating
High	1	9	6.3
	2	6	4.2
	3	0	0.0
	4	7	4.9
	5	4	2.8
	6	2	1.4
	7	9	6.3
	8	7	4.9
	9	3	2.1
	10	22	15.5
	11	23	16.2
	12	17	12.0
	13	1	0.7
	14	10	7.0
Low	15	22	15.5
Totals		142	99.8*

Mean rating 9.77; median rating 11.00.
* Round-off error.

time they were interviewed. And half that percentage, only seven (or 12 percent) of the 58 firms that originated from the large diversified technological company, had failed. Cooper and Bruno (1977) find a somewhat higher but still relatively low discontinuance rate of 29.2 percent after an average of ten years existence for 250 Silicon Valley high-technology firms. This matches the very highest termination rate experienced in any of the Massachusetts technological samples, five failures of 18 (or 28 percent) of one cohort evaluated seven years after their founding.

Low failure rate, however, does not characterize most new companies. The U.S. Small Business Administration (SBA) reports failures of as many as 90 percent of new companies during their first five years after founding. My study of consumer oriented manufacturers, carried out as a deliberate contrast to the technological firms usually researched, confirms the SBA findings. Due to business failure or other reasons for disappearance my assistant and I had been able to study in detail only twelve of the original 49 Massachusetts consumer products firms identified. And then, eight of those 12 had gone out of business within six years of founding, with an average life of 1.5 years. Moreover, 42 of the 49 had either probably or definitely gone out of business, giving a likely failure rate of 85 percent of the nontechnical consumer-products producers during the six years that followed their incorporation. Thus, the high survival rate of technology-based firms is noteworthy.

But note also in Table 9–1 that very few firms earn the highest ratings, including just 6 percent that score 1. Although these numbers are measures of company success, not of returns to their investors, this percentage credibly is only slightly higher than the relatively small number of venture capital investments found by Huntsman and Hoban (1980, p. 47) to have generated extraordinary returns, and is only slightly less than the folklore suggestion that one in ten investments will be "home runs". The remainder of this chapter is devoted to assessing the differences among the firms that lead to these marked differences between survival and success.

Whenever possible the elaborate measurement scheme described earlier is used in the analyses of results and is the basis for claims of success or failure in this chapter. For some of the studies much less elegant, less complicated, and less data-demanding success measures are employed. For example, the research on spin-off companies from a large diversified firm simply uses the average annual sales growth percentage and the fact of profitability in at least two of the last three years. Even more simple is the use of average sales growth over the past three years to measure performance of the young technical firms studied. These alternative sources of information, and their alternative measures of success, are pointed out when their results are cited in this chapter.

The Entrepreneur

The first set of factors examined for their influence upon company success are aspects of the entrepreneurs themselves. As indicated in Chapter 3 the research studies gathered data on family background, education, work experience, personality, and motivation, and now attempt to relate all of these to the later outcomes of the companies the entrepreneurs founded.

Family Background

Chapter 3 demonstrates strongly that a significantly disproportionate number of entrepreneurs had self-employed fathers, benefiting from the "entrepreneurial heritage" in their choice of careers. Similarly, those individuals who lacked self-employed fathers as role models but still became entrepreneurs were found to be disproportionately of Jewish background, reflecting what McClelland has identified as an "achieving" home environment. While these factors have affected the individual's decision to found his own business, neither religion nor other aspects of family backgrounds influences directly the later success of the enterprise. In only one study, that of the spin-offs from the large diversified corporation, is there even a weak link

($p = .11$) between new enterprise success and having a father who owned his own business. No statistical ties are found in any of the other samples. Thus, the home helps affect the career choice of entrepreneurship, but not the career results. To the extent, however, that this family background contributes to the individual's personality and motivation, company success may indeed still be affected, as discussed later in this chapter.

Education

The unusually high survival rate (compared with national survival percentages for all new companies) of 70 to 80 percent found for these technology-based companies probably relates in part to the high level of education that is characteristic of the technical entrepreneurs. The level of education is shown in Chapter 3 to be necessarily high, due to the nature of the source organizations at which these entrepreneurs are employed previous to starting their own companies. The overall high survival of these entrepreneurial firms might also be explained by a ready market existing for the technological outputs of the sampled companies. A nontechnical enterprise run by a less well-educated individual (as represented by the sample of consumer oriented manufacturers) might fail due to a lack of a ready market and not necessarily due to the founder's lower level of education. At any rate, within the technical entrepreneurs, whose median education level is an M.S. degree, it is difficult to determine what effect education has had on overall company success.

In two MIT lab subsamples the several companies founded by entrepreneurs without a college degree do not perform as well as those founded by higher-educated individuals. This relationship does not hold up when all of the MIT spin-off companies are aggregated into the same cluster for analysis. In one sample of 20 young technical firms data were collected on the education levels of all the co-founders and averaged as an education measure for each company. This average correlates favorably ($p = .10$) with early sales growth of those companies, and even more significantly (.016) with long-term sales growth.

For the most part, in the rest of the technological entrepreneurship studies, no statistically significant direct relationship can be found between education and success. Instead there appears to be somewhat of an "inverted U" relation between company performance and education level: Entrepreneurial successes seem to be dominated by what I label here a "moderate educational level", that is not more than a master of science degree. In contrast, the lower rated performers are even more highly educated, that is, holding Ph.D. degrees. The Ph.D.s as a group do not perform well as entrepreneurs. This is not to say that the additional education in itself is damaging. Rather the results suggest that the general temperament, attitude, and orientation of Ph.D. recipients are usually out of line with those necessary for successful technical entrepreneurship in engineering-based fields. A doctoral program involves, in addition to advanced study and research, a fair amount of regimentation, regulation, committees, and writing. The person who has attained a Ph.D. usually has a different set of motivations that the person who wants to get on with it, start putting ideas to work, and make something happen. (Having a Ph.D. makes me cautious in drawing these conclusions, especially since I certainly do know several very successful Ph.D. entrepreneurs, including Arthur Rosenberg, discussed in Chapter 1.) The data do not declare, "If you have a Ph.D., give up as an entrepreneur." They do call for especially thoughtful assessment by the Ph.D. as to whether he or she is really willing to sacrifice the time, energy, and other commitments to try to create a thriving successful organization. Indeed Van de Ven and colleagues (1984) find that Ph.D. founders of educational software companies outperform others during their start-up years (.05). Founders with Ph.D.s as a whole tend to form (1) single-founder firms, that are (2) more oriented to research and/or consulting, which are (3) initially funded by themselves and with low initial capitalization, all factors independently correlated with poor company performance, as is revealed later in this chapter.

The generally poor performance of Ph.D. founders may be different for the new generation of genetic engineering and biotechnology companies that are not included in my current samples. In fact, the exception within these samples to the poor results generally found for Ph.D. entrepreneurs is in the group of Massachusetts biomedical firms studied. There the 14 companies founded by holders of masters and Ph.D. degrees do significantly outperform (.10) the 11 firms formed by lesser educated individuals, although MD founders are associated with lower performing companies (.05).

Age and Experience

One of the most surprising findings of the study of the nontechnical consumer oriented manufacturers is their average age, 45 years, and experience, 24 years, at the time of company formation. One might hypothesize that this maturity would lead to higher probability of business success. That the opposite is true might in contrast be explained by saying that this older group lacks the ambition and drive characterized by younger people. In fact, only occasionally do the research subsamples provide weak evidence for the notion that younger technological entrepreneurs are more successful. The data as a whole do not support any link between age of founders and company performance.

Mixed results also arise for most attempts to relate specific aspects of the entrepreneur's work experience and later company success. In some of the studies the number of entrepreneur's publications and even more strongly the patents obtained while employed at the source organization are found to correlate with the success of his later new enterprise. Yet, most of the studies do not confirm this linkage. Given the findings to this point on family background, education, age, and general experience, it is not surprising that Sandberg and Hofer (1987) also find little impact on

new company performance of the *biographical* characteristics of the entrepreneur.

One set of work measures does relate to company performance, but
unfortunately has been gathered at only three MIT laboratories. In these
cases additional interviews were conducted with the former lab supervisors
of spin-off entrepreneurs, in which the supervisors evaluated their former
employees in regard to technical creativity, general technical ability, and
managerial ability. The obvious problem is whether or not these post hoc
evaluations are biased by the events that occurred after the entrepreneurs
left the MIT labs. In any event a composite evaluation formed by averaging
the three ratings significantly differentiates ($p = .10$) the high from the low
company performers, as shown in Figure 9–2 for the Lincoln Laboratory
spin-offs. Do note, however, that their supervisors evaluate 12 out of 27 or
44 percent of the entrepreneurs as having been high performers as employees, while our company performance measurement finds only five of
27 or 19 percent of their firms as high performing. Perhaps this evidences
a higher "hurdle" for technological entrepreneurs to overcome. Strong
correlation between supervisory ratings and company performance is also
found among the spin-offs from the MIT Electronic Systems Lab (.08),
where in fact all three individual supervisory ratings also correlate significantly. At the MIT Research Laboratory for Electronics, however, only the
technical creativity rating correlates with later company success (.05). It is
difficult to interpret the meaning of these ratings, especially given their
after-the-fact nature.

Figure 9–2
Supervisory Ratings as Predictor of Entrepreneur's Future
Company Performance ($n = 27$ Lincoln Lab spin-offs)

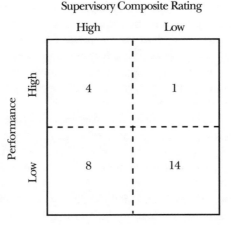

A more useful finding derives from the companies started by former employees of the electronic systems company, the only site at which the entrepreneurs' work experience information is coded for their possible prior role as supervisors, the role that was itself related to length of time at the lab (.01). In that sample set individuals who spent the largest fraction of their time in supervisory tasks later outperformed their former associates as entrepreneurs (.004). This seems reasonable from several perspectives. First, the personal characteristics that lead individuals into supervisory roles (i.e., that resulted in their excelling within the laboratory), also contribute directly to their later entrepreneurial success. Second, the supervisory role generates exposure to a spectrum of responsibilities and problems, including managerial ones that better prepare the individual for a wide variety of issues that are encountered in their new enterprises. Third, supervisors are better able to grasp the entirety of a technical project, rather than just one piece of it. This bigger picture enables the supervisors to transfer more technology (.03) of a meaningful nature to their new enterprises. Finally, supervisors are in a better position to become familiar with the customer, and are often involved in the marketing of laboratory technology. This increases the supervisors' knowledge of whom to approach and how to approach them, an understanding that helps when they enter their own businesses. Perhaps this latter point is the key. Of all the "spin-off" entrepreneurs from the electronic systems company, few had prior market contact and experience other than the supervisors. This perspective is supported by the in-depth study of 18 recently formed technical companies, where the extent of prior managerial experience of the principal founder is found to correlate strongly with later company success (.07).

Important ties are discovered between aspects of work experience and technology transferred, as described in Chapter 4. These are now seen to relate to later new company performance, without clear ability to separate the effects of the experience on success from the effects of the technology on success. For example, entrepreneurs from the MIT labs who then acquired commercial experience before establishing their own companies transfer less MIT technology into their new companies and are less successful. Similarly, entrepreneurs from the electronic systems company who acquired other commercial experience also performed poorly. The more companies they worked for, logically the lower their average length of employment per job (.006), and the poorer their entrepreneurial performance (.003). These entrepreneurs probably never stayed in one job long enough to acquire the technical knowledge needed and, therefore, transferred less technology, not only from the observed source lab but from their other employment as well.

Entrepreneurs who enjoyed their earlier technical work experience generate higher company performance (.02), with a similar tie between finding their lab work challenging and later entrepreneurial success (.01). This may well be a manifestation of the personalities of different people.

Some enjoy working and make it challenging, regardless of how dull it may seem to others. These individuals carry this challenging and enjoyable atmosphere with them to their own companies. The firm's employees and even its customers are then imbued with a sense of purpose and meaning for the new enterprise, helping the company to succeed. No relationship exists in my studies between the size or other characteristics of the source organization and later spin-off success. But Feeser and Willard (1989, p. 430) indicate that high-growth founders of computer equipment firms more frequently come from large and publicly held "incubators", and less often from universities and not-for-profit organizations. Their result disagrees with the clear dominance of MIT-related spin-offs in the Greater Boston area.

Personality and Motivation

The studies reported in Chapter 3 help to define the personalities and motives of technological entrepreneurs. No attempt is made to link personality with later company success. The examination of entrepreneurial motivations also gathered data on the sales growth of their companies. That sample, using Thematic Apperception Tests to measure the entrepreneurs' drives, finds statistically significant ties between need for achievement (n-Ach) and company performance. The companies led by entrepreneurs with high n-Ach exhibit an average growth rate of almost 250 percent higher than those led by entrepreneurs with a moderate n-Ach. As illustrated in Figure 9–3, 79 percent of the technological companies led by entrepreneurs with a high n-Ach enjoy a growth rate above the median for the total sample of entrepreneurs. The low scoring n-Ach group performs slightly better than the moderate group, although still significantly lower than the high n-Ach cluster. Smith and Miner (1984), with later confirmation from their extended studies (Miner, Smith and Bracker, 1989), also find that more successful technological entrepreneurs exhibit much greater task or achievement motivation than entrepreneurs who head slower growth firms. Substantial variations in performance do exist among the firms in each n-Ach category. For example, within the high n-Ach group alone, performance scores based on sales growth range from 0.14 to 2.10. Similar within-group ranges exist in the low and moderate n-Ach groupings.

Looked at independently, need for power (n-Pow) and need for affiliation (n-Aff) lack important statistical relationships with company performance. Clear linkages do exist among the three, n-Ach, n-Pow, and n-Aff, explaining in part the performance differences. Most important is the moderate n-Pow observed among the highest performing technical entrepreneurs in the high n-Ach group, with the lower-rated performers manifesting either high or low need for power. High need for power by the company founder, accompanied usually by an autocratic or authoritarian style of leadership, drives out the possibility that others may effectively contribute to the organization's growth and success, assuring that the hu-

Figure 9–3
Percentage of Companies Above Median Growth Rate, as
Function of Entrepreneur's Need for Achievement ($n = 51$)

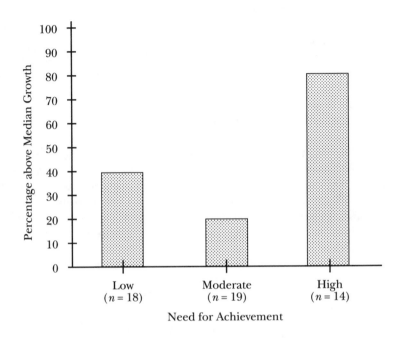

man resource capabilities of others in the company are underutilized. It forces a "one-man show", seldom a correlate of effective growth, as shown in the next section. Low n-Pow, on the other hand, symbolizes a founder who lets everyone in the organization do his or her own thing without regard for central goals. The low n-Pow entrepreneur is likely to create a firm in which little leadership and direction is being exercised—the laissez-faire in which every person is her or his own boss, the antithesis of focused management priorities, leading to dissipation of the scarce manpower and capital resources of the young firm. The middle of the n-Pow spectrum represents a mixed influence of the two extreme styles, linked strongly to what is best described as a participative organizational style. That entrepreneur is indeed the boss, formulating and maintaining clear central goals and objectives. He also shares responsibilities with colleagues and subordinates who have latitude to make their contributions as best they can. This combination of moderate n-Pow and high n-Ach characterizes the most successful technological entrepreneurs. Fombrun and Wally (1989) support the notion that high-growth technology oriented firms "foster greater autonomy and risk taking in their cultures, whereas firms pursuing a cost strategy ... place direct controls on employees that reduce autonomy" (p. 115), but find no overall linkage with these firms' performance.

Number of Founders

Beyond these characteristics of individual entrepreneurs one of the strongest and most consistent correlates of later success of the new technical enterprises is the number of original company founders. This shows up as a statistically important influence in essentially all of the subsamples studied over the entire twenty-five-year research program, however success is measured. For example, 14 percent of the multiple founder companies from MIT labs are in the highest performance group (ratings of 1 to 3), while only 3 percent of the single founder enterprises attain that level. The positive relationship between an increasing number of founders and increasing company success is significant as well ($p = .05$) among the spin-offs from the electronic systems company. Even during their early years, sales growth of technological companies correlates with initial number of founders (.09), as shown in careful monitoring of twenty young firms. Of those young companies, 63 percent of those with more than two founders perform better than average, whereas only 20 percent of the firms with one or two founders exceed average performance. Among computer-related companies, as another case, the mean number of founders among the low performers is 1.9, whereas the mean for the high performing group is 3.2 co-founders. Cooper and Bruno (1977) also find significant relations between multiple founders and high growth in their Silicon Valley studies, and Feeser and Willard (1990) confirm this in their study of computer-related companies. Van de Ven et al. (1984) describe an exception to this rule in finding higher performance for single founders among their 12 young educational software firms, citing frequent power struggles among multiple founders in their sample.

The success of multifounder companies is obviously not due merely to the presence of more than one entrepreneur, but to a number of advantages possessed by teams of founders. The greater variety and depth of talents, capabilities, initial capital (as shown in Chapter 5), and experience available to new firms as the number of founders increases should lead to higher performance. Indeed, in the study of recently formed companies, the founding teams that include both technical and managerial experience are the higher performers (.05). Although contrarians might argue that division of authority among several founders dissipates company efforts and effectively reduces performance. As shown in Chapter 6, larger founding groups work more total hours (.008), in particular in both engineering (.13) and sales (.006), with both total hours and each of these allocated effort levels correlating closely with early sales growth of these firms. (Analyses of the Kendall tau partial rank correlation coefficients determines that the number of founders and total hours worked are the principal influences on these effects.) Indeed, these benefits seem so obvious that the presumption of the importance of partners to business success may well have motivated the very establishment of many of the multiple-founder teams. Roure and Maidique (1986) also illustrate the importance of more complete founder teams in their sample of high-technology companies. The running of a business is an awesome responsibility for one person to handle, especially for the technical individual lacking in business experience. This is reflected in the fact that those individuals with the least prior business experience are most likely to join with others in the formation of the new enterprise (.01), presumably in part because they recognize the need to broaden the experience base of the company.

Furthermore, as suggested from the negative correlation found earlier between an entrepreneur's n-Pow and later company success, entrepreneurs with the desire to lead and achieve, rather than to control the actions of others, may well seek partners with whom they can share both aspirations and burdens. This tends to isolate the power-driven entrepreneurs heavily into single-founder firms, which then turn out to have far lower corporate success. Unfortunately, the data are not available to test this hypothesis in regard to executive style.

The advantages of beginning with more people who share the responsibilities carry over into the later operations of the firm. More founders can undertake more time consuming tasks that demand different sets of skills. Chapter 6 demonstrates that the number of founders relates statistically to the existence of three marketing oriented aspects of business operations: a marketing department, sales forecasting, and market analysis. As shown later in this chapter, all three of these elements significantly relate to high company performance. Furthermore, as Chapter 6 indicates, those companies starting with multiple founders are more likely to commence operations in some form of hardware production, also shown later as correlated with successful company performance. And, of course, as Chapter 5 shows, multiple founders start with significantly more capital financing (.002), which also turns out to correlate with company success. Later financing is similarly biased in favor of multiple founder firms, with venture capitalists among others (as shown in Chapter 7) preferring to invest in the larger founding teams.

The Technological Base

Rapid Technology Transfer

Chapter 4 explores the many factors that affect the degree of technology transferred from the entrepreneur's source organization into the new enterprise. That technological base of the new firm turns out to be far more than a nicety. It directly correlates with and is a causal influence upon the eventual degree of company success. Of 119 firms spun-off from four MIT laboratories and an engineering department, those that transfer the greater amounts of technology are consistently the most successful ($p = .02$). The same finding arises from the entrepreneurs who departed from the electronic systems company. Those companies that were founded on the basis of rapid and direct translation of advanced technology from the parent laboratory to the new enterprise are the ones most likely to succeed. Similarly, the degree of technology transfer is significantly greater for the high performing energy-related companies (.10) in one industry sample. The direct transfers in effect remove a major stumbling block from the company's path. A technology that has already been developed to varying degrees is being utilized, refined, or modified somewhat, and presented to the market. The companies that effect more technology transfer into their new firms have, almost by definition, less costly and less time-consuming "in-house" product development to accomplish prior to marketing a profitable product or service. The technology they transfer gives them a significant advantage over companies that have to "start from scratch". Although not examining the specific technology "transferred" to the new firm, Cooper and Bruno (1977) find that high growth firms are similar to their "parents" in both the technology utilized and the markets served, recently confirmed by Feeser and Willard (1989, p. 429, and 1990).

The most evident and negative effect of the technology factor is on those who wholly fail to utilize it. This condition of no source lab technology transfer applies, for example, to 11 of 103 companies in one sample. Only one of those 11 firms scores above the bottom two-thirds of the companies in performance. These entrepreneurs seem to lack appreciation of the potential usefulness of source lab technology. In some instances it might reflect fear of law suits from a previous industrial employer. In contrast, the electronic systems company entrepreneurs who report learning more at their lab as opposed to applying what they had previously known perform significantly better in their new firms (.04). One manifestation of that learning is the granting of patents to the individual entrepreneur, which also correlates with his later company success (.02).

As Chapter 4 shows, a critical influence on use of source organization technology is whether or not the entrepreneur engages in further work after leaving the source labs. Frequently, a prospective entrepreneur takes a follow-on job in industry in search of business experience that he presumes might contribute to his later entrepreneurial success. Unfortunately, that additional work experience tends to diminish the amount of technology transfer and in the MIT spin-offs correlates negatively with eventual company success (.10). Commercial work experience, in particular, has a dissipating effect on the transfer of advanced lab technology from the MIT labs and generates an even stronger negative relationship with company's performance (.01). If the intervening experience is longer than four years, the negative effect is striking (.05), and indeed the longer the delay the worse the eventual performance of the company (.001).

Partial correlation analyses indicate that the time lag, or added experience, is not crucial in itself. Rather, the dominant mechanism here is that increasing time lag deteriorates the degree of technology transfer, which is then responsible for lowering the performance rating of the firm. Figure 9–4 portrays conceptually the effect of the time lag between leaving the source lab and setting up the new firm on technology transfer and eventually on company success. Unique technical knowledge and/or skills constitute the strongest competitive advantage that a fledgling company has potentially available. If entrepreneurs lose this advantage while in the process of seeking business acumen, they no doubt lose their ability to compete with already established businesses.

Product-Oriented Businesses

Chapter 6 documents that among those who directly transfer technology from their source organizations is a large number of entrepreneurs who initially establish consulting and/or contract research and development companies. Unfortunately, few of those firms have become high performers, as measured by our sales growth and financially oriented measures. Only one company in my MIT laboratory samples that started and remains wholly in contract research and development work achieves the highest rating group. All the rest of the high performance cluster from the MIT labs and departments are engaged either solely in hardware production or are in hardware production in combination with other work. For example, only

Figure 9–4
Effect of Time Lag in Establishing New Firm on the Degree
of Technology Transfer to That Firm

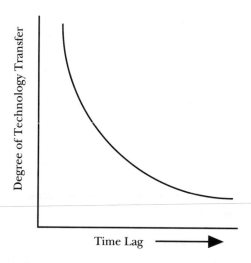

three of the eight companies started by faculty of the MIT Aeronautics and Astronautics Department began with a product orientation. Those three have the largest gross sales among the departmental faculty spin-offs. In other samples the highest successes include also software product companies, but not those engaged primarily in custom software or programming services.

Of course, many of the firms that are product-involved from the start do not achieve significant success. Therefore, product orientation alone does not assure success. But lack of such orientation almost certainly assures lack of high growth and high profitability. Thus, although direct technology transfer provides a competitive advantage to the entrepreneurial carrier of that technology, embodiment of the technology in product form is necessary for significant and leveraged company growth to occur. A related observation is that most of the highest performing companies (nine of 11 in one large MIT sample) have made noticeable attempts to become less dependent on government markets, seeking commercial applications of their technological bases, despite the initial primarily governmental funding of their technology source organizations. Far fewer lower performing companies have tried to move away from their original technical–financial governmental dependencies. Increasingly, new entrepreneurs are even finding ways of embodying their transferred technology into repetitive productlike services that also demonstrate potential for successful company performance. For example, the higher performing energy companies feel they knew of a new product or service not being adequately developed by their prior employers ($p = .01$).

The Impact of Regulatory Delays

The immediate and extensive movement of advanced technology into the marketplace thus constitutes a critical success factor for the spin-off companies described earlier. But what happens when governmental regulations both (1) relate in their severity of impact to the degree of technical innovativeness of a firm's products, as well as (2) prevent rapid commercialization of new technology? Both liabilities characterize the environment facing new companies in the biomedical products field. The extensive regulatory constraints imposed by the U.S. Food and Drug Administration (FDA) and their counterpart agencies in many other developed countries emerge as one of the most significant differences of the biomedical industry vis-à-vis other technology-based industries. The extent of this external interference and control of quality standards is overwhelming, including both the efficacy and the safety of the product (FDA 1976, pars. 59-515). The regulations also include directions about manufacturing and record-keeping procedures (par. 501), and labeling and advertising standards (par. 502). Both sets of standards are far more rigorous than standards that apply to nonbiomedical industries. Research by Birnbaum (1984) on the x-ray equipment manufacturing industry, and by the Congressional Office of

Technology Assessment (OTA) on the contact lens industry provide strong evidence that increased regulations have diminished biomedical product innovation, especially by smaller firms.

These issues led to my initiating research on technological entrepreneurship in the area of biomedical products, with specific concerns for the consequences of governmental regulations on new and young technical enterprises that are engaged in the development of risky but innovative products.[3] The risk arising from the new biomedical products was evaluated by the external medical experts separately in regard to explicit risk to the patient and degree of "patient invasiveness" of the product. For example, a medical device used for lab diagnosis is less risky on both dimensions than a pill swallowed by the patient. The extent to which "new or first-of-a-kind" technology is important to the biomedical products correlates statistically significantly with the evaluated risk associated with use of each of the first ($p = .10$), second ($p = .10$) and third ($p = .05$) products of the firms. While new products in nonbiomedical fields also often embody risk to the user, seldom is this risk subject to regulatory control. Growing concerns about product safety and liability issues outside of the medical world might well cause the spread of new product regulation to other areas of high-technology entrepreneurship.

As sensed by the 26 young biomedical companies studied, the FDA responds to this perceived risk by affecting the firms somewhat in proportion to the risk of their products. Table 9–2 shows that increased product risk (as assessed by the independent expert panel) correlates significantly with increased overall impact by the FDA (as reported by company executives and indicated as well by company FDA-related expenses). An apparent outcome is a slight negative correlation between the company's overall technological innovativeness and its annual sales during its first two years and a slight positive relation between innovation and its later average sales,

Table 9–2
Risk Associated With Use of the Firm's Products
and the Impact of FDA Regulations

The Firm's Products	Correlation of Risk With Overall Impact of FDA Regulations	Correlation With Expenses for the FDA Interface
First product	0.35*	0.20
Second product	0.29	0.32
Third product	0.14	0.51*
The firm (products average)	0.32†	0.47*

Table lists the Spearman and Pearson correlation coefficients.
Statistical significance of the correlations: *$p = .05$, †$p = .10$.

neither relationship being statistically significant. The failure to find a strong positive tie between this highly regulated medical innovation process and the company's success is disappointing at this stage, but I return to and clarify this issue in the next section.

Financing

Initial Capital

Another factor that is consistently linked to later company success is its financial base at the time of initiation. Table 9–3 displays the association of later company performance and the initial financing ($p = .006$) for 101 MIT spin-off companies. The performance rankings are in five groups, each consisting of three rankings, with 1 being best and 15 being worst company performance. Eight of the nine companies in the highest performing group received initial financing equal to or in excess of $50,000. Of the lowest performing group, only four of twenty-three received that much initial capital. Similarly, the electronic systems company spin-offs show strong correlation between their performance and their initial funding (.002). The special study of young firms also reveals a significant relationship between initial capital and sales growth (.026). And the firms seeking venture capital financing also demonstrate positive relations between initial capital and growth rate.

Table 9–3 also confirms that substantial initial financing does not guarantee success. Ten of the 23 firms that received more than $50,000 initially

Table 9–3
Amount of Initial Capital and Company Performance
($n = 101$ companies)

Initial Capital* ($ thousands)	Performance Ranking*					
	1–3	4–6	7–9	10–12	13–15	Total
< 1	—	4	—	11	8	23
1 – < 10	—	1	2	16	6	25
10 – < 50	1	2	6	16	5	30
50 – < 100	2	1	1	2	1	7
100 – < 250	1	—	1	4	2	8
250 – < 500	2	—	1	—	1	4
≥ 500	3	1	—	—	—	—
Totals	9	9	11	49	23	101

* tau = .201, $p = .006$.

are in the lower clusters of performance ratings (ratings of 10 to 15). Of course, lower performers occur with far higher incidence among those companies with lower levels of initial capital.

The earlier identification of the importance of a larger number of founders mentions the strong correlations between the size of the founding group and the initial capital. This poses a question as to whether the funding is a critical success factor in its own right or whether it might just be acting as a surrogate for the size of the founding team. Applying partial correlation analyses to several data samples determines that the initial capital relationship with sales growth and with overall company performance are independent of the associations between capital and the number of founders, meaning that the amount of initial capital is crucial in itself.

The importance of initial capital in company performance is best illustrated by the firms that began with a product. Table 9–4 displays the positive relationship (.025) for a subset of 38 of these companies. Each of the thirteen companies that began with a product and less than $10,000 in capital has a performance score higher than 9, where 1 is the best rating. In contrast, of the twelve product companies financed equal to or in excess of $50,000, half are ranked better than 6 with only four ranked worse than 9. Earlier in this chapter I point out that initial product orientation is a needed ingredient for company success. Now it becomes clear that greater initial capitalization facilitates the growth and success of product-based firms, and that product entrepreneurs who raise less initial capital have less likelihood of eventually succeeding.

The strength of these associations should indisputably indicate that initial financing is influential in affecting the success of technology-based

Table 9–4
Initial Capital and Company Performance for Companies
Begun With a Product ($n = 38$ companies)

Initial Capital* ($ thousands)	Performance Ranking*					
	1–3	4–6	7–9	10–12	13–15	Total
< 1	—	—	—	2	—	2
1 – < 10	—	—	—	8	3	11
10 – < 50	—	1	2	7	3	13
50 – < 100	2	—	—	—	—	2
100 – < 250	—	—	1	2	2	5
250 – < 500	2	—	1	—	—	3
≥ 500	1	1	—	—	—	2
Totals	5	2	4	19	8	38

* tau = .259, p =.025.

companies. However, one other fact still needs to be considered. Apart from the amount of initial financing, all companies do not begin equal. Prospects for success for some companies appear uncertain even at their beginnings and for others they are even rather dim. Other companies offer good prospects for success to the potential investor. It is likely that the latter more attractive-appearing companies are financed more heavily at the outset than the other companies, as indicated by Chapter 7's discussion of venture capitalist decision making. To the extent this is true, the amount of initial financing depends on the prospects of a company's success, especially as seen by outsiders. Perhaps initial capital does not actually influence success but is merely symptomatic of a company's chances for success, as influenced by other factors. This notion is supported by the observation that of the nine highest performing companies in Table 9–3, all but one received initial financing from sources other than the founders themselves or their families and friends. The other research samples confirm that the successful firms are financed by sources different from those used by unsuccessful companies. For example, among the energy-related firms those entrepreneurs that started with only personal equity do not perform as well as those that also generated outside equity or loans at the outset ($p = .05$). I indicate in Chapter 5 that the outside financial sources also supply the largest sums of initial capital. This couples with the fact that when these sources are initial investors, they tend to invest in companies that appear, at least to them, to have good prospects for success. Bruno and Tyebjee (1985) make the same observation after finding the association between outside investment improves company sales and growth. But let us not make too much of this argument and draw the probably inappropriate conclusion that initial capital does not influence later company performance. Companies with good prospects might easily have performed poorly if they had not been backed by substantial funds. Other companies might have performed better if they had received larger initial capitalization. While many entrepreneurs attest that lack of capital was not a major problem for them or their firms, only a few argue that limited funding was good for them in terms of forced carefulness and induced discipline in managing their young companies.

The Financing of Biomedical Companies

The section on Technology briefly describes my study of young biomedical companies, as well as the failure to find a definitive link between their technological innovativeness and their economic success. In general, as mentioned in Chapter 5, the financial attributes of those biomedical firms are comparable to other technology-based enterprises. A look in detail at the distribution of their initial financing turns up two outliers, each with over $850,000 in initial funding. That financing threshold turns out to be critical to keep these innovative firms going during the long time, enforced by the FDA-regulated product evaluation procedures, that they are unable to generate product sales revenues. Accordingly it is essential to test the linkage between technology and corporate economic performance, without the possible influence of these especially well-financed companies. The results in Table 9–5 are eloquent: The previously ambiguous relations between the several indicators of technological innovation and company success (as measured here by average annual sales over the preceding three years) become explicit and significantly negative for the 20 biomedical firms that could not mobilize the necessary initial threshold financing. Larger initial capital has just been shown to be associated with significantly better later performance for technology-based companies. Now, even more strongly, the results show that unless a biomedical firm is adequately financed at founding, its technological innovativeness is in general detrimental to its own future health and well-being! This finding from simple bivariate correlations is also supported by more detailed multivariate regression analyses (Roberts and Hauptman, 1987, pp. 390–391).

Subsequent Financing

The effect of company performance on the acquisition of later capital is unclear. Similarly unclear is the role of later financing in causing company success. One might logically infer that the company that is successful early in its life might both need and be able to obtain additional capital. Therefore, high success implies high additional capital, and not necessarily the reverse. In support I show in Chapter 8 that more successful companies (selling "the steak") encounter a far easier and less costly process in going public, one important mode of additional financing, than younger companies that are selling "the sizzle". Without indicating the direction of causality here, data from the electronic systems company spin-offs indicate that the occurrence of secondary financing correlates well (.08) with company performance, with tertiary financing even more strongly correlated (.025). Similarly, a positive and significant correlation (.08) exists between company success and the magnitude of additional capital for those twenty firms

Table 9–5
Technological Innovativeness and Biomedical Company
Success (excluding well-financed firms) ($n = 20$)

Measures of Technological Innovativeness of Firm's Products	Correlation with Company Success
New technology or first-of-a-kind	– 0.47[*]
Special specifications or purpose	– 0.44[*]
Calibre of product or personnel	– 0.60[†]

Table lists the Pearson correlation coefficients.
Statistical significance of the correlations: [*]$p = 0.05$; [†]$p = 0.01$

studied in regard to their proposals to venture capital companies. Sales growth also correlates (.07) with subsequent capital for that group. And the more successful energy technology companies more frequently consider going public (.05).

As displayed in Table 9–6, subsequent financing does occur for firms that end up at all points in the performance spectrum, even from the various sources of investment. Chapter 7 shows that very few companies receive additional funds from personal savings or from family and friends, presumably because these sources can no longer provide the larger sums needed subsequently. Most of the subsequent capital comes from sources new to the company.

For seven of the 63 companies in the Table 9–6 group public issuance of stock is the source of additional capital, split approximately 50-50 between high and low performers, no doubt including some that had sold the sizzle and had never performed well and others whose eventual performance may have declined from their time of going public. Other sophisticated sources of capital—venture capital funds, private investors, and nonfinancial corporations—also support both good and poor performers. It is interesting that essentially all those high-technology firms that rely for subsequent funding on the personal funds of the founders or their families and friends end up as low performers. Either their poor prospects or their poor performance made them unable to raise money elsewhere, or the lack of sufficient subsequent financing contributes to their eventual poor performance, or both.

Using information from one of the larger samples I had hoped to find indications of changing investor risk or "decreased failure rate" as a function of stage of investment. The results shown in Table 9–7 are not overly

Table 9–6
Source of Subsequent Financing and Company
Performance ($n = 63$ companies)

	Performance Ranking					
Source	1–3	4–6	7–9	10–12	13–15	Total
Personal savings	—	—	2	2	1	5
Family and friends	—	—	—	2	2	4
Private investors	3	2	3	8	5	21
Venture capital funds	1	2	2	3	1	9
Nonfinancial corporations	—	1	2	3	2	8
Commercial banks	—	1	1	4	3	9
Public stock issues	4	—	1	1	1	7
Totals	8	6	11	23	15	63

Table 9–7
Failure of Capital Investments in Technological Firms
by Stage

	Percentage of Failed Investments[*]		
Source	First Stage	Second Stage	Third Stage
Personal savings	46	40	50
Family and friends	50	100	—
Private investors	63	38	22
Venture capital funds	13	22	0
Nonfinancial corporations	0	18	40
Public stock issues	0	14	37

[*] Company performance ratings ≥ 11.

persuasive that investors do better by delaying until the technology-based company matures. Private investors, who account for the largest number of subsequent investments itemized in Table 9–6, do invest in fewer failures as they shift their attention from start-ups to second round and then tertiary investments. No other investor group is consistent in that regard. Indeed both nonfinancial corporations and the public market demonstrate the contrary result of an increasing failure rate as the stage of their financing advances downstream. The results here are also clearly contrary to the expectations of investors that their risks of failure decline substantially as they avoid the earliest stage investments (Dean and Giglierano, 1989; Ruhnka and Young, 1987; Wetzel, 1983). These results are consistent with my early studies of venture capital portfolio performance, which found best performance by those funds that focused on young technology-based firms.

Marketing Orientation

A wide array of founder and company characteristics and activities testify to the critical role of marketing in generating success of the technology-based enterprise. These evidences are in sharp contrast to the technologically arrogant assertion, "If you build a better mousetrap, the world will beat a path to your door." For example, data from the sample of eighteen young firms support the importance of sales experience among the founding group. Sixty-seven percent of the enterprises with above the mean of 2.6 years of cumulative prior sales experience by the founders have sales growth exceeding mean performance; only 25 percent of the companies with low founder sales experience perform well. As might be expected greater founder sales experience correlates strongly ($r = .62$) and significantly with greater percentile effort devoted by the founders to early sales activities, which in turn is also tied to better company growth ($p = .05$). Specifically, 55 percent of the companies with above average sales effort have high performance, whereas only 14 percent of those with below average sales effort do as well. Related to this is the early recognition of need for dedicated marketing personnel. For example, the more successful technology-based energy product companies studied tend to add marketing people at the start of their firms (.05).

Company–Market Interface

The importance of accurate perception of customer needs for company success is demonstrated repeatedly by the research. In probing more deeply into the early relationships between the companies and their prospective customers, entrepreneurs in the 18 young firms were questioned on three subfactors relating to their first three years of existence: (1) frequency of customer contact; (2) purpose of customer contact; and (3) customer helpfulness.

Frequency of customer contact by itself does not relate to sales growth. When the purpose of customer contacts is broken down into eight different categories, only the frequency of contacts aimed at determining customer needs correlates significantly (.05) with sales growth. Of the companies that utilize less than 25 percent of their customer contacts for need determination, only 22 percent have better than average growth in sales, while 45 percent of the companies with greater than 25 percent needs determining contacts perform better than average. Contacts with customers so as to better understand their needs is an essential demonstration of a marketing orientation.

The entrepreneurs' ratings of their customers' helpfulness in determining product or service specifications positively relate to sales growth (.01). These ratings may be objectively valid or may reflect primarily the entrepreneurs' attitudes toward being customer oriented. Of the seven companies that perceive customer help as low, only one achieves above average sales growth, while six of 11 firms that perceive customers as more helpful achieve better than average sales performance. In addition the companies that claim they use market-connected sources, such as customer requests or suggestions, for their new product ideas, as opposed to relying just on their founders, have significantly greater early sales (.025) and sales growth (.09). Forty-five percent of the firms using customer sources perform above the group average, while only 20 percent of the companies not reliant on customer ideas perform well. Here the perception may be even more important than the reality. Seeing the customer as helpful and as the source of product ideas is another reflection of a market oriented point-of-view. Van de Ven's (1984) more successful educational software companies also involve their customers more actively in product planning and market niche assessments. Neil Pappalardo of Meditech says that half of its new products and services come from learning from customers what pleases them and what bothers them; the other half, he asserts, come from Meditech's own ideas of new features or systems. In particular, he points out that you cannot ask customers to guide you in areas where they have no prior experience, such as color terminals or "window" screens for Meditech's hospital information systems.

Competitor awareness and assessment are not widespread among young technical firms, and yet strongly relate to company success. The small sample of eighteen young firms was divided into two groups based on their awareness of competition. Normalized average sales for the two groups are shown in Table 9–8, along with the statistical significance of their differences. These differences increase meaningfully as the firms' relative sensitivity to competition (with presumed responsive actions) has effect over time on company sales. Obviously, those companies with competitive awareness significantly outperform their less market perceptive cohorts.

For the past decade especially an orientation toward international markets might also be presumed to be of importance. McDougall (1989) shows that high-technology companies do differ strategically, especially in their distribution and marketing strategies, as a function of their international perspective, but she finds no ties to company performance. In contrast Feeser and Willard (1990) find that high growth firms derive significant revenues from foreign sales.

Internal Marketing Activities

The importance of market oriented factors also shows up symptomatically by several internal marketing measures. For example, the very existence of a separate marketing department strongly correlates consistently with the later success of the company. (The significance level is $p = .002$ for the firms derived from the large electronic systems company, .007 for the companies formed out of the large diversified technological corporation, .01 for the

Table 9–8
Normalized Average Sales Based on Awareness of
Competition ("unaware" in year 1 = base of 100)
($n = 18$ companies)

	Sales in			Growth Rate in		
	Year 1	Year 2	Year 3	Years 1–2	Years 2–3	All 3 Years
Aware	295	667	1257	372	590	481
Unaware	100	176	229	76	53	60
Significance	.18	.05	.02	.06	.05	.04

sample of recently formed technical companies, and .03 for the large number of MIT spin-off companies in the data base.) To some extent this is both a cause and an effect of success. As is shown previously, the existence of a market orientation early in a firm's life bodes well for its future performance. In addition, those companies that do perform well can generally be expected to establish separate marketing departments as they grow and prosper. Overall 46 percent of the MIT spin-offs have created marketing departments. Although they exist in firms doing all types of work, they particularly characterize the higher performing companies. For example, 78 percent of the top two categories of performers have marketing departments, whereas only 35 percent of the lowest two categories have them.

The same linkage is found between sales forecasting and company success, even though here most of the technological companies engage in this marketing activity. Eighty percent of the enterprises within the two highest performing groups do sales forecasting, with but 51 percent of the two lowest groups. The use of sales forecasting correlates with success (.08) for those companies from the large diversified technological corporation. Analysis of market potential, a key dimension of market research, is carried out by over 70 percent of the highest performers in contrast to 33 percent or less of the companies in all the other performance groups. Companies that do formal market analyses are considerably more successful ($p = .03$) than their counterparts who either feel this is unnecessary or lack the resources to conduct such studies.

Managerial Orientation

While less pervasive than the support for marketing orientation as a key to company success, the studies do provide frequent indicators that an orientation toward business perspectives and management is an important contributor. I noted earlier the correlation of supervisory and other managerial work experience with eventual company success. In addition, detailed analysis of the early years of 20 technical companies shows significant association of the business experience of the founding teams with early sales growth of those companies. The amount of administrative effort expended in these firms correlates significantly with their early sales success (.025). More successful entrepreneurs seem to balance their initial effort allocation well among engineering, sales, manufacturing, and administrative aspects of their young firms (.025), without allowing their strong technical backgrounds to generate an overemphasis on technical activities. Slevin and Covin (1987) support this general view of the importance of managerial capabilities in finding that high-performing high-technology companies implement their tactics effectively: "Doing things right" is as important as "doing the right things", they conclude (p. 94).

Moreover, several of the other research data clusters indicate that the more successful companies have, postfounding, recognized the importance of managerial skills and have brought in senior staff for the explicit purpose of handling some of the management considerations and activities. For example, ten of eleven high-performing firms among the Lincoln Lab spin-offs hired trained business people, whereas only six of 27 lower performers did the same (.01). The same pattern is observed among the spin-offs of the electronic systems company (.06). More significant than just the hiring of management staff is the finding by Tushman, Virany, and Romanelli (1985) that changes in the chief executive officer followed by strategic reorientation of the company led to success in a number of high-performing minicomputer firms. This dimension is discussed further, along with more data, in Chapter 11.

In a similar spirit, management consultants are used to supplement in-house managerial talents in 45 percent of the technological companies, with most of this concentrated in the higher performing firms (.009). Van de Ven et al. (1984) also find that better performing young educational software companies more rigorously follow a program planning model in developing their firms and bring in professional consultants to help them. But the obvious issue here, as in a number of prior areas, is which is cause and which is effect. Does the recognition of the importance of management-related skills and the follow-up hiring contribute to success, or does success cause the recognition of managerial aspects of the firm? Low performers protest that they cannot afford the luxury of management staffs and certainly not management consultants. And high performing companies get to see, due to their own successful growth, aspects of organizational issues and needs that may well lie outside the interests and capabilities of their technology-based founders. Being able to afford specialized managerial support staff is likely to mirror company performance, at least as much as it contributes to future company success.

One critical dimension of managerial sensitivity is toward the costs of doing business. Many founders do not understand accounting or the need for conscientious cost control. Some of these entrepreneurs, as well as other founders, take naive approaches to product pricing and do not assure large enough margins for the firm to become significantly profitable. The data gathered from the "early years" study show that companies that are conscious of their general and administrative (G&A) cost structure in their first few months, and account for it in their product costing, achieve greater early sales growth success ($p = .01$). The same result occurs relative to sensitivity to product development costs and the need to recover them in product sales (.025).

Entrepreneurs were asked to identify the areas of their major business problems. Not surprisingly, the primarily technical founders appear to have their principal difficulties in nontechnical areas (that is, personnel, marketing, finance). In the MIT samples the only problem area that correlates significantly with company performance is "personnel"; that is, those entrepreneurs who regard people as one of their key problem areas are among the most successful ($p = .03$). In contrast those entrepreneurs who describe other kinds of problems, but do not include people as a key problem area, are usually in the lower performing category. I suspect that the larger more successful companies do not really have more people problems. Rather, successful entrepreneurs are more likely to manifest their concerns for their employees as the principal productive element of their technical organizations. As shown in the earlier discussion of motivations, the high-achievement oriented technical entrepreneur who performs best is one who shares power with his employees, trying to provide challenge and satisfaction to his employees as well as himself. Bruno and Leidecker (1988) find that the perceived problems that lead to company failure in a small sample of Silicon Valley firms are essentially the same in the 1980s as in the 1960s, including a number of product–market issues, financing questions, and managerial–key employee reasons. No single issue dominates the failures.

Summary and Implications

Success of the technology-based company is obviously multi-faceted and attributable to multiple causes, as proposed in Figure 9–1. For most of my research studies I have chosen to measure success with an index based on average corporate growth rate, modified to account for how long the company has been in business, and incorporating information on the profitability of the firm. This carefully designed performance measure correlates strongly with several other common corporate assessment indices.

The Entrepreneur

Many presumptions to the contrary, family background variables show no direct effect on company success. Many aspiring entrepreneurs will be delighted to know that successful high-technology entrepreneurs are made, and not born.

Age of founders also has no effect on success. But the more successful companies are primarily founded by entrepreneurs with what is labeled in the samples "moderate educational levels", that is not more than an M.S. degree; lower performers tend to be even more highly educated—that is, Ph.D.s generally perform poorly, except in biomedical companies.

Entrepreneurs who leave university labs to acquire commercial experience before starting their firms transfer less advanced technology into their companies and are less successful. The number of years of their work experience does not relate to later company success, but supervisory and other prior managerial work experience shows up as an important predictor. Prior founder sales experience also correlates with enterprise success.

Entrepreneurs motivated by high need for achievement perform significantly better than founders motivated in other ways. The best performers combine this high need for achievement with moderate personal need for power.

Companies formed by multiple founders achieve greater success. In general, the more initial co-founders the higher the eventual company performance.

The Technology Base

The greater the degree of technology transferred from the entrepreneur's source organization to the new firm, the greater the success of the resulting corporation. Poorest performance is found among those companies with no lab technology transfer.

The longer the delay between leaving an advanced technology source lab and setting up a new firm, the lower the technology transferred and the worse the resulting company performance. Despite the admirable intent, gaining industrial work experience after departing from an MIT lab results in poorer, not better, entrepreneurial outcomes. Rapid technology transfer importantly contributes to company success.

Companies started with products or with products under development far outperform companies that start as consultants or contract R&D organizations. Prospective entrepreneurs should carefully consider whether "hedging their bets", by starting out as consultants, is indeed in their best interests.

Financing

Significant evidences in the MIT spin-offs and in several other research samples link success with the amount of initial company capitalization. This relationship is true for all companies considered together, and is similarly strong when only product oriented firms are examined (thus removing the sample bias of including consulting and contract R&D organizations). These results are independent of the number of company co-founders. This finding may be symptomatic of the fact that start-ups that are initially more attractive to outside investors raise greater amounts of initial capital and achieve greater eventual success. An even stronger case is made for firms that encounter extended regulatory delays prior to product revenue generation, such as biomedical companies, for which initial funding needs to exceed certain critical threshold levels for later corporate success to be achieved.

Less consistent ties are found between subsequent financing and company success. Success often generates the need for the further financing of company growth, although additional funding obviously also facilitates that growth. The lack of strong correlation is reflected in all forms of later financing. Indeed, half of those that raise their later financing through public offerings perform poorly. All entrepreneurs who raised their later financing primarily from personal resources or from their families and friends perform poorly, suggesting the total inability of firms with low appeal to persuade more professional investors of their merits.

Prospective investors should note (with surprise) that there are no consistent evidences of "learning", or lower risk of failure, as investors move out from first- to second- to third-stage investments in these technology-based companies. In terms of percentage of investments that fail, seed-stage investments are no worse than much later-stage investments in high-technology enterprises.

Marketing Orientation

Many marketing-related factors are associated with the later success of emerging technology-based companies, beginning with elements present at the time of company formation and later evidenced by organizational developments and practices postfounding. Founder prior experiences in sales activities and intensity of early company sales efforts both correlate with later success. Company contacts with customers that are aimed at determining customer needs, and the perception of customer helpfulness in determining product specifications also relate to success. More successful companies more often use market-connected sources, such as customer requests or suggestions, for new product ideas, rather than relying on founder ideas alone. Explicit awareness of competitors also characterizes the higher performing firms.

Whether a cause of success or one of its by-products, a separate marketing department shows up far more frequently among the high performers. This is also true for the practice of sales forecasting as well as for carrying out formal market analyses.

Managerial Orientation

Despite apparent professional investor preferences, the prefounding business and management experience of new company entrepreneurs appears to be only somewhat related to company success, less so than the marketing and sales backgrounds discussed earlier. More successful companies do demonstrate early recognition of the importance of skilled managers through their early hiring practices. This, too, may be more a result of early (unmeasured) success than an underlying cause of the later measured success. Finally, founder-heads of the more successful corporations most frequently perceive of "people problems" as their key issues, reflecting in my judgment a heightened sensitivity to the potential importance of people in attaining their organizational success.

Notes

1. The overall performance measure is developed for each company in three steps.

A. Average Sales Growth

A least squares regression analysis is utilized to determine the "best fit" to the time series of sales for each company. The purpose of this regression line is to get some uniform representation of sales growth patterns for each company. Each of these lines is normalized to the lower left hand corner of an arithmetic grid and graphed. In somewhat arbitrary fashion the quadrant in which the graphed lines fell is divided into groups by a 45° line, a 22.5° line, and a 11.25° line. The sales growth group numbers assigned to each company is the section of the graph on which its average sales growth line falls, as shown in Figure 9–5. If a company went out of business or had a declining average rate of sales, it is classified as group V.

B. Years in Business

The companies are then categorized in accord with the number of years they have been in business. Those in business under three years are felt to be too young to be considered as yet clearly on the way to stable success. Those in business from three to five years are seen as in what is often called a gestation period. A company in business over five years is considered to be a going and relatively stable concern. Although arbitrary, these classifications have been demonstrated to have at last some practical real-life justification. Combining

Figure 9–5
Sales Growth Groups

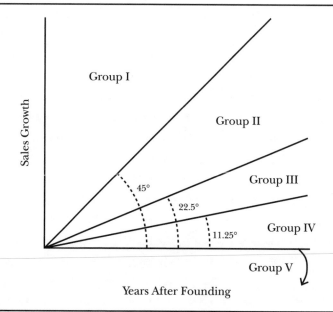

the age with the sales growth produces performance ratings of 1 through 15 as shown in Table 9–9, an index that measures the persistence of company growth.

Table 9–9
Rating Scheme for Overall Performance Measure

		Years in Business		
		Under 3	3–5	Over 5
Sales Growth Groups	I	3	2	1
(from Figure 9–5)	II	6	5	4
	III	9	8	7
	IV	12	11	10
	V	15	14	13

When company is not profitable, adjust rating as follows:
   If in business under three years, make no adjustment
   If in business 3–5 years, move rating down one vertical entry in Table 9–9 (e.g., 8 would become 11; 11 would become 14)
   If in business over five years, move rating down and to the left one group in Table 9–9 (e.g., 7 would become 11; 10 would become 14; 13 would become 15).

C. Profitability

Complete and reliably comparable detailed profit data are not available for all firms, requiring that the companies be classified simply as being profitable or not. The profitability information is integrated into the performance rating scheme as described in Table 9–9, penalizing older unprofitable firms more than younger ones.

2. The intercorrelations among the five success measures are shown in Table 9–10.

Table 9–10
Relationships among Five Performance Measures
($n = 84$ MIT spin-off companies)[*]

	Overall Performance	Average Sales	Sales Growth in Dollars	Adjusted Actual Sales	Projected Sales
Overall performance	1.0000				
Average sales	0.6750	1.0000			
Sales growth in dollars	0.6715	0.7602	1.0000		
Adjusted actual sales	0.6468	0.8600	0.7789	1.0000	
Projected sales	0.5425	0.6661	0.8164	0.7161	1.0000

[*]Values in the table are Kendall taus corrected for ties. All relationships are significant at the .0000 level (one-tail).

3. My sampling process aimed at collecting data on all new firms formed over a seven-year period in Massachusetts (for data-gathering convenience as well as

consistency with most of the other firms in the entrepreneurial research studies) that had the explicit integrated purpose of developing, manufacturing, and marketing biomedical and/or pharmaceutical products. The time frame preceded the rapid emergence of genetic engineering and biotechnology firms, whose somewhat unique characteristics do not, therefore, affect the findings of this study. Usable comprehensive data were obtained from in-depth structured interviews with entrepreneur-founders of 26 firms. In addition to company data sources, evaluations of the risk associated with the use of the 62 new products released as these companies' first, second, and third products were collected independently from three external medical experts. Details on data collection methods and specific measures are provided in Roberts and Hauptman (1987).

References

P. H. Birnbaum. "The Choice of Strategic Alternatives under Increasing Regulation in High Technology Companies", *Academy of Management Journal*, 27, 3 (1984), 489–510.

A. V. Bruno & J. K. Leidecker. "Causes of New Venture Failure: 1960s vs. 1980s", *Business Horizons*, November–December 1988, 51–56.

A. V. Bruno & T. T. Tyebjee. "The Entrepreneur's Search for Capital", *Journal of Business Venturing*, 1 (1985), 61–74.

A. C. Cooper & A. V. Bruno. "Success Among High-Technology Firms", *Business Horizons*, 20, 2 (April 1977), 16–22.

B. Dean & J. J. Giglierano. "Patterns in Multi–Stage Financing in Silicon Valley", in *Proceedings of Vancouver Conference* (Vancouver, BC: College on Innovation Management and Entrepreneurship, The Institute of Management Science, May 1989).

P. Dubini. "Which Venture Capital Backed Entrepreneurs Have the Best Chances of Succeeding?", *Journal of Business Venturing*, 4, 2 (March 1989), 123–132.

H. R. Feeser & G. E. Willard. "Incubators and Performance: A Comparison of High– and Low–Growth High–Tech Firms", *Journal of Business Venturing*, 4, 6 (1989), 429–442.

H. R. Feeser & G. E. Willard. "Founding Strategy and Performance: A Comparison of High- and Low-Growth High-Tech Firms", *Strategic Management Journal*, 11 (1990), 87–98.

C. J. Fombrun & S. Wally. "Structuring Small Firms for Rapid Growth", *Journal of Business Venturing*, 4, 2 (March 1989), 107–122.

Food and Drug Administration. *Federal Food, Drug, and Cosmetic Act (as amended) October 1976* (Washington, D.C.: U.S. Government Printing Office, 1976).

B. Huntsman & J. P. Hoban, Jr. "Investment in New Enterprise: Some Empirical Observations on Risk, Return, and Market Structure", *Financial Management*, Summer 1980, 44–51.

P. P. McDougall. "International vs. Domestic Entrepreneurship: New Venture Strategic Behavior and Industry Structure", *Journal of Business Venturing*, 4, 6 (1989), 387–400.

J. B. Miner, N. R. Smith, & J. S. Bracker. "Role of Entrepreneurial Task Motivation in the Growth of Technologically Innovative Firms", *Journal of Applied Psychology*, 74, 4 (1989), 554–560.

E. B. Roberts & O. Hauptman. "Financing Threshold Effect on Success and Failure of

Biomedical and Pharmaceutical Start–Ups", *Management Science,* 33, 3 (March 1987), 381–394.

J. B. Roure & M. A. Maidique. "Linking Prefunding Factors and High-Technology Venture Success: An Exploratory Study", *Journal of Business Venturing,* 1 (1986), 295–306.

J. C. Ruhnka & J. E. Young. "A Venture Capital Model of the Development Process for New Ventures", *Journal of Business Venturing,* 2 (1987), 167–184.

W. R. Sandberg & C. W. Hofer. "Improving New Venture Performance: The Role of Strategy, Industry Structure, and the Entrepreneur", *Journal of Business Venturing,* 2 (1987), 5–28.

D. P. Slevin & J. G. Covin. "The Competitive Tactics of Entrepreneurial Firms in High– and Low–Technology Industries", in N. C. Churchill et al. (editors), *Frontiers of Entrepreneurship Research, 1987* (Wellesley, MA: Babson College, 1987), 87–101.

N. R. Smith & J. B. Miner. "Motivational Considerations in the Success of Technologically Innovative Entrepreneurs", in J. A. Hornaday et al. (editors), *Frontiers of Entrepreneurship Research, 1984.* (Wellesley, MA: Babson College, 1984), 488–495.

M. L. Tushman, B. Virany, & E. Romanelli. "Executive Succession, Strategic Reorientations, and Organization Evolution: The Minicomputer Industry as a Case in Point", *Technology in Society,* 7 (1985), 297–313.

A. H. Van de Ven, R. Hudson, & D. M. Schroeder. "Designing New Business Startups: Entrepreneurial, Organizational, and Ecological Considerations", *Journal of Management,* 10, 1 (1984), 87–107.

W. E. Wetzel, Jr. "Angels and Informal Risk Capital", *Sloan Management Review,* Summer 1983, 23–34.

CHAPTER 10

Product Strategy and Corporate Success

Formed by three engineers who had done advanced computer systems development in a large electronic systems firm, Computer Technologies, Inc. (CTI) initially developed calibration machines for the production of magnetic tape storage devices, sold directly to large computer manufacturers. Encountering limited opportunity the company then developed its own read/write heads for tape machines, sold as components directly to its manufacturing customer base to satisfy a new user functionality. CTI subsequently developed a third and unrelated product line, attempting to enter the microcomputer software business with a proprietary operating system for a popular microcomputer. In a following product release the firm also developed small business applications software. The two software products entailed major changes from CTI's earlier user functionality, customer groups, and channels of distribution, selling these packages through office systems dealers and computer retail stores. None of the products became profitable and, after six years of struggling, Computer Technologies ran out of funds that had been provided by the founders and some private investors. The founder–CEO thinks they would have made it if only they had more capital. I think the fundamental problem was in their product strategy.

Deciding what products to make and how to make them is a constant challenge to the management of technology-based firms. Companies operating in areas such as computers, electronic components, optics, medical devices, telecommunications, lasers, and biotechnology are frequently and profoundly affected by rapid advances in their respective product technologies. In these industries where the rate of new product introduction is high, a stagnant research and development effort can be disastrous. Even good ideas are insufficient; firms must turn them into successfully marketed products. Given that technology-based companies must continue to innovate their products (and perhaps their manufacturing processes as well) to survive and grow, they must make fundamental choices about their technology and market strategies.

This chapter draws heavily on two articles that I previously co-authored with Marc H. Meyer (1986, 1988).

This chapter is divided into three portions that treat the formulation and implementation of product strategy for the high technology firm. First, I present some of my strongly held convictions, which both influenced the related field research studies and were enlightened by them. Next, I describe a framework that evolved during the research for envisioning a company's product history in a manner that may reveal its de facto product strategy. This method was used in four separate studies to gather data on a number of New England computer-related companies. The framework is presented along with three case studies from the research. The third section presents some of the statistical results from the field studies, which support the importance of "strategic focus" in new product development to achieve success in the technological firm.

Some Perspectives on Corporate Strategy

During the 1970s, especially, corporations were urged to develop diverse product portfolios to grow and prosper. The leading U.S. strategy consulting firms created techniques, such as the "market share/ market growth" matrix of the Boston Consulting Group, to help managements visualize product lines as pieces of a financial investment strategy, often premised on the diversification of risk. Buying and selling product lines and businesses was considered a pathway to achieving optimal portfolio mix. Intensifying or diminishing the internal investment in a business was a function of whether the product line was a "star" or a "cash cow". The technologies associated with those businesses were often considered only peripherally in the making of these decisions, and rarely viewed as a separate strategic issue—they were most often just lumped in as an amorphous entity that came or went with a business-unit portfolio change. As a result, acquisitions and divestitures often preceded major reorganizations of a company's R&D effort. For managers, this resulted in an unstable engineering resource pool and often ineffective new product development programs.

In the 1980s, the business community generally came to appreciate that these earlier perspectives were both naive and wrong. The public as well as the management of many firms began to see that company growth and prosperity depend upon "excellence" at something that the marketplace values, be it a stream of products or the delivery of certain services (Peters and Waterman, 1982). Today, the underpinning of excellence in a product's performance is more clearly understood to lie largely with its technology, which had better be planned and managed effectively.

In planning the development of new products, management has three basic choices in regard to technology and three comparable choices re-garding market application. In terms of technology a company may pursue a *focused* strategy of building a critical mass of technological skills for a closely related product portfolio, believing that the distinctive competence achieved in its *core* technology will become the basis of long-lasting competitive advantage. Ketteringham and White (1984) argue for the importance of key core technology in strategic analysis. A second option, that of *evolving* the technology, once again stresses internal technology development, but targets multiple and perhaps unrelated technologies. A firm creates a diverse set of products that does not depend on the continuing importance of a single core technology. Third, a company may generate a diverse portfolio of products through an *unfocused* or *diversified* strategy of acquisition—buying into new technological fields by acquiring other technology-based companies, or at least their technologies, and avoiding the long-term effort of building the needed technological expertise internally. The growth of "strategic alliances" among firms frequently reflects one or both of the strategic partners adopting this technology strategy. A firm can combine the third strategy, in varying degrees, with either of the first two.

Similarly, in terms of market applications, a company can adopt a *focused* market applications strategy, pursuing a single product/market area with stable selection of distribution channels. The firm's products offer solutions for the same set of problems, are applied to a single set of customers, and are sold in one basic fashion for the life of the company. Alternatively, the firm can follow a *leveraged* market strategy, releasing products that address different customer groups, typically sharing the same basic functional need, often reached through the same distribution channels. Leveraged products that are sold to different, yet related, customer groups also tend to be based on a single key technology that the firm then customizes to specific niche requirements. The third strategy pattern in the market dimension is a *diversified* one, characterized by products that contain changes in all three market-related parameters of customer needs, end-user customer groups, and channels of product distribution. Again companies may use acquisitions and/or alliances to implement this third strategic choice in regard to market applications.

Combining the several-product technology strategies with the several-market applications strategies leads to a wide range of optional approaches to product development and/or acquisition and sales. Which of these is most beneficial to a company? The answer no doubt depends on many factors specific to the company and its industry. Product diversity and acquisition may have been attractive growth strategies in corporate America and may be effective for some large companies. My instincts, however, strengthened by the evidence presented later in this chapter, strongly indicate that they are ill advised for emerging technology-based start-ups. The best opportunities for rapid growth of a young firm come from building an internal critical mass of engineering talent in a focused technological area, yielding a distinctive core technology that might evolve over time, to provide a foundation for the company's product development. Those products should be targeted at a focused set of customer needs, sold to gradually broadening groups of end-users through single channels of sales and distribution.

Companies observed in field research studies that attempt to build an overly diverse portfolio of products (through either internal development or acquisition) eventually find themselves with technologically mediocre products and diffuse marketing. Companies that concentrate on the internal development of a single technology or a closely related set of technologies, and that focus on related market applications, achieve both technological product excellence and deep understanding of their customers. These results agree with Cooper's findings (1984, 1985) from survey research on new product strategies by Canadian companies. Without a defensible core technology, typically, the technological venture has difficulty assuming a leadership role in its target markets and finds itself playing catch-up with competitors. In contrast, companies that develop a strong core technology show the ability to develop new products faster, with greater reliability and quality, than unfocused companies. With a core technology, these technological "winners" are more capable of responding to competitive events and in many cases are able to assume industry leadership by virtue of an exciting new product strategy. The more successful companies stay close to a single set of customers, using their technological advances to capture increasing market share, and gradually broaden their base of customers into related fields. The evolving horizontal integration is achievable with essentially a single type of selling process and stable channels of distribution. From a human resource management perspective, the company can create more readily a close-knit cadre of talented engineers and is adept at hiring and training new engineers for its R&D group. The firm can also recruit and manage sales and field service personnel in a smoother manner.

Beginning with strongly held convictions based on personal experience and the indirect observation of many technological firms is a potentially dangerous way for researchers to proceed. Yet my graduate students and I exercised great care in the development of a data collection framework and in the actual information gathering and analysis needed to test our hypotheses.

Developing a Framework

New product decision making in the technology-based enterprise addresses four basic issues:

1. What are the basic needs or user functions that the firm will satisfy with its products and services?
2. What are the groups of customers that share these needs or functional requirements and to whom products and services will be sold?
3. What technology will be used to build the products or deliver the services, and what is the source of that technology?

4. What distribution mechanisms will be employed to bring successfully developed products to the marketplace?

A company that finds a set of consistent answers to these questions, supported by a track record of company actions taken in support, has a firm basis for a product strategy. Product strategy obviously encompasses two key dimensions: the technology embodied in the products, reflecting both the personal skills and techniques that achieve a physical manifestation in any given product; and the market applications of the products, including the intended functionality of the product from a user's perspective and the specific customer groups to which the product is marketed.

In searching for a framework to identify a company's product strategy Chandler's lead (1962) is helpful. He traces the evolution of seventy large American companies over approximately 20 years of their growth, emphasizing their shifts in strategy and organizational structure. In turn, my approach is to trace the evolution of high-technology companies over their life spans, emphasizing the changes in the products they develop and market. To study technological content of the products specific tangible levels of change are identified between successive products created and sold by a given company, building from concepts first presented by Johnson and Jones (1957). Similarly, to study market change specific shifts are identified in the three marketing oriented parameters of product functionality, end-user customer groups, and distribution channels.

The Technology Dimension of Product Strategy

Every product made is based on an identifiable engineering skill set, or what might be called a technology. Most products are in fact composed of multiple technologies, some of which are created within the company's R&D group, while others are licensed from outside sources or purchased as components. Assessing technology strategy requires investigation in depth of the internally developed technologies used in products. These technologies evolve within companies over time, finding their way into successive products. As each new product emerges, the cumulative body of the company's technology experience expands. That broadened experience becomes the base for evaluating the "incremental newness" of the technology embodied in the next new product. My concentration is on the changes made in the key core technology that provides the firm with a proprietary, competitive edge and differentiates it from other companies making similar or substitute products. This can usually be distinguished from other "base technologies," also used by the firm in its products, but more commonly available in the marketplace as components. A firm that is in an industry characterized by rapidly advancing technology typically concentrates on one or possibly two specific key core technologies and, by packaging or integrating its core with a variety of component base technologies, generates its final products. The key core technology becomes the basis for the "value added" of the firm.

Tracking the evolution of technology in a company's products involves assessing the degree of improvement in or additions to the technology over time. This level of technological change runs along a conceptually continuous range of expended resources and effort. However, the research studies used four discrete levels of change or newness in product technology to evaluate more than 200 products developed by twenty-six companies. The first and "smallest" level of technological change identified is a *minor improvement* to the company's existing product technology. This level of change is illustrated by one of the printer manufacturers that, having produced a series of 80-column dot-matrix printers for microcomputers, developed a 132-column printer. The project took less than six months and was introduced easily into the company's manufacturing and sales operations. Minor improvements can also include efforts as marginal as repackaging existing technology or customizing a product in response to customer requests. For example, a terminal manufacturer developed a series of equipment that contained new communications and terminal "emulation" capabilities so that it could more readily be tailored for use with computers produced by Digital Equipment, Data General, Burroughs (now Unisys), and others. Often, new products that embody minor technological improvements simply correct known problems. Not surprisingly, this is a common type of "new product" among software companies which seem continually to release new versions of a basic product line with more "bug fixes" than genuine new features.

The second level of technological change is a *major enhancement* to an existing product technology, incorporating a substantially larger effort in the improvement or advancement of a technology in which the company has previously developed expertise. A firm often achieves major enhancement through the addition of new base technologies to a product line, frequently requiring substantial development effort. By adding new components or subsystems, the firm can leverage its existing key technologies into new product/market areas without having to develop new core technologies of its own. Companies that can continually succeed with major enhancements often become the "standard setters" in an industry. For example, one of the photocomposition systems developers pioneered the application of color-imaging technology in the 1970s and now sells high-ticket expensive systems to magazines, newspapers, and other publishers as a state of the art production facility. A more recent new product allows the user to define extensive graphics "libraries" so that, for example, a digitalized photograph of a sailboat can be augmented with a "prestored" digital female figure, the designer's favorite bathing suit and sunglasses, and other graphic "objects" such as a dog, a beach ball, and a bottle of fine Chardonnay.

Major enhancements tend to be sequenced in intervals of three to five years within specific product lines, although this pace of technological change has been accelerating in recent years under intensified foreign competition. For example, one manufacturer, which has focused on high-speed line printers, privately labeled for resale by a large number of computer manufacturers, has over the years upgraded its printing head technology from early rotating "drum" devices in the late 1960s, to "linked chain" printing heads in the mid-1970s, to soldered "band" technology in more recent years. Terminal manufacturers, as another example, have developed high-resolution graphics terminals, more recently with color capability, as an extension of long-standing alphanumeric display technology. None of these major enhancements to an existing product technology took less than nine months in R&D, and some required two to three years of concentrated effort. At the same time, however, the companies achieve both of these first two levels of technological change with a stable cadre of engineers, augmented periodically with new talent at the junior level, within the company's evolving core-technology skill set.

The third defined level of technological newness occurs when a company develops an entirely new core technology that is integrated with an existing company technology in the final product. Here is an example. One of the terminal manufacturers makes transaction-processing terminals used by bank tellers. The smaller-than-usual terminals are loaded with communications software. In a move to expand on both its technology and its customer base, the company created an automated teller machine. While its previous terminal screens and transaction communications software are employed directly for the screen displays of the automated teller machine, the company's engineers had to develop two additional technologies: the electromechanical technology for the cash withdrawal and deposit safebox inside the machine, and all the applications software for handling the dialogue with the bank user. At first the company employed the services of a software R&D contractor but, finding that approach too unreliable, was forced to hire a number of software engineers. In subtle ways these software applications engineers represent a different culture or style than the company's traditional R&D group and present a new challenge to management in terms of integration and control. When new technology is combined with existing company technology in this way, the third level of technological change is involved, here labeled *new, related technology*. Another example is a software company that had developed as a core-product technology a version of the Unix operating system for personal computers. It then created a new product, a data base management system that ran on its Unix operating system. Again, while some of the initial operating system engineers were shifted onto the data base project, within a year a half-dozen new engineers were hired who had specific skills in data base storage, query languages, and screen interfaces for users. The skill set required for development of the commercial data base management system clearly separates it from operating systems work. Yet, since the product is designed for use with the earlier operating systems offering, for this company the data base management product is a new, related technology effort.

The fourth level of technological change encompasses new core technology that is not combined with existing product technology in the company. This *new, unrelated technology* is the "highest" level of change in a company's technology evolution, a major departure from technological focus. Why do companies undertake the risk associated with such diversity? One reason may be corporate survival. Several companies in the sample introduced first products that failed commercially and, rather than cease business operations, management tried a new product technology for a different application. For example, one company initially implemented a cable television network for a local municipality. Today, its cable business no longer exists but the company has become a leading supplier of plastic card scanners used by banks for automated teller machines and by corporations and residential complexes for access control. An unfocused technology strategy may also be the result of engineering oriented management that continually seeks "new hills to climb." One photocomposition company (whose founders are also MIT professors) has developed and sold optical character recognition devices, a computer-based camera and image composition system, and a multiuser text-composition system, all for use in the newspaper industry. While the first two products are sometimes delivered as a single system to newspaper companies, the third is a stand-alone product, entailing the new core technology of the text-composition applications software. Large-scale additions of different types of engineers were necessary to implement these new products.

These four levels of technological change—minor improvement and major enhancement to an existing company core technology, and the development of new technology that is either related or unrelated to existing technology—can be used to assess the technological diversity of any new product. The framework can also be used to develop a portrait of a company's technological evolution over its entire history. Obviously, by using measures of marketing change between successive products, the same assessment can be made of a company's product-marketing history.

The Market Applications Dimension of Product Strategy

As mentioned earlier the market applications framework has three parameters, adding distribution considerations to the product usage and customer groups vectors used by Abell (1980) for evaluating business opportunities. The first of these, *product functionality*, is the general set of customer needs that a product satisfies. It is clearly distinct from the technology that is embodied in the product: Functionality is the goal of a product, whereas technology is the tool for delivering that functionality. The same functionality may be delivered by different technologies, perhaps by a process of technological substitution. Conversely, a single technology or group of technologies may be extended to different sets of functionality, if the earlier technology can be stretched to satisfy needs that are different from those addressed by earlier products of the firm.

End-user customer groups is the second criterion used for measuring market applications change between successive products. Industrial classification codes, common organizational environments, and levels of user experience are criteria employed to segment markets into customer groups. Abell and Hammond (1979) suggest additional factors that distinguish customer groups: "Customers may differ in their needs for information, reassurance, technical support, service … and a host of other "nonproduct" benefits that are part of their purchase" (p. 48).

The third facet of the market applications dimension is the *distribution channels*. Distribution channels for the technology-based firm include:

1. Direct sales
2. Original equipment manufacturer (OEM) reselling
3. Nonmanufacturing value-added resellers (VARs)
4. Nonmanufacturing, nonvalue-added resellers
5. Mail order

In the first category, direct sales, the firm's own sales force sells the product directly to product end-users. The company typically assumes responsibility for customer support, which may include training and equipment maintenance, and sometimes for the integration of the product with other vendors' products that are required by the end-user. Technology-based firms frequently employ the next distribution means listed, the OEM channel. Microprocessors, software packages, terminals, printers, peripheral storage devices, and even entire computer systems are commonly distributed through large manufacturers for integration with the manufacturer's own product line. In the third channel of nonmanufacturing value-added resellers (VARs), the firm distributes its manufactured products through systems integrators that specialize in particular vertical market niches. VARs bring together a number of different components, only one of which is the firm's product, and tailor these components to provide complete or "turn-key" systems to end-users. Electronic Data Systems, now a subsidiary of General Motors, is a large VAR that has combined and customized outside vendors' software and peripherals with its own software packages for application to IBM mainframe environments, successfully penetrating market segments that include insurance, banking, and government agencies. The fourth distribution channel, nonmanufacturing nonvalue-added resellers, are more usually called distributors, and offer lower levels of support to end-users than the previous channels, typically selling a range of products from different suppliers. In the area of low-end computer products the microcomputer store is this type of reseller. Independent sales representatives are a component of this channel of distribution. Finally, the firm may decide to undertake mail order distribution by advertising in publications read by their prospective customers or by direct mail campaigns. "Direct mail" and "direct sales" are appropriately at opposite ends of the spectrum, involving dramatically different commitments of company resources in contact with and support of its customers.

Adopting any one of the five channels identified does not preclude the use of any other channel. Similarly, as firms grow they may shift channels or add new channels of distribution to those employed for earlier products. For example, starting with the development of a microcomputer version of a popular graphics and statistics package used at MIT, Mitch Kapor developed a graphics package that was compatible with the then popular Visicalc "spreadsheet" package. That product, labeled "Visiplot", was sold as an OEM product through Apple Computer. Then Kapor teamed up with Jonathan Sachs and organized Lotus Development Corporation to develop an integrated system, combining his graphics software with their own "spreadsheet" and simple text editing software into the pathbreaking "1-2-3" product. A distribution agreement was signed with a large nonvalue added reseller that brought the product to hundreds of retail computer outlets. With additional financing, Lotus expanded its market and captured the margins previously sacrificed to distributors by creating its own direct linkages to retail stores. Finally, direct selling to large corporate accounts has also been used more recently by the firm.

Using these three market oriented parameters, a matrix for the measurement of market applications change can be constructed, as shown in Table 10–1. The first level of market applications change is no change, that is, when all three parameters remain unchanged from the previous product release. If only one of the three parameters changes, either a new user functionality or a new customer group or the adoption of a new channel, the product is assessed as being at the second level of market change. Similarly, a change in any two of the three parameters brings about the third level of change. Finally, when all three parameters change the product is measured at the fourth and "highest" level of market applications newness.

Table 10–1
Levels of Change in the Market Applications Dimension

	Customer Groups	Usage Functionality	Distribution Channels
Level 1	Same	Same	Same
Level 2	A new group	Same	Same
	Same	A new function	Same
	Same	Same	A new channel
Level 3	A new group	A new function	Same
	A new group	Same	A new channel
	Same	A new function	A new channel
Level 4	A new group	A new function	A new channel

The three different product market applications strategies described in this chapter's introductory section on product strategy are associated with these various levels of product market change. The *focused* product market strategy reflected in a series of Level 1 product releases is illustrated by the high-speed line printer company described in the technology section. The firm has always sold its printers, designed for high-speed data processing use, through OEMs. The *leveraged* market applications strategy with new products that address different customer groups, but usually satisfying the same basic need and distributed through the same channel, is employed by the access control systems vendor. Its magnetic card readers are found in bank ATM machines, computer facilities of large corporations, and residential complexes. In more recent product offerings, the company has developed a set of applications software for "time-in, time-out" management, selling turnkey systems where dozens of its card readers may be attached to a microcomputer. The firm has also recently enhanced its direct sales channel to include sales representatives who cover particular geographic areas and vertical markets. The third and last strategic pattern of *diversified* market applications, distinguished by changes in all three dimensions, is epitomized by one firm in an initial pilot study, identified here as Computer Technologies, Inc. (CTI) and described at the outset of this chapter.

The Product Innovation Grid

The two dimensions of new product strategy can be integrated into the Product Innovation Grid shown in Figure 10–1. The term *innovation* reflects the perspective that innovative activities in the realm of new product strategy are not confined to technology development alone, but also encompass the market applications dimension of the products. A firm's historical product portfolio can be plotted on the grid, where each product is measured along both the technological and the market applications dimensions for its level of change compared to the product developed before it. Abetti and Stuart (1987) have conceptually modified this two-dimensional grid into three dimensions, separating out as the third axis the product functionality aspect of the market applications considerations.

Four generic labels of product strategy characterizations are shown on the Figure 10–1 grid, representing "average" levels of combined technological and market newness for a firm's entire product sequence. The "Highly Constrained" pattern is one where the company chooses to perform only minor enhancements to a single core technology and sells its products to a particular market niche for one usage and with an unchanging sales mechanism. The "Focused" pattern is marked by major enhancements in technology that are leveraged into products for several customer groups. The firm aggressively employs new component technologies to provide new levels of functionality to its users. The third pattern, called "Mixed", involves a strategy where the firm has ventured into new product areas by

Figure 10–1
The Product Innovation Grid

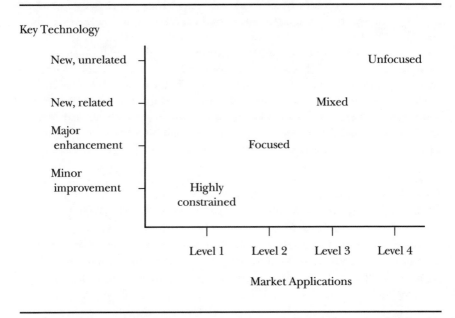

Key Technology

developing new core technology and integrating it with existing core technology. New functionality, different customer groups, and different distribution channels are encountered in such efforts. Other firms in this product strategy group may be companies that have tried various product development approaches before settling down into a more focused strategy. Finally, the "Unfocused" strategy represents wide diversity along both dimensions in the firm's product sequence.

It is important to realize that a high level of interproduct change is not synonymous with overall aggressiveness, even with regard to technological advance. Focused companies that exhibit low or moderate levels of change in product technology are hardly stagnant. Remaining competitive in dynamic technological fields required equal if not greater amounts of research and development on the part of the companies studied as venturing into new and different technologies. The successful technologically focused company demonstrates a combination of aggressiveness and "working smart" to build a distinctive competence and generate a strong core technology.

Three Case Studies

In the course of carrying out the research reported in this chapter the process of developing and displaying a plotted presentation of a company's product history provided useful managerial perspectives. The resulting

Product Innovation Grid can be used to provide a "snapshot" or portrait of a firm's product activities at any point in time. Three companies from the data base are discussed in detail, both to expand on the methods used in product assessment and to illuminate the arguments for a focused product strategy.

FastPrint. Figure 10–2 displays the product sequence of a printer manufacturer that has followed a clear technological and market focus: It has developed a strong core-technology capability, has developed and reinforced its primary marketing approach, and competes effectively against Japanese as well as American companies.

Let's call the company FastPrint. Notice that in Figure 10–2 the lowest number on the grid is 2, which represents the company's second product. In this methodology the first products of companies are not scored on the grid, but instead are used as the baseline to evaluate the newness of the second and subsequent products. FastPrint has released a total of eighteen products since its founding in the late 1960s. It was started by several MIT professors whose first product was, of all things, one of the first electronic-gambling systems for a Las Vegas casino. Requiring inexpensive printing stations for the gambling systems and unable to find them on the market, these entrepreneurial academics then made one of the first small dot-matrix printers; it was the company's second product. Thus, product 2 is positioned on the grid as (relative to product 1) having new, related technology and Level 4 market applications change, meaning concurrently new customers, new user functionality, and new channels of distribution. In the products listing in Figure 10–2, product 2 is given a Technology Newness Score of 3 ("new, related") and a Market Applications Newness Score of 4. All the later products are positioned on the grid and scored in a similar manner.

From this point on FastPrint's product strategy was focused on printing technology and its applications in the minicomputer and later the microcomputer marketplace. FastPrint scored its biggest success in its mid-life by making the first popular desktop dot-matrix printer, which was widely sold through retail stores along with the first popular Apple microcomputer system. The company's technology development has been continually aggressive, with repeated major enhancement efforts designed at providing faster speed and better dot-matrix printing at lower cost. The technology descriptions associated with the product numbers in Figure 10–2 demonstrate this pattern. To develop the ratings shown, minor improvements were differentiated from major enhancements by working with the vice president of engineering to assess the time and resources allocated to each product. Major enhancements that went into one product were often consolidated later with minor improvements in new product releases, either to reduce production cost or for repackaging. On other occasions, when FastPrint wanted to go into a new technological area, such as building a higher-speed line printer, it

Figure 10–2
The Products of FastPrint: A Focused Strategy

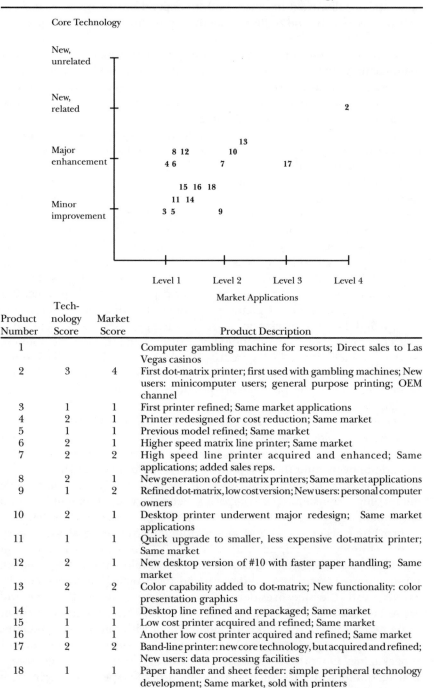

Product Number	Technology Score	Market Score	Product Description
1			Computer gambling machine for resorts; Direct sales to Las Vegas casinos
2	3	4	First dot-matrix printer; first used with gambling machines; New users: minicomputer users; general purpose printing; OEM channel
3	1	1	First printer refined; Same market applications
4	2	1	Printer redesigned for cost reduction; Same market
5	1	1	Previous model refined; Same market
6	2	1	Higher speed matrix line printer; Same market
7	2	2	High speed line printer acquired and enhanced; Same applications; added sales reps.
8	2	1	New generation of dot-matrix printers; Same market applications
9	1	2	Refined dot-matrix, low cost version; New users: personal computer owners
10	2	1	Desktop printer underwent major redesign; Same market applications
11	1	1	Quick upgrade to smaller, less expensive dot-matrix printer; Same market
12	2	1	New desktop version of #10 with faster paper handling; Same market
13	2	2	Color capability added to dot-matrix; New functionality: color presentation graphics
14	1	1	Desktop line refined and repackaged; Same market
15	1	1	Low cost printer acquired and refined; Same market
16	1	1	Another low cost printer acquired and refined; Same market
17	2	2	Band-line printer: new core technology, but acquired and refined; New users: data processing facilities
18	1	1	Paper handler and sheet feeder: simple peripheral technology development; Same market, sold with printers

licensed products from other companies and refined them for its own purposes. This occurred in products 6, 15, 16, and 17.

From a marketing perspective FastPrint's initial printer, product 2, had the highest level of change, shifting from direct sales of its earlier gambling machine to sale through OEM computer manufacturers for use in mini-computer printing and by small data processing facilities. Subsequent products show a consistent focus on OEM channels, with minor occasional market applications changes. For product 7, for example, FastPrint added independent sales representatives to its distribution channels. With product 10 came the new user group of microcomputer users. A new user function-ality of color printing was served by product 13, and a new customer group, the high volume data processing facility, was reached by product 17.

FastPrint is a clear example of a company that is both technologically and market focused. Its distinctive core technology, developed over years by a fairly stable corps of dedicated engineers, has been a key factor in the company's leading market position. Its long-term focus on relating to and selling through OEM computer manufacturers has generated customer relationships based on mutual understanding and shared dependence, as well as ready access to high volume sales opportunities.

Techlabs. A contrast to this focused strategy is the case of a newspaper composition systems company that pursued many technologies and, though sticking to the same newspaper organizations as customers, attempted to satisfy varied needs of different individuals, often through different sales channels. The product history of this company is shown in Figure 10–3.

Founded also by an MIT professor, the company, here called "Techlabs", created one of the first "raster display" graphics terminals in the late 1960s, thus permitting time-shared minicomputers to have graphic displays. The initial product was sold directly to universities and other scientific institu-tions. Soon, however, Tektronix released its own (and now industry stan-dard) raster display graphics terminal and has since come to dominate the marketplace. Techlabs responded not with another terminal, but with a graphics tablet that could be attached to engineering workstations. This new technology was marketed exclusively through a large computer-aided design systems manufacturer. Techlabs then used the cash generated from this product to venture into yet another technological field, developing a text-editing workstation in the mid-1970s, complete with hardware and applications software. In addition to direct sales, the company sought to contract with distributors to sell this product. In subsequent products Techlabs undertook costly hardware projects, in a sense pioneering micro-computer architectures for its own text-editing product line. With limited success the company then focused on its text-editing software, releasing a series of packages aimed specifically at small newspaper companies. Its more recent products, for example, include packages for managing classified advertisements, newswire communications, and text composition. Outgunned by its various competitors in the domestic marketplace, Techlabs

Figure 10–3
The Products of Techlabs: An Unfocused Strategy

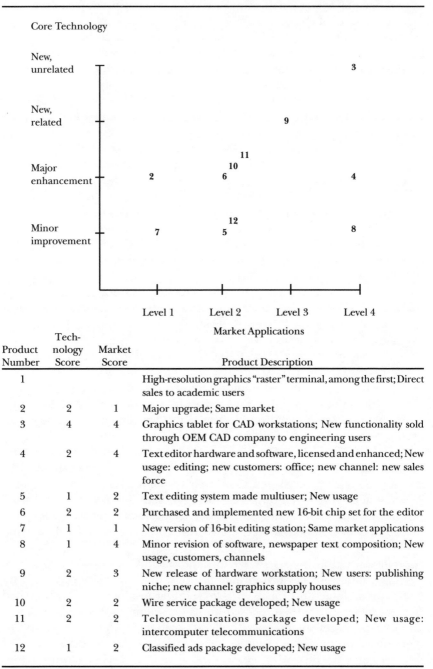

Product Number	Technology Score	Market Score	Product Description
1			High-resolution graphics "raster" terminal, among the first; Direct sales to academic users
2	2	1	Major upgrade; Same market
3	4	4	Graphics tablet for CAD workstations; New functionality sold through OEM CAD company to engineering users
4	2	4	Text editor hardware and software, licensed and enhanced; New usage: editing; new customers: office; new channel: new sales force
5	1	2	Text editing system made multiuser; New usage
6	2	2	Purchased and implemented new 16-bit chip set for the editor
7	1	1	New version of 16-bit editing station; Same market applications
8	1	4	Minor revision of software, newspaper text composition; New usage, customers, channels
9	2	3	New release of hardware workstation; New users: publishing niche; new channel: graphics supply houses
10	2	2	Wire service package developed; New usage
11	2	2	Telecommunications package developed; New usage: intercomputer telecommunications
12	1	2	Classified ads package developed; New usage

has recently sought to exploit the European marketplace through distributors that include graphics supply houses in various European countries.

With such diversity in technology (requiring major enhancement efforts in both hardware and software) the company cannot be clearly identified by a single core technology. Its engineering pool has undergone numerous transformations in terms of skill content and emphasis. Furthermore, the company's diverse products, each targeted to different types of customers for widely varying uses, has yielded multiple distribution channels and marketing programs. When Techlabs managers were interviewed recently, they were clearly struggling with this complexity. The company is experiencing little growth and its cash flow cannot sustain current operations.

BestScreens. A company's product strategy can also change dramatically. Companies that were once highly focused and successful can dissipate their core technology and, with a commensurate lack of market focus, find themselves very quickly in financial straits. A third case description illustrates this. "BestScreens" had risen to approximately $50 million in sales by supplying a highly reliable yet inexpensive family of alphanumeric terminals that could be used efficiently with a range of computer manufacturers' protocols, including those of Digital Equipment and Unisys. These terminals were sold through OEMs and dealers. BestScreens had also produced a very popular graphics terminal that could, at the same time, be used as an alphanumeric terminal. Thus, its product strategy had been classically focused, major enhancements to a single technology with market adaptation for a series of related customer groups.

Then, as a result of ambition (or greed) and not desperation, BestScreens' management changed its orientation and sought to become a full-fledged computer company through both internal R&D and technology acquisition. BestScreens first acquired a small company that had made a portable microcomputer. Management established limited retail distribution for the new product. The product was a costly failure, especially after IBM and Compaq, among others, released comparable products. Still maintaining its success with its long-standing terminal product line, management decided to have another go at diversification. BestScreens proceeded to develop in-house a multiuser desktop minicomputer based on the then new Intel 80286 chip. While designing and manufacturing the new computer internally with the best of its existing hardware engineers, the company also had to hire a number of operating systems software specialists needed to integrate the Unix operating system that the company had licensed from AT&T. The new computer was aimed at the value-added resellers distribution channel and, compared with its previous products, targeted new applications. BestScreens' second venture into diversification had a more telling impact than the previous one. This publicly traded company went into a tailspin, and within two years BestScreens sought legal protection from its creditors.

Empirical Evidence Supporting Focus
in Product Strategy

The three previous examples illustrate how product focus in terms of both technological and market applications dimensions figures into the success of high-technology companies. Are these observations merely flukes, or are they representative of an underlying truth that is generally applicable to technology-based companies? To find an answer the Product Innovation Grid framework was applied systematically to evaluate product change in a sample of 262 products from twenty-six New England companies.

The Strategic Focus Hypothesis

The main hypothesis is that firms with a high degree of strategic focus in their product innovation will over time outperform less focused companies. This hypothesis is pictured in Figure 10–4. Examining the limits of the hypothesis suggests one refinement. At one extreme, the bottom right of the diagram suggests failure for an organization that pursues an ultimately unfocused strategy, implementing for each new product a new unrelated core technology, and targeting new functional uses, different customers and distribution channels. Cooper's (1979) findings support this reasoning. His "High Budget, Diverse" firms, whose products have unrelated tech-

Figure 10–4
The Strategic Focus Hypothesis: Strategic Focus
versus Performance

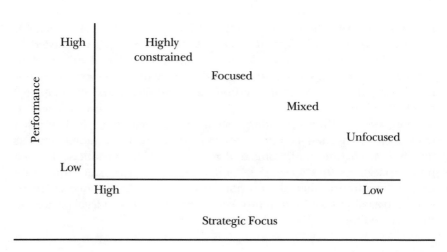

nologies and are scattered in market orientation, contain the weakest performers, in line with my expectations.

But the other extreme, the top left of Figure 10–4, suggests greatest success for the firm that undertakes for all its products only minor improvements to its single initial core technology, to be sold exclusively to one set of customers for one specific function through one stable channel of distribution. At first glance this may seem the least risky of strategies because the firm takes no chances in exploring new technological areas or market applications for its products. However, a dogged faith in the continued viability of a single technology/customer set could, over time, prove to be a very risky course of action. By labeling this strategy "Highly Constrained" I have already set it up by implication to mean "Too Highly Constrained".

Indeed several reasons suggest that the "Focused" strategy should assume the top position with respect to expected performance. The environment of rapidly changing technologies mandates that firms keep abreast of technological change through well-timed major enhancements to internal core technology. Similarly, a company often cannot be satisfied with a single customer group for the life of its entire product line. A specific customer group may be limited in size or the object of greater competition as time progresses. New markets for technology products tend to evolve into more well defined subgroups, and products targeted for the initial market undergo needed "differentiation" to satisfy better the requirements of the emerging market niches. In addition, new markets for technologies are continually born, and may present attractive opportunities for the firm to leverage its core technology into new functional applications and customer groups.

The company that performs periodic major enhancements to its product line and aggressively pursues new customers is very different from the firm that relies on a single, familiar customer set with successively repackaged and customized technology. A "Focused" versus a "Highly Constrained" strategy is also potentially more successful because the firm seeks new related growth opportunities. Thus, the revised hypothesis is that the most effective product strategy is one that focuses on some level of highly directed change in either the technology or market applications dimension, pictured in Figure 10–5 as a bell-shaped curve skewed to the left.

Methods and Measurements

The sample evolved over the course of four studies of New England firms into small but consistent clusters in four computer oriented industrial groups: terminal manufacturers, printer manufacturers, systems houses making newspaper composition systems, and software companies. All of these groups have experienced high levels and widely different patterns of complex product innovation, but, of course, may not be representative of issues encountered in other technological fields. (One area of possible difference between technology-based industries is the importance of gov-

Figure 10–5
The Revised Strategic Focus Hypothesis

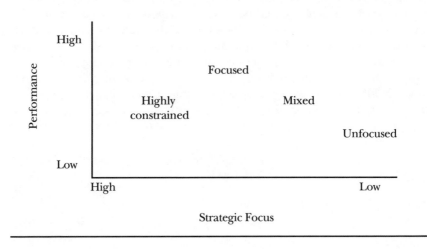

ernment regulation, discussed in Chapter 9 in regard to biomedical products.) Rather than conduct a telephone or mail-based canvasing of several hundred companies, which could yield only a superficial level of product strategy information, the students and I worked within company locations to examine closely 262 individual products, including 26 initial products, constantly probing for evidence to measure degrees of technological and marketing change.

All the companies studied are small or medium sized. We gathered extensive product and financial data for each company since its founding year up to the time of the interviews. Usually the founder-CEO or senior technical and marketing managers provided the information, requiring on average about four hours in each company. As might be expected in these volatile industrial sectors, a broad range of success in overall performance is encountered. The largest company in the sample has sales approaching $200 million, while several of the smallest companies have become bankrupt since the time of data collection. The range conforms to my attempts throughout the overall entrepreneurship research program to develop sample compositions that include both successful and failed companies, allowing clearer differentiation of policies that worked from those that did not.

Relying on careful joint determination with interviewees, all products are assigned technological and market applications newness scores of 1, 2, 3, or 4, based on the four-level typologies explained earlier. The level of newness is measured relative to all product development activities undertaken by the firm before the specific release of a given new product. There-fore, the base against which both technological and market newness are measured grows with each successive product. Then three different quanti-tative indices of product focus are calculated for each company: techno-logical focus and market applications focus, looking separately at each dimension of interproduct change; and overall product focus, generated by combining the two separate measures.[1] These indices are based prima-rily on the average level of change for each company's products, and secondarily on a measure of the consistency of change (a simple math-ematical variance). Overall performance for each firm is assessed by using company sales data, the most readily available and consistent performance indicator that relates to the entire product history of this sample of compa-nies, normalized (to produce a time series of current sales per year of company existence) and then averaged over the life of the firm.

Company Data

The processed data for the 26 firms are shown in Table 10–2, including rank orders for each company in regard to Technology Focus, Market Focus, and Overall Product Focus for all the firm's products, and Sales Performance. The companies range in age from four to 17 years and in sales from $150,000 to $167 million in the last complete year prior to data collection.

Figure 10–6 shows the distribution of levels of technological and mar-ket innovation among the 236 subsequent products from the 26 firms. The frequency of 55 percent first level marketing changes, almost twice that of second-level changes (29.8 percent), is somewhat as expected. However, the relatively greater number of major enhancements to minor improve-ments on the technological scale underscores the degree of rapid and substantial technological change embodied in these companies' products. While the issue of strategic focus in R&D and marketing activities still remains to be explored, the data show the firms as strong technological achievers. Self-assessment by the firms as to the levels of technological accomplishments in the first products of these firms, with respect to the state of the art of the industry at the time of product introduction, also reflects a high level of asserted technical aggressiveness. Table 10–3 presents the majority claim for the first products as being "highly distinctive", with only three of the 26 assessed as a "major breakthrough", no doubt due to modesty on the part of many of the entrepreneurs! In fact, Tushman and Anderson (1986) show that technological "discontinuities" which destroy the competences of an existing industry are typically introduced by new firms, opening up a product class to a wave of new entrants (p. 460).

Statistical Results

The principal statistical results relating to the strategic focus hypothesis have already been suggested by the various rank order tabulations in Table

10–2, where company rank according to sales growth performance can be compared visually with the ranks according to degrees of technological, market applications, or overall product focus. For example, the top-ranked company according to overall focus of its products, case S (a dot-matrix printer manufacturer), ranks second in the sample for performance (now exceeding $200 million in annual sales). The second-ranked company in product focus, case I (a producer of complex composition systems for large newspapers), ranks third in performance. Conversely, the last-ranked

Table 10–2
Companies Ranked by Focus Indices and Performance

Company	Description	Tech-nology Focus	Market Focus	Overall Product Focus	Sales Perfor-mance
A	Airline reservations terminals	6	12	12	17
B	Electronic funds transfer terminals	12	19	16	19
C	CAD/CAM and medical imaging terminals	11	11	10	10
D	Infrared factory control terminals	24	26	26	26
E	Hand-held process control terminals	10	17	13	23
F	General purpose terminals	1	14	3	4
G	Lottery systems terminals	18	4	4	6
H	General purpose terminals	9	13	15	11
I	Newspaper composition systems	4	10	2	3
J	Newspaper composition systems	15	22	21	18
K	Newspaper composition systems	25	9	19	9
L	Graphics composition systems	21	24	22	16
M	Image scanners	13	25	23	21
N	Color photocomposition systems	16	15	17	5
O	Dot-matrix printers	5	7	5	1
P	Color ink-jet printers	20	18	18	13
Q	Letter-quality impact printers	23	21	24	14
R	High-speed line printers	22	5	14	8
S	Dot-matrix printers	2	6	1	2
T	Mainframe spreadsheet programs	8	8	6	12
U	Graphics programs for microcomputers	17	1	7	20
V	Mainframe data base management system	7	3	8	7
W	Unix data base management system	26	23	25	25
X	Mainframe data base management system	19	2	11	15
Y	Language compilers	14	20	20	24
Z	Microcomputer Unix operating system	3	16	9	22

Figure 10–6
Distributions of Levels of Technological and Market
Change in All Products in Entire Sample ($n = 236$ products)

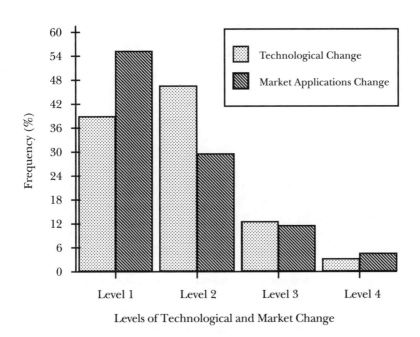

Levels of Technological and Market Change

company in overall product focus, case D (a producer of infrared factory control terminals), is also last in performance. Perhaps panic reaction to imminent company failure causes a flailing lack of product focus, rather than the other way around. No doubt strong positive feedback relationships exist between focus and performance. Many less clear matches are also

Table 10–3
Distribution of First Product Technology
($n = 26$ companies)

Technology Relative to Competition	Frequency	Percentage
Undistinctive	2	8
Somewhat distinctive	8	31
Highly distinctive	13	50
Major breakthrough	3	12

shown in Table 10–2. While exact matching in rank order would be beyond belief, a clear pattern does emerge. Based on Spearman rank-order correlation coefficients, shown in Table 10–4 (where 1.0 represents a perfect match), the products' overall focus is demonstrated to couple closely with the overall growth performance of the companies.[2] An independent analysis of 42 computer firms (Feeser and Willard, 1990) confirms the significance of product focus in achieving high growth.

Separating the overall product focus measure into its two-dimensional components permits examining the importance of each direction of product change. Technology development is assumed to be a less difficult resource to manage for relatively small technology-based firms than their market applications resources. Many high-technology firms achieve outstanding technical feats, but often fail to demonstrate comparable ability to implement effective sales programs for their products. One reason for this is likely to be the engineering backgrounds of most entrepreneurs of technology-based companies, as shown in Chapter 3, as well as the relative lack of marketing people within the founder groups. Similarly, new technology development may often be achieved with a relatively small number of talented engineers. Whereas the implementation of sales programs for the market applications of new products requires participation of many groups of individuals, some of whom are usually external to the company. The range of activities needed includes end-user documentation, development of marketing materials and advertising, implementation and maintenance of sales programs, and the creation of effective product support mechanisms. By implication, then, market applications diversity should be most difficult for a small technology-based firm to manage effectively, even more so than the development of multiple core technologies. The data analyses

Table 10–4
Correlations of Overall Product Focus and Performance:
Technology and Market Applications Combined
($n = 26$ companies)

Product Cluster	Number of Companies	Rank Order Coefficient	Statistical Significance
Composers	7	.943	.05
Printers	5	.900	.05
Software	7	.750	.05
Terminals	8	.881	.01
Entire sample*	26	.646	.01

*The entire sample was sufficiently large to permit calculation also of a Kendall rank coefficient = 4.151, with significance = .0005.

indicated in Table 10–5 support this hypothesis. In three of the four clus-ters, and confirmed by the product sample as a whole, product focus for the market applications dimension correlates more strongly with perfor-mance than technological focus. The stronger correlation of performance with high levels of relatedness in product market applications than with technological focus is also supported by an analysis of variance.[3]

The importance of strategic focus in products is reinforced by the absence of statistically significant relationships between overall company performance and the technological aggressiveness of first product launches (see Table 10–3 for the underlying data). Nor does company performance correlate with the rate of new product releases per year. Each of these areas is postulated by others as a possibly critical determi-nant of success for the technology-based company. (Feeser and Willard, 1990, confirm no relationship between technological pioneering and high growth for their sample of computer firms.) In fact, the additional analyses provide specific support to the arguments on behalf of focused product innovation. For example, no significant relationship exists be-tween the technological aggressiveness of the first and subsequent prod-ucts, indicating that firms that begin with technical leaps are able to exploit their advantages through a continued set of minor improvements and timely major enhancements, rather than a series of continuous am-bitious technological jumps. The innovation intensity of the product strategies (as measured by the mean rate of technological change em-bodied per year in the company's new products) correlates negatively and significantly with sales performance for the entire sample, bolstering the concept that somewhat lower levels of average technological change per year are preferable.

Table 10–5
Comparison of Product Focus in Technology and
Market Applications (Spearman Rank Correlation
Coefficients) ($n = 26$ companies)

Product Cluster	Technology Rank Coefficient	Market Applications Rank Coefficient
Composers	.028	.886
Printers	.800	.600
Software	.428	.679
Terminals	.429	.786
Entire sample*	1.986	3.828

* Kendall rank correlation used for entire sample.

Strategic Conclusions

Companies that historically show product strategic focus perform substantially better over extended periods than companies that implement multiple technologies and/or seek market diversity. A quick telephone follow-up with the sampled firms shows that this hypothesis is still on target. The ten top companies in terms of product focus have an average product-related sales level of approximately $56 million. This contrasts sharply with the bottom ten companies, again ranked in terms of product focus, whose average sales are approximately $3 million. The research demonstrates that managing wide-scale product diversity is, at the least, a most difficult endeavor for the small- or medium-sized technology-based company. Conceivably, larger firms or less technology-dependent ones might be better able to handle greater product-line diversity, although the strategic advice of "stick to your knitting" (Peters and Waterman, 1982) and earlier studies of diversification (Rumelt, 1974) and acquisition strategy (Ravenscraft and Scherer, 1987) also generally support the importance of focus for large companies.

The data do evidence, however, an "inside limit" to the strategic focus concept. Not surprisingly, companies that show a total lack of technological aggressiveness, undertaking only minor improvements to their core technology, do not perform as well as companies that over time make major enhancements to their core capabilities. This is true also in regard to better performance being achieved by firms that seek market expansion through steady introductions of new functionality, moving toward related customer groups, and adding distribution channels to exploit fully the available marketplace.

Notes

1. The mathematical formulae used for calculating the three focus measures are shown here, as calculated for each company. First, the focal point for each dimension and their combination are calculated. Next, the variance for each dimension and their combination are computed. Finally, the overall strategic focus index are determined for each dimension and for their combination.

Focal Point

$$FP(T) = \frac{\sum\limits_{i=2}^{N} |\Delta T_{P_i}|}{N-1}$$

$$FP(M) = \frac{\sum\limits_{i=2}^{N} |\Delta M_{P_i}|}{N-1}$$

Variance

$$V(T) = \frac{\sum\limits_{i=2}^{N} |\Delta T_{P_i} - \Delta T_{P_{i-1}}|}{N-1}$$

$$V(M) = \frac{\sum\limits_{i=2}^{N} |\Delta M_{P_i} - \Delta M_{P_{i-1}}|}{N-1}$$

$$FP(TM) = \frac{\sum\limits_{i=2}^{N} \left| \Delta T_{P_i} \times \Delta M_{P_i} \right|}{N-1} \qquad V(TM) = \frac{\sum\limits_{i=2}^{N} \left| \Delta T_{P_i} - \Delta T_{P_{i-1}} \right| \times \left| \Delta M_{P_i} - \Delta M_{P_{i-1}} \right|}{N-1}$$

Strategic Focus:

$$SF(T) = FP(T) \times \sqrt{V(T)}$$

$$SF(M) = FP(M) \times \sqrt{V(M)}$$

$$SF(TM) = FP(TM) \times \sqrt{V(TM)}$$

where: FP = focal point; T = technology dimension; M = market applications dimension; TM = combined technology and market applications dimensions; ΔT = level of technological newness; ΔM = level of market applications newness; N = total number of products; p = a product; V = variance; and SF = strategic focus.

2. These results are reaffirmed with statistical confidence through a variety of additional analyses, including: (1) tests of the sensitivity of the findings to possible shifts between rank orders of pairs of firms within each cluster, (2) recalculation of the combined focus measure using alternative formulas, and (3) computation of the performance index based on nonnormalized sales data.

3. To carry this out the 26 companies are divided into four groups in terms of their measured degrees of focus upon each of the two dimensions. The four quadrant data are shown in Figure 10–7. The F-statistic for the overall model is most significant, at 21.699, with effective probability of error of 0.001. The F-statistic for the technological dimension is 3.640 (p=.07) and 8.254 for the market dimension (p = .009), with no significant interaction effect observed between the two variables.

Figure 10–7
Sample Divided into Product Strategy Quadrants

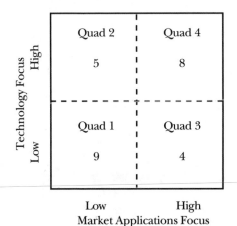

References

D. F. Abell. *Defining the Business: The Starting Point of Strategic Planning* (Englewood Cliffs, NJ: Prentice-Hall, 1980).

D. F. Abell & J. S. Hammond. *Strategic Market Planning* (Englewood Cliffs, NJ: Prentice-Hall, 1979).

P. A. Abetti & R. W. Stuart. "Product Newness and Market Advantage", *High Technology Marketing Review*, 1 (2) (1987), 29–40.

A. D. Chandler. *Strategy and Structure* (Cambridge, MA: MIT Press, 1962).

R. G. Cooper. "The Dimensions of Industrial New Product Success and Failure", *Journal of Marketing*, 43 (Summer 1979).

R. G. Cooper. "New Product Strategies: What Distinguishes the Top Performers?" *Journal of Product Innovation Management*, 1 (September 1984), 151–164.

R. G. Cooper. "Overall Corporate Strategies for New Product Programs", *Industrial Marketing Management*, 14 (August 1985), 179–193.

H. R. Feeser & G. E. Willard. "Founding Strategy and Performance: A Comparison of High and Low Growth High Tech Firms", *Strategic Management Journal*, 11 (1990), 87–98.

S. C. Johnson & C. Jones. "How to Organize for New Products", *Harvard Business Review*, May–June 1957, 49–62.

J. Ketteringham & J. White. "Making Technology Work for Business", in R. Lamb (editor), *Competitive Strategic Management* (Englewood Cliffs, NJ: Prentice-Hall, 1984).

M. H. Meyer & E. B. Roberts. "New Product Strategy in Small Technology–Based Firms: A Pilot Study", *Management Science*, 32, 7 (July 1986), 806–821.

M. H. Meyer & E. B. Roberts. "Focusing Product Technology for Corporate Growth", *Sloan Management Review*, 29, 4 (Summer 1988), 7–16.

T. J. Peters & R. H. Waterman. *In Search of Excellence* (New York: Harper & Row, 1982).

D. J. Ravenscraft & F. M. Scherer. *Mergers, Sell–Offs, and Economic Efficiency* (Washington, D.C.: The Brookings Institution, 1987).

R. P. Rumelt. *Strategy, Structure, and Economic Performance* (Boston: Division of Research, Harvard Business School, 1974).

M. L. Tushman & P. Anderson. "Technological Discontinuities and Organizational Environments", *Administrative Science Quarterly*, 31 (1986), 439–465.

CHAPTER 11

Super-Success

What does it take to go beyond the better and become the best? How does a technology-based firm that has achieved some degree of success go on to the realm of "super-success"? This chapter tries to identify the strategic actions needed through an intensive investigation of high-technology companies in the Greater Boston area that had already survived for at least five years and had attained sufficient sales to be deemed by many as successful. The evidence supports the notion that to find super-success most high-technology firms must transform themselves toward a marketing oriented strategy.

Perspectives from Prior Strategy Research

Strategic aspects of high-technology companies are perhaps the least developed area of academic entrepreneurship research. The studies carried out thus far go beyond the nonstrategic correlates of success discussed in Chapter 9: demographic and personal characteristics of the entrepreneur, venture capital and other financing considerations, sales/ marketing activities of the young firm. For example, the analyses of product strategy described in Chapter 10 fit into this overall strategic dimension. Other recent strategic research on technology-based entrepreneurial firms are of several types: overall corporate strategies or marketing strategies; organization structures; decision making processes; and executive influences. Tushman and Romanelli (1985) present a useful overview of many of the issues treated, linking the disparate literatures on organizational evolution, executive leadership, and strategic reorientation.

Romanelli (1987) and Eisenhardt and Schoonhoven (1989) note the persistence of early strategies among high-technology firms. Sandberg (1986) finds that early-stage entrants in an industry need to employ different strategies from later-stage entrants. Smith and Fleck (1987) find some high-technology firms lacking explicit long-term plans but behaving consistently as highly specialized niche-market players, trying to preserve founder financial control. Slevin and Covin's (1987) comparison of high- and low-technology industries identifies few significant differences in competitive tactics of high- and low-performing firms and concludes that effectiveness of implementation rather than tactics alone may explain performance.

Miller and Friesen with Mintzberg (1984) characterize 32 percent of their sample of large organizational transformations as "entrepreneurial revitalizations". They find that new CEOs in those firms pursue "new market opportunities, ... become more aggressive and innovative in dealing with competitors and more imaginative in meeting the needs of customers", increasing both "proactiveness and product-market innovation" (pp. 133, 134). Bahrami and Evans (1988) argue that high-technology entrepreneurs try to design organizational structures that "emphasize fluidity and flexibility, while retaining cohesion across interdependent functional and technological activities" (p. 3).

Eisenhardt and her co-authors (1989; Bourgeois and Eisenhardt, 1987, 1988; Eisenhardt and Bourgeois, 1988) discuss decision styles and consequences in "high velocity environments", demonstrating that "fast decision making" can be carried out with good use of data and careful consideration of alternatives. Tushman and his students find that new CEOs (and other top managers), when also accompanied by multidimensional strategic change, lead to performance improvement in these companies (Tushman, Virany and Romanelli, 1985, 1987). Furthermore, top management characteristics "shift over time—in their early years, hiring a large portion of executives with engineering expertise and shifting their recruiting emphasis over time toward sales and marketing" (Virany and Tushman, 1986, p. 264). The research by Tushman and colleagues, in particular, relates closely to the point of view as well as some of the findings that are described in this chapter.

In Chapter 6 I indicate that within a few years of their founding many technology-based firms begin transitional evolution from a primarily inward orientation focused on internal technical inventiveness into more balanced operations, increasingly devoting their attentions to customers and market. I now hypothesize that in search of ultimate success the technological enterprise must complete this transformation. It can no longer be primarily an exploiter of its technical origins and hopefully continuing strengths. It must become a servant of its customer's needs, practicing what might be regarded as true marketing oriented management. This is consistent with Peter Drucker's classic perspective: "Marketing is ... the whole business seen from the point of view of its final results, that is from the customer's point of view" (Drucker, 1973). Of course, technological innovation must continue to play a key competitive role for the still relatively small firm, differentiating it from its larger rivals in providing product performance in servicing its customers' priorities. This broad strategic hypothesis is explored in the data analyses now reported.

Methodology and Measurements

Sample Development

To investigate this broad hypothesis of the need for complete market oriented transformation my graduate students and I carefully developed a sample of 21 Greater Boston firms in two high-technology Standard Industrial Codes (SIC), electronic computing equipment (3573) and medical instrumentation (3811), that were from five to about 20 years old and that had already attained sales of over $5 million. The 21 participants and the 13 that also met the study criteria but declined to participate are listed at the end of this chapter.[1] A three-person interviewing team conducted structured interviews at all 21 participating firms, gathering data from an average of four persons in each firm, including the CEO plus vice presidents of marketing, finance, and corporate development or their equivalents.

Degrees of Success

Figure 11–1 shows the revenue distribution of these firms. At the time of the study, with the median firm thirteen years old, six had already achieved current year sales of over $100 million. No significant differences were found here or on any other key variable between the clusters of firms from

Figure 11–1
Revenue Distribution of Sampled Firms (n = 21 companies)

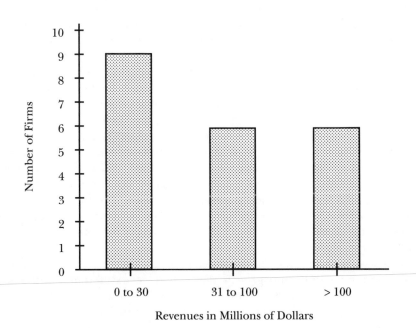

Revenues in Millions of Dollars

the two different electronics oriented SICs. No correlation exists between the age of the firms and their sales revenues. Yet, size alone is not a sufficient measure of success. Overall worth of the firms, as determined perhaps by their public stock valuation, might have been desirable as an indicator of success, but unfortunately was not adequately available.

Presurvey participants, as well as the interviewed corporate executives, agree that although the personal goals of the technological entrepreneur vary widely as a function of the individual's motives, the financial goals of the high-technology firm are generally seen as twofold: growth and profitability. Both are essential to maximize the shareholder's short- and long-term returns. The growth dimension is measured using the annual average compound increase in sales. Profitability is measured as the average annual return on equity (ROE).[2]

Some of the analyses require "standards" for the degrees of growth and profitability that constitute success. Although some firms hope for and even realize higher growth, and other firms target lower growth, the consensus of the presurvey and the company interview participants is that 30 percent growth is a good target that, with effort, can be controlled and managed. Similarly, they perceive 15 percent return on equity as a successful standard of continuing profitability over a several-year period, although returns in any given year, especially of high inflation, might be considerably higher.

These success measures divide the participating companies into four clusters, as shown in Figure 11–2. Type 1 includes the seven most successful firms, with both high growth (averaging 60 percent) and high return (averaging 25 percent). The five type 2 and 3 companies are reasonably successful, but more on one dimension than on the other. Based on the criteria established for this special study of super-success, the seven type 4 firms are not successful overall, with both less growth (20 percent) and negative average returns (–9 percent). Two privately held companies in the sample could not provide adequate financial time series for the several-year averages used in setting up this matrix, but their omission does not appear to introduce any additional bias in the data. Thus, although all 21 companies have reached at least $5 million in sales, assessing their success in terms of growth rate, essentially a projection of their likely future development, and their return on equity, a measure both of financial solidity and potential returns to their stockholders, one third of the sampled firms appear more to be "survivors" than successes. Statistical analyses of the firms' current sales against these financial success measures show no significant correlation.

The logical approach to developing a single simple measure of success is to combine ROE and growth rate. Their very different mean values would weight the growth parameter too heavily if just a simple addition is used. Based on the data, success is, therefore, defined as:

Figure 11–2
Success Matrix for Sampled Companies (n = 19 companies)

Growth

	> 30%	< 30%
Return on Equity > 15%	+ +: Type 1 7 firms Average Growth: 60% Average ROE: 25%	+ –: Type 3 3 firms Average Growth: 23% Average ROE: 21%
< 15%	– +: Type 2 2 firms Average Growth: 57% Average ROE: 10%	– –: Type 4 7 firms Average Growth: 20% Average ROE: – 9%

$$\text{SUCCESS} = 2\ (\text{ROE}) + \text{GROWTH}$$

where ROE and Growth are expressed in percentages. Experimentation with several alternative equations for defining success, using other weightings of return on equity and growth rate, supports the use of this approach.[3] Throughout the rest of this chapter I use this measure of success to look more closely at these firms.

In Search of Success

In their drive for corporate success high-technology entrepreneurs manage a complex process of continuous growth and change. They develop and deploy strategies and tactics affecting all aspects of their firms, sometimes thoughtfully and deliberately, sometimes by happenstance. I explore here the actions of the sampled high-technology companies in four broad areas: marketing, technology, finance, and human resources. Before closing I also examine the special problems that are linked to the ouster or resignation of the founder/CEO and/or to related critical events affecting overall corporate transformation.

The Marketing Side of the Technological Enterprise

In "Marketing Myopia" Levitt observes "that the top-heavy science-engineering production orientation of so many electronics companies works reasonably well ... because the companies are in a position of having to fill, not find markets; of not having to discover what the customer needs and wants, but of having the customer voluntarily come forward" (1960). This may indeed still be true during the early days of a technological firm's existence, when the enterprise is bringing new technology to the marketplace to serve new needs or to better serve old ones. As the company grows in sales, satisfies its initial niche, and begins to encounter competitors, it tends to face a substantially different market. The CEOs and various vice presidents interviewed among these 21 firms that have been around for an average of thirteen plus years identify a number of critical issues that have emerged in the last decade that challenge their future success. These concerns include:

1. Shorter product life cycle
2. Increased competition, both domestically and especially internationally
3. Difficulty in maintaining state of the art technology in all areas of business
4. Decreased product differentiation
5. Shift to nonengineering customer base
6. Problems in maintaining a growth atmosphere in their companies

To evaluate the marketing perspectives and activities of these high-technology companies, the CEO and/or the vice president of marketing/sales in almost all cases provided the data reported later on issues related to: corporate objectives and growth strategies, market planning and research, market change, and product line structure. I assess each area statistically against the overall financial SUCCESS measure, using five year compound averages for the ROE and GROWTH variables.

Corporate Objectives and Growth Strategies. Corporate objectives define a firm's business domain, generally in market-related terms such as market share, growth in sales, or profitability. All but one of the companies studied identifies sales growth as one of their key objectives. Lack of growth leads to loss of entrepreneurial engineers and managers, leaving the door open to technical obsolescence, a fate second in fear only to bankruptcy in the eyes of the executives interviewed. Real growth estimates for the firms, adjusting for their estimates of inflation, range from zero to over 50 percent per year and are shown in Table 11–1. This growth objective correlates significantly with SUCCESS ($r = .488$, $p = .05$).

All but three of the companies are attempting to pursue some form of integrative growth strategy. In a *horizontal integration strategy* the company

Table 11-1
Real Growth Objectives
($n = 21$ companies)

Real Revenue Growth (% / yr)	Number of Firms
0–10	4
11–20	4
20–30	0
30–40	6
40–50	4
>50	3

seeks ownership and/or increased control over its competitors in one broad product line that meets increasing portions of its customers' needs. Only six of the 21 firms claim they have an aggressive policy of horizontal integration, and yet use of this policy correlates significantly ($p = .10$) with overall success. Of course all 21 of the companies seek continuing growth through product line and/or market expansion. When asked to prioritize their sources of future growth, four companies anticipate their primary growth as coming from new products through acquisitions, eight firms plan mostly to develop new products internally, and nine companies intend to emphasize expansion of their existing markets.

Many more of the firms follow *vertical integration strategies*, attempting *backward integration* by seeking increased control over their suppliers, and/or *forward integration*, by trying to gain ownership or control of their distribution systems. Conventional wisdom dictates that high market share businesses tend to be more vertically integrated. They "make" rather than "buy" their components and they attempt to control their products' access to their customers. Yet, the firms in the sample are clearly split on adopting vertical integration strategies, as shown in Figure 11-3. Only five of these high-technology companies are following aggressive backward integration policies, while eight of 21 are pursuing aggressive forward integration policies. These two approaches are quite different in orientation and different outcomes might be expected.

The research reveals two related strategic motives for integrating backwards. Some firms perceive that the number or quality of certain key suppliers is declining. Other firms feel that a certain component is the key to their competitive advantage. The perceived benefits from backward integration include tighter quality control and decreased effort in maintaining vendor relations. In many areas of the computer field final product performance is seen as especially dependent on critical components or semiconductor chips. As these are increasingly supplied by vertically integrated

Figure 11–3
Growth Strategies: Vertical Integration Policies
($n = 21$ companies) (Level 1: No Integration; Level 2: Some
Integration; Level 3: Aggressive Integration)

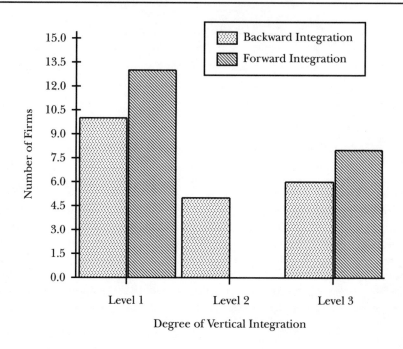

overseas competitors, local companies have growing concerns about availability and price of supplies in the future. On the negative side, however,
backward integration ties up more of the firms' assets in bricks and mortar
and makes them more dependent on a particular technology. Utterback
(1987) has argued that in an age of rapid technological changes this may
be more of an albatross than a blessing. The statistical analyses support this
negative view of backward integration, those companies with lesser degrees
of backward integration achieving greater success (.05).

Forward integration, on the other hand, allows a firm to seek higher
value-added and correspondingly higher profit margin products through
closer ties with the ultimate user. It requires that a firm's products be
differentiated from its competitors', whether achieved through product
characteristics, service, or image. Not a mere coincidence, forward integration reflects a marketing oriented perspective of getting closer to the customer, whereas backward integration tends to focus on a technological
point of view of securing the technical base for building a product. Consistent with my overall hypothesis that market oriented transformation leads

to ultimate company success, the statistical results show that aggressive forward integration is closely tied to financial success (.05).

Most of the firms see the external strategic environment as undergoing dramatic change. For example, thirteen of the companies sense both the number and activity of their competitors as increasingly relative to five years ago; only four firms see competition as diminishing, with four perceiving a stable competitive setting. Interestingly those companies that see increasing competition are also likely to be more successful (.05), suggesting that competition is attracted to rapidly growing market opportunities. This argument is supported by the close link between company growth in revenues and an increasing sense of competition ($r = .539$, $p = .02$).

In line with this shift is repeated testimony that sales used to be much easier to make. Several executives attribute this change to more buying of total solutions, placing higher emphasis on cost–benefit relationships. As shown in Figure 11–4 the companies believe that their competitive advantage five years ago was primarily in the area of technological innovation and product quality. Today, these same firms see their competitive advantage as having shifted toward price–performance and customer service. This should not be construed as a lessening of the importance of high-technological

Figure 11–4
Shift in the Perception of Competitive Advantage
(Level 1: Technical Innovation, Quality;
Level 2: Price/Performance, Service)

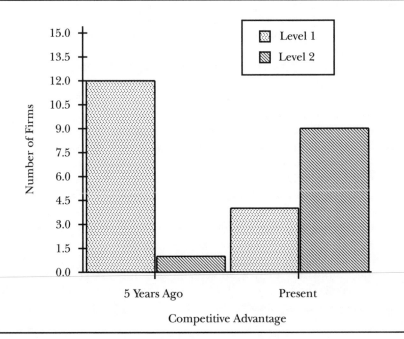

quality but rather as a signal that today's customers have a wider selection of products and are, thus, becoming more interested in price and service. No doubt this shift in company perception of customer priorities also reflects the aging and growth of these firms over the past five years, beyond the initial market niches they may have "owned" outright into more general competitive arenas. This phenomenon is indicative of a market with declining product differentiation, such as occurs in the mature stages of most product life cycles, more representative of today's situation for these high technology firms than for their condition in years past. Some firms moan that their products have moved to commodity status.

All but one of the companies have developed written strategic plans, featuring a wide variety of thoroughness of coverage as well as a wide disparity in the length of the planning horizon. This contrasts with Smith and Fleck's (1987) observation of no written plans in British high-technology firms, perhaps due to their generally younger and smaller status. For the 20 firms preparing plans, four durations were observed: one year (three firms), three years (seven firms), five years (nine firms), and ten years (one firm). Self-assessment of the adequacy or accuracy of the plans is hard to elicit from the company executives. When pushed for replies, about half think their strategic planning is very accurate; the other half are divided evenly between thinking their plans are grossly inadequate, admitting that their performance surpasses their most optimistic predictions, or offering no comment. The quality of these twenty strategic plans was not independently measured and no correlation is found between the firms' planning horizon and their computed success measure. As part of my overall research program I did carry out a separate pilot longitudinal research study on the strategic planning of eight other technology-based firms. That study demonstrates little correspondence between initial expectations and what actually occurs during the life of a new high-technology firm and no correlation between planning quality and later overall company success. However, Chapter 7 does point out strong links between good initial business plans and a company's ability to get venture capital funding.

Market Planning and Research. In a more specific vein, technology-based companies in the sampled fields of semiconductors, computers, and biomedical instrumentation have widely accepted the concept of the product life cycle as characterizing the distinct changes in the sales history of their products. Except for a few products, such as semiconductors, most of those interviewed see the maturity phase of their product's lives as extremely short, with the decline phase setting in with a rapid drop in sales. Logically this product change environment might be expected to engender strong appreciation of the need for market planning, a process of market oriented goal setting, competitor analysis, strategic positioning, market/product opportunity analysis, and associated programs, budgets, and controls. The interviews divide the sample firms into three levels of adoption of market planning:

1. An informal system of discussions among top management;
2. A formal planning system tied in with the sales forecasts and budgets; or
3. A formal system integrated with the strategic planning process of the firm, with formality measured crudely by whether the plans are committed to paper.

Note from Figure 11–5 that less than 50 percent of the companies coordinate their marketing activities with their strategic planning process. As expected, firms with shorter product life cycles do less market-strategy integration, but not significantly so.

In the five firms at level 1 with regard to market planning, often the only input is from the CEO. Interviews suggest that they maintain a "we know best; the customers do not know what they want" attitude. As no written plan ever appears, the companies argue they can move with the state of the art of technology, and are not tied down by inflexible plans. Level 2 companies make some effort to incorporate data from their employees about future product needs. Generally level 2 executives speak as if planning involves managers all the way down the line. In essence, however,

Figure 11–5
Formality of the Market Planning Process
(Level 1: Informal Discussions; Level 2: Formal Planning
Tied to Sales Forecasts; Level 3: Formal System Integrated
with Strategic Planning)

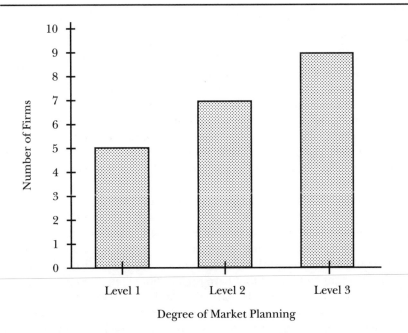

Degree of Market Planning

the market plan is generally written by one person, usually the VP of sales, with inputs from a few key salespeople. Little evidence suggests that research and development or engineering or manufacturing personnel are involved. Level 2 plans tend to be generated once a year, with no formal system for interim updates to reflect new competitor data or economic trends.

In contrast the planning process in level 3 companies usually involves people representing each of the key functional areas: R&D, engineering, manufacturing, marketing/sales, finance, and personnel. Their market planning is an integral part of the three- to five-year strategic plan and is updated yearly. A formal system, such as special forms/reports or departmental meetings, encourages employee participation throughout the organization. Periodic meetings to review any new material or revised information are held regularly on a monthly or quarterly basis. Those companies that do formal market planning, as capstoned by level 3 practices, are generally higher performers by my SUCCESS measure ($p = .10$) than those that do not. Companies that are experiencing a declining product life cycle have significantly stronger focus on market planning (.10), probably propelled by pressures from the market changes.

Not surprisingly adopting market planning also strongly correlates ($r = .725$) with formalizing market research in the companies. Yet most of the companies have not wholly committed themselves to rigorous practices of understanding who are their customers and what they need. Six firms carry out no or only informal market research. In some ways, despite their age and size, they hold tightly to the belief that superior technology products sell themselves. Half of the sample, ten companies, purchase outside data on market size, growth potential, and/or industry trends. Some of these companies assert that they are using the purchased data as phase one in their formulation of a more elaborate market research scheme, as financial resources permit. Others in this group suggest they had only acquired the data to see what information their competitors might have used in their goal setting. Only four companies have instituted internal market research departments, believing that their own groups are best suited at keeping up with the rapidly changing technological and market environment. Some report that purchased reports are often outdated as soon as they are published. The more formal the market research, the more successful the company (.10).

Consistent with these specific aspects is the more general issue of whether differences between marketing and sales are appreciated in the firm. To probe this the interviewers examined whether marketing was identified on the organization chart and whether a separate marketing budget could be found, distinct from a sales budget. The firms divide into three levels: eight companies that have no identifiable marketing activities— only sales; nine firms in which marketing and sales activities function together, with no separate personnel or budgets; and four companies that have separate marketing and sales organizations, which usually report to one vice president. As might be expected the more that firms separate the marketing role, the more they engage in both market planning and market research (.01) and, more importantly, the more they succeed ($r = .417$, $p = .10$). The unanswerable question is whether these symptoms of market point of view contribute to company success, or do they merely result from the success. With growth of organization size some elements of functional separation might come about purely as a way to manage larger numbers of specialists, and might not reflect managerial attitude. I have heard successful technical entrepreneurs cursing about the continuing escalation of marketing staff in their own companies, as if they can't do anything to alter the situation.

All of these dimensions of formal marketing—separate organization, formal planning, and formal market research—also correlate significantly (.05) with increasing intensity of competition, perhaps the driving force behind adopting a heavier marketing presence.

Market Change and Product Line Structure. As high-technology firms grow most begin to serve distinctly different market segments of customers with distinctly different needs. As Chapter 10 indicates this poses severe problems for the young technology-based company that is usually better able to absorb technical change than market oriented change. Figure 11–6 shows the range of market segmentation in the 21 sample firms. Most of the firms

Figure 11–6
Market Segmentation ($n = 21$ companies)

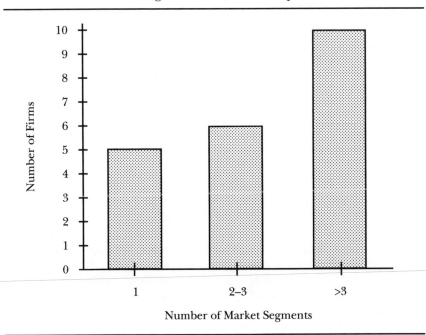

Number of Market Segments

under $150 million in sales concentrate on several market niches, each in the range of $5 to $20 million in sales per year, in the oft-repeated hope that the narrow markets will cause the "big guys to leave us alone".

Logically the number of market segments served might be assumed to relate to some extent to the number of main product lines that the company produces. A main product line consists of a family of products with the same technology base, that perform basically the same function, but perhaps for somewhat different applications or customers. For example, a word processor designed for analytical laboratories (one market segment) is in the same product line as a word processor designed for office workers (a second and different market segment). Limiting the definition to main product lines that are more than one-year-old and account for at least 5 percent of revenues, only five firms have more than three product lines, eleven have two or three, and five firms still have only one main product line. To my surprise the number of products and the number of markets do not relate statistically. Indeed, the number of products most strongly relates only to the age of the firm ($p = .05$), suggesting product line proliferation as the firm gets older, a problem given the desirability of focus shown in the previous chapter.

One change that occurs as the companies age is in their customer base. Earlier in their lives most of these companies' revenue streams tend to be dominated by original equipment manufacturers (OEMs). But increasingly a shift toward end-users takes place. Eight of the companies still rely primarily on OEMs, with four firms now focused directly on the other extreme, the end-user. The bulk of the sample, nine companies, now serve a mix of OEMs and end-users. The shift is requiring different sales techniques (no longer engineers just selling to engineers) and increased emphasis on quality control issues such as debugging (new customers being less tolerant of equipment operating problems). Several firms say the combination of these changes has led to higher sales costs and increased R&D spending, the latter claim being supported significantly (.05) by the data.

None of the elements described in this section—the number of market segments or product lines, or the customer base—relates statistically to the financial success of the companies.

Technology and Product Development

The sampled firms are all high technology product developers and manufacturers. Their roots are usually technological as is the training and experience base of most of their founders. To learn more about the technical aspects of these firms, the interviewers gathered data on research and development expenditures as well as on the process of new product development. The R&D data proved difficult to use, given the wide variance in reporting practices of the companies. Some firms do much of their R&D on government contracts, which often does not show up in the R&D figure quoted in their annual reports. Different practices in allocating field and manufacturing research also cause concerns. In the end the only useful measure seems to be whether the absolute R&D spending over the past five years has been increasing (13 firms) or decreasing (eight firms). Satisfying to me is the analytical result that companies with increasing R&D expenditures are the most successful ($r = .66$, $p = .01$), but the dilemma of separating cause from effect is especially marked here.

The areas of concentration of R&D spending fall into three clusters:

1. Redesign of existing products (minor changes such as in physical attributes);
2. Different versions of existing products (technical changes such as range or alteration of specs so that the product can be used by a new market segment);
3. Totally new product concepts (new technologies utilizing a firm's expertise, such as a move from computer graphics to robotics).

Figure 11–7 shows how the firms are distributed in regard to the thrust of their R&D expenditures. As Chapter 10 leads us to anticipate, this orientation of R&D spending by itself does not relate to overall financial performance. That chapter documents the need for strategic product focus, rather than just high R&D expenditures, for the young technical firm

Figure 11–7
Primary Thrust of R&D Expenditures ($n = 21$ companies)
(Level 1: Redesign of Original Features; Level 2: Different
Versions of Same Product; Level 3: Totally New Products)

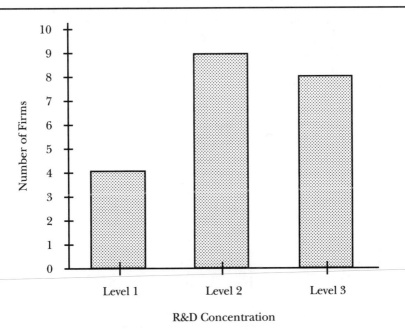

to continue its growth. The new data on the 21 older and successful firms continue to rebuff the assumption that technological companies ought to obsolete their base businesses with wholly new products. Those companies that are emphasizing totally new products serve significantly more market segments and have a higher degree of forward integration, both forces demanding new products to satisfy new customers.

The final elements of product development that seem of strategic interest are the sources of direction and control. Who decides what products the firm develops? A focus on the informal process rather than on organization charts determined the existence of two key control points: (1) the person or group who controls new product ideas and (2) the one who controls the new product development process. In each case one of three players has been dominant: top management, usually the CEO and an occasional right hand person; the key engineering decision maker; or the key marketing decision maker. In several companies the key decision maker is not the vice president or department head but rather a senior engineer or the "old timer" in the organization. The "idea control" dimension emphasizes the "valve" that permits or restricts the flow of new product ideas to the company from both internal and external sources. The "process control" dimension emphasizes the source of resolution of conflicts when product development negotiations between departments stalemate. At both stages the role is informal and had to be discovered through open discussion with the interviewees. In line with my overall hypothesis on the need for market oriented transformation these concepts led to documentation of the distribution of the firms along a loosely defined six-stage progression from total technologically oriented product development (stage I) toward wholly marketing oriented product development (stage VI) within a technological firm. Table 11–2 depicts this progression and the breakdown of the sample firms.

Half of the firms are still in stages I or II, with the CEO alone or

Table 11–2
Stages in New Product Control
($n = 21$ companies)

Stage	Idea Control	Process Control	Number of Firms
I	CEO	CEO	3
II	CEO	ENG	8
III	ENG	ENG	3
IV	ENG	MKTG	3
V	MKTG	ENG	2
VI	MKTG	MKTG	2

together with engineering in full control of the new product development cycle. The other firms respond that new product ideas and process control are shared roles between engineering and marketing. Probing into the informal structure elicited where the dominance is. Firms in stages I and II are clear in identifying (and not at all apologetic about) the strong control exercised by their CEO. Both in terms of influence on the idea generation process ($p = .05$) and the control over the product development process ($p = .01$), those firms with greater marketing orientation have achieved greater success.

Financial Management

An analysis of overall capital structure of the companies was carried out by averaging each firm's financial ratios and then computing averages for the firms in each of the four type performance categories displayed in Figure 11–2. Table 11–3 presents those data for our 19 sampled companies along with equally weighted industry averages for 25 electronics firms available from the Compustat Annual Industrial Tape.

Comparison of the total sample with the industry averages reveals two general trends. The sample of Greater Boston high-technology companies have a significantly higher degree of financial leverage than their industry counterparts, as indicated by all three measures of leverage in the table: long-term debt/equity, total debt/equity, and current ratio. This runs counter to the conventional wisdom that young technological firms should be financed mainly by equity (Brealey and Myers, 1981, p. 394). This apparent contradiction may be resolved by the second observable trend, that the Boston high-technology sample operates far more profitably than the

Table 11–3
Financial Ratios for High-Technology Companies

	Type 1 (+ +)	Type 2 (+G, – ROE)	Type 3 (– G, + ROE)	Type 4 (– –)	Total High Technology Sample	Electronics Industry Average
Sales Growth (GROWTH)	0.60	0.57	0.23	0.20	0.39	
Return on Equity (ROE)	0.25	0.10	0.21	–0.09	0.10	0.05
Standard Deviation ROE	0.17	0.21	0.26	97.18	35.93	
Long-Term Debt/Equity (LTDE)	0.77	0.48	1.30	0.19	0.60	0.50
Total Debt/Equity (TDE)	2.2	1.5	2.6	2.6	2.3	1.07
Current Ratio (CR)	2.5	2.1	2.8	2.4	2.5	3.17
Times Interest Earned (TIE)	17.7	10.3	6.4	4.0	10.1	5.22
Return on Assets (ROA)	0.20	0.15	0.20	0.11	0.16	0.04
Earnings/Sales (E/S)	0.07	0.04	0.07	0.00	0.04	0.01

industry, whether measured by times interest earned, return on assets, or earnings/sales, even including in the sample the seven firms that were judged to be not successful. This profitability generates a high level of confidence for investors who are considering lending money to these firms.

Those firms with higher return on equity are especially likely to be highly leveraged through debt, as reinforced in Table 11–4 where the high-technology sample is divided to reflect the differing returns on equity. What is not clear is the cause and effect. Are firms with higher debt to equity ratios able through leverage to increase the returns to their stock-holders? Or are more profitable firms better able to obtain debt? Corporate managers reassured us in discussion that an iterative process is at work. Initially the high-technology firm is essentially equity funded until some profit record is established, at which time investors become willing to debt finance the company, enabling debt-based investments in further growth of sales and profits.

The executives also revealed that the choice of capital structure is not simply based on the financial capability of the company. Indeed, it often reflects personal convictions, also a contradiction to the modern financial theory that asserts that all managers have the same utility function and accept risk to the extent the stockholder is compensated for it. Some entrepreneurs simply do not want any debt, feeling their high degree of business risk does not allow them also to accept the increased risk associated with financial leverage. Some, like Neil Pappalardo in Chapter 1, feel rather moralistic that a policy of no debt or only short term debt forces an appropriate pay-as-you-go situation in which present management cannot encumber future management's actions or responsibilities. Others match the theoretical expectations and have sought to obtain the maximum leverage possible to increase the potential returns to equity owners. Note that Smith and Fleck (1987) observe that British entrepreneurs seek to minimize outside equity to maintain control, a behavior also frequently observed in U.S. technological entrepreneurs.

Only one of these 21 hardware firms has ever issued a dividend, reinvesting all their earnings in anticipation of future growth and the capital gains treatment of an increased stock value. Successful software and service

Table 11–4
Return on Equity (ROE) and Use of Debt

	High ROE		Low ROE	
	Type 1	Type 3	Type 2	Type 4
Long-term debt/equity	0.77	1.3	0.48	0.19
Total debt/equity	2.2	2.6	1.5	2.6

companies, however, both types omitted from the sample, often generate high cash flow without comparable need for equipment and inventory investments, and therefore may follow quite different financial policies. The elimination of preferential treatment of capital gains occurred after the data collection and, therefore, does not figure into the behavior recorded here.

Unfortunately, despite these interesting differences between the capital structure of the high technology companies studied and the electronics industry as a whole, no consistent relationships between capital structure variables and overall corporate success are uncovered except for the rather obvious ones mentioned earlier. Financial success strongly correlates to return on equity, return on assets, earnings/sales ratio, and times interest earned, all more-or-less in line with the definition of the success measure. The strongest correlation of the financial variables to success is the negative correlation ($r = -.71$, $p = .01$) with the standard deviation of the return on equity ratio. This is not really managerially interesting: Less successful firms somewhat obviously have greater variability over the years, generating both positive and negative returns in various years. In contrast, the most successful companies produce only positive returns from year to year, resulting in much lower variability and lower calculated standard deviation.

Overall Corporate Development: The Human Side

The continuing search for critical success factors brought us to a number of softer dimensions on the human side of the firm: the role of the board of directors, the overall management of human resources, and the evolution of the CEO and the senior management team. The importance of the board of directors to the success of the firm receives comments in just about every interview. Its significance to high technology companies is only now beginning to be recognized in the academic community (Rosenstein, 1988). The interviewers collected data on board size, composition, role and changes in these. The typical board has six or seven members and all have both inside and outside members. As shown in Table 11–5 the outside board members most heavily include someone from the financial community, someone from business, with consulting, academia, and law filling the other positions. Although the variety of backgrounds on the boards is fairly consistent among the firms, the ratio of outside to inside members varies widely, splitting about evenly into three clusters: 30 to 50 percent outsiders, 50 to 80 percent outside members, and more than 80 percent. After the fact most of these entrepreneurial CEOs claimed that they considered their boards successful when they have always been active and influential, although one must wonder whether the thirteen original CEOs no longer in those positions would agree with this assertion. Again, however, the key finding is that no statistically significant relationship exists between any of the board-related variables and the company's success.

Table 11–5
Representative Professions on
the Boards of Directors

Profession	Number of Firms
In-house	18
Finance	
Venture capital	11
Banking	5
Private investing	3
Insurance	1
Business	
Company-related	8
General	5
Consulting	12
Academia	6
Law	2

Despite turnover rates of 20 to 25 percent for both engineers and managers being cited as usual, and despite strongly voiced appreciation of the criticality of creative and entrepreneurial human resources in these organizations, I cannot find any personnel-related factor that correlates with success among these 21 companies. Their perquisite packages are different from each other in varying ways, their performance assessment and reward systems also show unique qualities, and their orientation approaches for new employees reflect differing levels of formalization. But none of the differences relates quantifiably to financial success.

"Founder's Disease"

Prominent in the folklore on entrepreneurship is the identification of "founder's disease", the "diagnosed" inability of the founding CEO to grow in managerial and leadership capacity as rapidly as the firm's size and further potential grow. In many cases the "disease" is "cured", for the firm at least, by the founder stepping down or being ousted by outside board members who inevitably replace the founder with a new CEO, usually brought in from the outside. One colleague claims that the stock price inevitably jumps upward when such an ouster occurs, but I have no evidence of this. Eisenhardt (1989) reveals several instances of CEO changes in her sample of eight microcomputer companies. Tushman et al. (1985, p. 308) indicates that ten of sixteen high-performing minicomputer firms had experienced executive succession, but senior managers in addition to the CEO are included in that count. My 21 company sample contains much evidence of the instability of the chief executive's job in high-technology firms. Only eight of the founders are still in office at the time of the data collection, and those primarily in the older firms for some inexplicable reason. Ten companies had changed CEOs once and three of the companies had more than one change of CEO.

This sample, the overall studies of high-technology entrepreneurs, and my past 25 years of experience, especially since being involved in venture capital investments, reveal many different founder's diseases. All of them are terminal in the presence of outsider control of the board of directors, or even when multiple founding partners get frustrated with their initial co-founding leader. The easiest version of the malady to identify among high-technology entrepreneurs is the problem of the technical founder who really is inadequate to run the company. In these somewhat obvious cases the person who started the firm, therefore, becomes its president, but is sufficient only to run a technical development program, which often occupies the first year or so of the company's existence. Frequently, even that initial product development is misdirected, lacking appropriate marketing inputs despite possibly state of the art technology. When initial product development is over, the new issues facing the company quickly overwhelm the founder.

A generalized version of that ailment affects the founder who is suitable for the first stages of company life but not capable of handling the later stages. This issue may arise in regard to sheer size of firm, with some founders able to manage effectively when they can be in personal touch with all people and activities of the firm. As the company grows and effective management requires measurement and control through information systems, some founding executives fall by the wayside. A variant of this disease occurs with the founder who is uncomfortable in delegating authority and responsibility, as discussed in regard to the Chapter 9 discussion of founders with high need for power. That entrepreneur may be effective in total personal control as long as the firm is relatively small, but falls apart if the company grows beyond his grasp. Alternatively, some founding presidents may do well when able to operate in the programmed or structured mode that is often dominant in the early aspects of the company, such as writing the business plan, carrying out systematic fund raising, moving forward on product development, introducing a first product to the market. Such persons may be frail when exposed to unprogrammed or unstructured decision requirements, such as responding to competition, handling an ineffective employee, turning around a bad situation, conditions likely to arise later in the company's history.

Very different symptoms are prevalent when the disease arises from growing disparity between founder views and outside board desires. This is most likely to happen when the company is not performing well or in accord with prior plans. Of course, some time must elapse postfounding for such a disparity to be created and sensed and the response undertaken. In attempting to rescue "their" company, the board replaces the founder with a subordinate or more frequently an outside executive. The company "sickness" leading to the action determines the specific character of the remedy, such as a more aggressive executive to carry out more rapid growth, a marketing oriented manager to shift from a "technology push" toward "market pull" approach, a controller to cut excessive costs. Indeed, Gorman and Sahlman (1989, p. 240) find this occurrence to be a regularity in the life of a venture capitalist: "The mean (in the statistical sense) venture capitalist has initiated the firing of three CEO/Presidents, or one CEO/President per 2.4 years of venture investing experience."

Statistical analyses of the 21 Massachusetts high growth firms show no relationship between the number of CEOs during a company's life and the overall company financial performance: The surviving original founder-CEOs as a group perform as well in generating even ultimate corporate success as do their replacement CEOs. How these surviving founders achieve their success is not knowable from the data. But the folklore clearly needs to be adjusted. The data indicate that for every two stories of a Steve Jobs being replaced by a John Scully in search of continuing onward and upward company performance at Apple Computer there is about one story of a Ken Olsen maintaining his founder-CEO role into the period of greatness of his firm.

Careful examination of the evolution and size of senior management similarly produced lots of interesting stories but no relationship with success of the quantified variables.

The "Critical Event" and Corporate Transformation

In analyzing the data an unanticipated phenomenon was noted that I label here the *critical event*. This is defined as a period of time during which a series of actions occur that bring about comprehensive changes in management structure, the financial, marketing, and planning processes, and eventually for many of the affected firms in their corporate success. Discovery of the critical event did not come about directly, but rather as a result of the "war stories" and discussions of how the firm got to where it is today. The critical event is distinctive because it is promulgated by outside stimuli, such as outside directors, outside management from an acquiring company, an act of God. Several different patterns are apparent. Most frequently the critical event happens when one or two of the outside directors on the board instigate an ouster of the CEO and then bring in an experienced manager to become the new chief executive. For example, this occurred to Arthur Rosenberg at Tyco Laboratories, as discussed in Chapter 1. In five cases an acquisition triggered the changes—either the company acquired another firm, or was acquired, or in one situation reacquired itself. In one company an act of God is responsible for the critical event when the founder-CEO was killed in an accident. Of great interest is that Tushman et al. (1985) also identify *environmental discontinuities*, brought about by external forces, as promulgating executive succession and strategic reorientation in minicomputer firms.

Applying this definition of critical event to the data collected, 16 of the 21 firms are classified as having critical events. Company scenarios preceding the critical events are far from uniform, including cases of steadily increasing revenues, erratic revenues, stable and/or declining revenues, all patterns that are also observable in the companies that do not experience the so-called critical events. In terms of company age at which the events occur, Figure 11–8 shows the event timing as a percentage of total company age to date, indicating a roughly even distribution between 30 to 80 percent of the companies' current age.

Three of the six largest companies in the sample, all over $100 million in sales, have experienced critical events; all of the five smallest companies, all with sales under $10 million, have also experienced critical events. Only one of the five oldest companies in the sample, all older than fifteen years, has experienced a critical event. All five of the companies in the sample that have not yet experienced a critical event still have their founder as CEO, as well as three of the sixteen critical event firms. Most of the new CEOs, brought in from the outside, have strong marketing backgrounds; most of the displaced CEOs have heavy technical backgrounds.

Figure 11–8
Timing of Critical Event Relative to Present Company Age

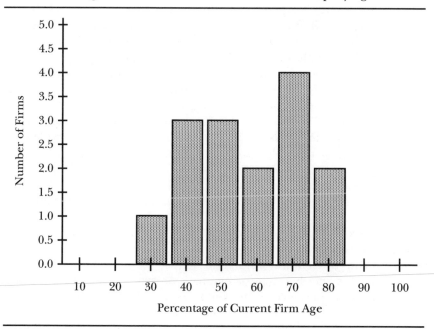

Within one to three years after every critical event a dramatic change in management structure is completed. Of the approximately 100 senior management positions traced in the 16 critical event companies, only seven of the positions have been filled by promotions from within since the event occurred. Part of this is no doubt due to the increased departure of managers and technical personnel after installation of the new CEOs.

Strategic planning processes tend to be installed where none existed before the critical event, gradually extending in planning horizon. For those companies that already had a planning process, changes occur through broadened lower level participation, extended forecasting periods, greater levels of detail, and more frequent monitoring and updates. Shifts in strategy tend to follow, most frequently changes in product line or growth strategy, often including increased horizontal integration.

Table 11–6 contains a synopsis of many of the typical firm's key characteristics before and after it underwent a critical event and the ensuing changes. Across the board restructuring and reorientation usually result, in many ways reflecting dramatic shift from a more technological to a more marketing-oriented strategic perspective. This matches Miller and Friesen's description of the "entrepreneurial revitalization" transformation: "A new CEO attempts to revive his enterprise by increasing innovation, pursuing new market opportunities, and devising more adaptive strategies" (1984, p. 133). Of most importance is that the occurrence of the critical event divides the sampled companies into two groups, with significantly higher financial success for the 16 event firms as a group ($r = .399$, $p = .10$). Along with this correlation are several supporting findings, including strong links between overall success and the creation of a new management team postevent ($p = .05$) as, well as the hiring of new senior management staff ($p = .02$). I will soon explore the implication that the event causes the company success.

Many of the managers interviewed suggest that a firm experiences two distinct phases of organizational development. Some refer to these as the shift from an entrepreneurial to an organizational phase. Post hoc the stages seem more to reflect the technology and marketing stages of a firm's life. The technological phase is the period when the firm is developing its core technology or technologies, from founding (as discussed in Chapter 4) through early successful product growth (as discussed in Chapter 10). The firm is learning the capabilities of its technologies, their applications, strengths and weaknesses. The horizons of discovery are primarily internal to the firm.

The marketing phase occurs for most firms after a critical event that shakes up the company or forces it to change. The firm does not necessarily lose its core of technical innovation. Rather the critical event modifies the company's somewhat singular emphasis on state of the art technology to include the importance of marketing as well. In all the critical event companies in the sample, control of the flow of new product ideas as well as the new product development process shifts toward marketing, but in no case does the product line shift to a new base technology. The horizons of discovery become predominantly external during this phase, secondarily internal to the firm. The shift in product development control can, however, become the leverage point for gradual erosion of technological competi-

Table 11–6
Characteristic Critical Event Phenomena

Factors	Before	After
Growth strategy	None/backward integration	Horizontal/forward integration
Market planning	Nonexistent	Formalized and integrated into strategic planning
Market research	None/bought but unused	In-house department
Control of new product development	CEO or engineering control	Marketing or marketing and engineering control
Separation of sales and marketing functions	No marketing or marketing combined with sales	Independently staffed marketing and sales functions
R&D concentration	Redesign, new product development, process optimization	New product development
Intensity of competition	1 to 3 competitors in one market	Many markets with one to many competitors in each
Market segmentation	Unsegmented or few	Segmented/hierarchical
Product life cycle	Up to 10 years	3 to 5 years
Number of products	1 to 3 main product lines	More than 3 main product lines
Backward integration	Frequent	None/infrequent
R&D expenditures	Varied patterns of change	Increasing over time
Customer base	OEM	Distributor and end-user
Planning horizon	1 to 5 years	3 to 10 years
Ratio of inside to outside directors	1:1	1:5
Management team	Founders	Managers hired from outside
CEO	Founder	Manager with marketing expertise

tiveness, marked in Table 11–6 by lack of appreciation in the R&D concentration of such elements as process optimization, despite almost uniform initial increases in R&D expenditures. The firms in which more than one CEO change occurs are too few to test whether later problems might arise eventually, following the initial critical event, demanding another altered focus resulting in and/or produced by a second critical event.

Within the sixteen critical event companies statistical analyses sought to differentiate the more successful from the less. They primarily reaffirm the factors described previously in this chapter, adding support to the initial findings. The only important new correlates of the more successful firms within the critical event subgroup are greater customer orientation toward end-users (.05) and shorter product life cycles (.10), both reflections of increased marketing sensitivity, and a CEO with a marketing background (.10), perhaps the explanation of that sensitivity.

Repeating the analyses for the subcluster of just the 13 CEO-change companies, all of whom are included in the 16 company critical event subsample, produces few important findings, with the key exception being that a marketing oriented CEO background ceases to be significant statistically. This says that a new CEO need not have a marketing background to be successful.

Combining this new finding with the more important earlier total sample outcome that the eight surviving original CEOs, with mostly technical backgrounds, perform as well as a group as the new replacement CEOs now sharpens the strategic question. Is a critical event truly necessary for producing super-success? Obviously for some of those firms, the critical event leads to failure, or at least to less than spectacular success. For many companies, however, super-success is indeed achieved after the critical event occurs. Some companies clearly achieve financial success without an externally induced critical change or without all the dimensions of transformation suggested earlier. And some of the two thirds that encountered critical change might have achieved stellar success without the changes, but we can never know this.

Four alternative strategic explanations for self-generated success (without a critical change event) seem plausible. (1) Some technology-based firms may have the right market orientation from day 1, and have no need to change. (2) Other companies might change toward the hypothesized needed marketing perspective through internal gradual evolution of their original management team. Chapter 6 clearly evidences that this evolution begins for many firms during their earliest years. Their CEOs may be the exceptions who do eventually make it from founding to greatness through their own personal growth and development. Other CEOs recognize their limitations and bring in and nurture strong and capable subordinates to manage the strategic transformation, while the founder-CEO reorients his own role as chief executive. I do not in any way suggest that my consulting firm, Pugh-Roberts Associates, is one of the super-successes. But I do note from personal experience that Henry Weil's evolution from first employee to eventual president of our firm, while I remained as CEO from the founding until we became a division of PA Consulting Group, exemplifies this intended internal progression of change. I believe that many other surviving founder-CEOs have consciously practiced corporate transformation through delegation and development. (3) Still other founders may have a technological orientation from the outset, sell largely technological advances perhaps primarily to OEMs, and build continually growing and profitable firms with that singular and unchanging strategy. Some might indeed even label that strategy as market oriented, once the very different needs of the specific OEM customer are properly taken into account. (4) Alternatively, Virany and Tushman (1986) demonstrate that "particularly visionary executives" lead the highest performing companies in the minicomputer industry, are not ousted from their roles as CEOs, and implement internal changes as needed to continue corporate growth and success. Their visionary leadership may have had little to do with either technological or market orientation per se. As Slevin and Covin (1987, p. 94) suggest, outstanding performance of the company may reflect "doing things right" rather than just "doing the right things". Unfortunately, neither the Virany and Tushman data set nor the eight company sample of no CEO change nor the five company sample of noncritical event companies is large enough to generate further meaningful explanations of company performance or to provide empirical support for these alternative pathways to greater corporate success.

Summary and Implications

This chapter attempts to advance the strategic understanding of high-technology firms by examining longitudinally a cluster of 21 Boston-area firms that had already survived for at least five years and had already attained at least $5 million in annual sales. The broad hypothesis that a market oriented strategic transformation is needed for super-success led to in-depth probes of many dimensions of managerial change. Backward integration toward self-sourcing a firm's components generally detracts from successful performance, while aggressive forward integration toward the firm's ultimate customers strongly correlates with success. Competitive advantages have shifted over time from technological uniqueness toward price/performance and customer service dimensions, no doubt reflecting both the aging and the growth of the firms, their markets and their core technologies. Formal market planning, integrated with strategy, formal market research, and formal organization of marketing are all significant factors in success. Most firms are gradually shifting from OEMs as their customers to end-users, with attendant complications in selling, servicing, and even engineering. New products are vital for these firms and increasing R&D expenditures as well as marketing oriented control of new product idea flow and the resultant development process all contribute significantly to financial success.

Financial managers of these high-technology firms generate higher financial leverage than their electronic industry counterparts, perhaps due to the greater profitability of the sampled firms. Dividends are almost never issued in these hardware companies, all earnings being reinvested toward future growth. No aspect of capital structure helps to explain overall corporate success.

Thirteen of the 21 founding CEOs had already been displaced by new CEOs, the replacements usually being brought in after some set of externally generated "critical events". But the eight surviving original founder-CEOs as a group perform as well as the replacement CEOs. The new CEOs tend to have marketing backgrounds, in contrast with the engineering backgrounds of the first-generation CEOs. The new CEOs dramatically transform their firms toward the marketing-orientation, frequently achieving corporate success by means of that transformation. No additional insights explain the comparable degree of success achieved by the relatively few super-successful surviving founder-CEOs. Their entrepreneurial accomplishments remain as a deep mystery for later researchers to fathom.

Notes

1. A number of prominent Greater Boston high-technology firms are not included in this sample primarily because their formal SIC code was not 3573 or 3811, or because the company was too old. For example, Digital Equipment Company, involved heavily in the overall entrepreneurial research program as an MIT spin-off, is omitted from this special analysis because it had been founded more than 20 years before the time of study.

A. Firms That Participated in the Research

Analog Devices	Dynatech Corporation
Analogic Corporation	GCA Corporation
Applicon, Inc.	GRI Computer Corporation
Block Research and Engineering	Haemonetics Corporation
Division, Bio-Rad Laboratories	Helix Technology Corporation
CL Systems, Inc.	Modicon, Division of Gould
Computer Devices, Inc.	Prime Computer, Inc.
Computervision Corporation	Semicon, Inc.
Damon Corporation	Silicon Transistor Corporation
Data Printer Corporation	(BBF Inc.)
Data Terminal Systems	Xylogics, Inc.
Datatrol, Inc.	

B. Firms that Declined to Participate

Alpha Industries	Entwistle Company
American Science and Engineering	Ionics
Cambex	Inforex
Centronics	Intertel, Inc.
Compugraphic	Micro Communications Corp.
CSP, Inc.	Sigma Instruments, Inc.
Data General	

2. A problem arose in calculating average return on equity for those companies whose cumulative losses had caused the equity balance to become negative. In these few cases the ratio of a negative return over a negative equity balance generates a positive computed number, generally a large one since the negative equity is usually small. To avoid the problem, return on equity for companies with negative equity is calculated by dividing average annual loss by the original equity investment.

3. Once the assembled data set from the company interviews was correlated with this measure of financial success, SUCCESS = 2(ROE) + GROWTH, the sensitivity of the correlations was checked with other weightings of the two components:

$$A. \quad 3.5 \ (ROE) + GROWTH$$
$$B. \quad 3 \ (ROE) + GROWTH$$
$$C. \quad ROE \ + GROWTH$$
$$D. \quad ROE \ + 2 \ (GROWTH)$$

These alternative weights were selected from inspection of the ROE and GROWTH data. The means of ROE and GROWTH are 10.78 and 37.68, respectively, with standard deviations of 28.2 and 27.3. The approximately 3.5:1 ratio of the means suggests equation A as an alternate formulation of SUCCESS. Further inspection of the data reveals one ROE data point to be an outlier. Removing it generates a new ROE mean of 14.97, with a new standard deviation of 17.10. The resulting new ratio of the mean ROE to the mean GROWTH is now closer to 3:1, suggesting equation B. Equations C and D give more weight to growth and also seem to be viable alternatives to test.

Fortunately, correlation analyses performed with these four alternative financial success measures agree in general with the results determined from the initial formulation of SUCCESS = 2 (ROE) + GROWTH, except for some shifting among the .01, .05, and .10 significance levels. Within these levels 22 of the 48 coded variables show statistically significant correlations with SUCCESS. In light of the small sample size, the slight changes among the significance levels are considered trivial. The consistent results with respect to all variations of the success factor add confidence to the use of the original success factor equation. In addition, SUCCESS also correlates significantly with each of its components, ROE and GROWTH, as well as with return on assets and profitability, as measured by the earnings/sales ratio, in the entire sample of 21 firms as well as in several smaller subclusters that were studied, lending still further support to this choice of a success measure.

References

H. Bahrami & S. Evans. "Stratocracy in High-Technology Firms", *I.E.E.E. Engineering Management Review*, 16, 4 (December 1988), 2–8.

L. G. Bourgeois & K. Eisenhardt. "Strategic Decision Processes in Silicon Valley: The Anatomy of a 'Living Dead'", *California Management Review*, 30 (1987), 143–159.

L. G. Bourgeois & K. Eisenhardt. "Strategic Decision Processes in High Velocity Environments: Four Cases in the Microcomputer Industry", *Management Science*, 34 (1988), 816–835.

R. Brealey & S. Myers. *Principles of Corporate Finance* (New York: McGraw Hill Book Company, 1981).

P. F. Drucker. *Management: Tasks, Responsibilities, Practices* (New York: Harper & Row, 1973).

K. M. Eisenhardt. "Making Fast Strategic Decisions in High Velocity Environments", *Academy of Management Journal*, 32 (1989).

K. M. Eisenhardt & L. G. Bourgeois. "The Politics of Strategic Decision Making in Top Management Teams: A Study in the Microcomputer Industry", *Academy of Management Journal*, 1988.

K. M. Eisenhardt & C. B. Schoonhoven. "Organizational Growth: Linking Founding Team, Strategy, Environment and Growth among U.S. Semiconductor Ventures (1978–1988)". Unpublished paper, June 1989.

M. Gorman & W. A. Sahlman. "What Do Venture Capitalists Do?", *Journal of Business Venturing*, 4, 4 (1989), 231–248.

T. J. Levitt. "Marketing Myopia", *Harvard Business Review*, October 1960.

D. Miller & P. H. Friesen with H. Mintzberg. *Organizations: A Quantum View* (Englewood Cliffs, NJ: Prentice-Hall, 1984).

E. Romanelli. "New Venture Strategies in the Minicomputer Industry", *California Management Review*, Fall 1987, 160–175.

J. Rosenstein. "The Board and Strategy: Venture Capital and High Technology", *Journal of Business Venturing*, 3 (1988), 159–170.

W. R. Sandberg. *New Venture Performance: The Role of Strategy and Industry Structure* (Lexington, MA: Lexington Books, 1986).

D. P. Slevin & J. G. Covin. "The Competitive Tactics of Entrepreneurial Firms in High- and Low-Technology Industries", in N. C. Churchill et al. (editors), *Frontiers of Entrepreneurship Research, 1987* (Wellesley, MA: Babson College, 1987).

J. G. Smith & V. Fleck. "Business Strategies in Small High-Technology Companies", *Long Range Planning*, April 1987.

M. Tushman & E. Romanelli. "Organizational Evolution: A Metamorphosis Model of Convergence and Reorientation", *Research in Organizational Behavior*, 7 (1985), 171–222.

M. L. Tushman, B. Virany & E. Romanelli. "Executive Succession, Strategic Reorientations, and Organization Evolution: The Minicomputer Industry as a Case in Point", *Technology In Society*, 7 (1985), 297–313.

M. L. Tushman, B. Virany & E. Romanelli. "Effects of CEO and Executive Team Succession: A Longitudinal Analysis". Unpublished paper, July 1987.

J. M. Utterback. "Innovation and Industrial Evolution in Manufacturing Industries", in Guile & Brooks (editors), *Technology and Global Industry* (Washington, D.C.: National Academy Press, 1987).

B. Virany & M. L. Tushman. "Top Management Teams and Corporate Success in an Emerging Industry", *Journal of Business Venturing*, 1 (1986), 261–274.

CHAPTER 12

Technological Entrepreneurship: Birth, Growth, and Success

This book has attempted to explain the origins of high-technology entrepreneurs and the causes of their successes and failures. Data were developed from over forty integrated research studies carried out over a twenty-five year period of time, covering many aspects of high-technology enterprise. This chapter reviews my findings, identifies what still needs to be learned, and contemplates the future of technological entrepreneurship.

Birth

New technological enterprises are given birth by engineers and scientists deciding to become entrepreneurs, developing or adopting the technological bases for their new firms, investing or raising the financial resources needed to get going. Chapters 2 through 5 cover these elements.

Becoming Technical Entrepreneurs
Though not quantified by Chapter 2 the complex environment of the Route 128 region surrounding Greater Boston clearly has had tremendous influence on many decisions to become an entrepreneur. The positive influence of MIT on its own faculty, lab employees and alumni has been particularly noteworthy in this regard. And the resulting early formation of high-technology companies from MIT and its surrounds has gradually produced a strengthening positive feedback effect on future prospective entrepreneurs. The associated infrastructure of venture capital and knowledgeable bankers, accountants, lawyers, as well as the technical industry support structure of specialty vendors, machine shops, contract assemblers has grown to be a vital aspect of persuasive influence on those considering entrepreneurship.

The most striking predecessor influence on technical entrepreneurs is their family background: Between half and two-thirds share the "entrepreneurial heritage" of having a self-employed father. This is even evident

339

from the anecdotal cases presented in Chapter 1. Compared with the general population or more importantly with the "controlled population" of engineers and scientists employed at the key "source labs", these research findings significantly demonstrate one or more of the effects on the entrepreneurial career choice of home environment, parental nurturing, inherited "need for independence", and/or lessened fear of starting a business. Although large numbers of founders are "first born sons", they merely match in percentage the first borns among their employed former technical colleagues. When they do not come from homes of the self-employed, other elements of family background, notably an achievement oriented religion, have some measurable influence on the entrepreneurial choice, leading to relatively more Jews and fewer Catholics as high-technology entrepreneurs.

Technological entrepreneurs are generally young when they start their first company, averaging in the mid-30s, but with wide age variation from mid-20s up to 65 all encountered within the data. They are younger than their average technical colleagues in the organizations they leave. Few are over 40 when they begin their first companies, including none of the examples in Chapter 1. This relative youthfulness reflects a compromise between all the typical "preparation time" for becoming an entrepreneur— university education, graduate school, a decade plus of work experience, including nearly a year to get their specific firm underway—and the clear long-term "itchiness" to be their own boss, often identifiable even during childhood. Almost a minimum is the completion of a four-year technical undergraduate degree, with the representative entrepreneur adding a master's degree, usually in engineering rather than science.

Varied patterns of prior work experience also characterize the high-technology entrepreneur community from MIT and its environs. But the mean timing is 13 years, heavily dominated by experience in one key technical organization that I have labeled the source organization. In their prior work activities most of the entrepreneurs carry out applied development work, not research, focusing their creative talents and technical skills on real but advanced technical problem-solving projects. This reflects possible earlier inclination by these preentrepreneurs to do applied rather than theoretical work, and/or shows the positive influence that doing such work produces on the likelihood of becoming an entrepreneur. What is especially impressive is how good the entrepreneurs had been as engineers. The data significantly demonstrate that they dramatically outproduce their former technical colleagues in papers published, even more in patents filed and issued, and frequently in the performance evaluations of their former supervisors. Many had already risen to managerial levels before departing. I wonder whether more responsive organizations or a less encouraging regional environment might have led these productive technologists toward positions of senior leadership within their source organizations, instead of outward into entrepreneurial paths.

Beyond the indicated long-felt desire for their own business (remember from Chapter 1 that Morris can identify even childhood longings), our probes at discovering the technical entrepreneurs' personality and motives come away somewhat empty-handed. To be sure the founders attest to heavy orientation toward independence as well as a continuing search to meet and overcome challenges. Most express only moderate concern for financial rewards at their time of start, many becoming more financially oriented long after their companies are underway. Psychological testing reveals a wide range of entrepreneur motivations in regard to needs for achievement, power, and affiliation. Although the entrepreneur is more extroverted than his rather introverted technical colleagues, the high-technology entrepreneur still emerges looking more like an inventor than any other unique role.

The Initial Technology

The initial technology base of the Greater Boston high-technology firms comes primarily from the source organizations that had previously employed the entrepreneurs, especially from the employer at the time the new company was conceived. This is true for all the cases in Chapter 1. Yet this is not true for all high-technology start-ups. Thirteen percent of the new companies reflect no technology transfer from the prior source that my assistants and I had checked, although these new enterprises may have been dependent on other source organizations. (One deficiency of the studies is that I do not positively identify the key technological source for each enterprise.) It is important to state that few entrepreneurs leave their sources with a product that had been developed at work. The technology they move into their new companies is usually advanced knowledge, previously learned, and ready to be newly applied in a new setting.

The prior work that leads to the transferred technology is development, not research, indicating a double filter in effect: More development oriented individuals become entrepreneurs, and more development oriented entrepreneurs transfer technology immediately into their own firms. Longer years of work experience at a lab lead to more technology transfer to the new company, especially if the firm is started on a moonlighting basis, with the entrepreneur still employed full time at his prior job. A long delay between leaving a technology source organization and setting up a new high-technology firm dramatically diminishes the technology flow, with a four-year delay essentially fully eliminating technical influence of the prior work.

Personal characteristics of the entrepreneurs influence this technology base. Advanced education, but less than doctoral-level, increases technology transfer, but advanced age somewhat reduces it. Positive attitude toward their past employ relates to more technology ties, perhaps symbolized by Ken Olsen's warm feelings toward MIT Lincoln Lab. Indeed those who see an explicit opportunity to use some source technology in a business application are the ones who depart most quickly and who transfer the most technology to their new enterprises.

Initial Capital

The Greater Boston area is well known for its long venture capital history and large pools of investment funds. Perhaps surprisingly the data clearly demonstrate that most technology-based companies start with money provided by the founders themselves (74 percent of the firms) and in quite small amounts (almost half with less than $10,000). This dominant personal source and small initial amount have been more-or-less the same for two decades. Most technological entrepreneurs start on a shoestring. When larger capital is provided at the outset, outsiders do provide it, but more likely so-called angels, informal individual investors or small groups, than venture capital firms are the seed or first round financiers.

More money is directly invested, and also raised from outsiders, by larger teams of co-founders. Statistically these larger amounts and the founding teams are tightly coupled with having specific plans for their businesses, and focusing on a product from the start, rather than doing consulting or contract development.

Transition and Growth

Market Perspectives

The high-technology company is often a puzzlement as to its initial and early market focus. Many founders testify they were unsure as to who would become the initial customers for their products and services. Often they change their minds within the first weeks of "being in business". Forty percent of the firms started solely by offering their personal technical skills and know-how for consulting and/or contract research and development work with large government and industrial organizations. Another 20 percent began with the intent of combining this personal technical service work with some form of hardware or software product. A large fraction of this 60 percent somehow hoped to define new products and markets after their businesses got going, producing some income for their founders in the meantime. Only 40 percent of the companies began with a clear product/market focus, and then usually with no more than past technical exposures as the basis for market assessment. Chapter 6 documents this initial market orientation (or its lack) and the early transition as the new technological enterprise moves forward. Chapters 7 and 8 focus on financial growth dimensions of the company.

Within the first few years of existence 22 percent of the companies dropped consulting/contracting as their sole business and developed some product base, leaving just 16 percent of the technological firms surviving without a hardware or software product orientation. Marketing oriented activities become more widespread as the companies develop, but over half still lack a formal marketing department after five to seven years of life. Multifounder firms, described earlier as having more product orientation and more capital at initiation, are also more market oriented from their beginnings. This reflects in part the greater tendency of larger entrepreneurial teams to include someone experienced in sales or marketing. Entrepreneurial teams devote a larger fraction of founder time to marketing and sales than sole founders, with relatively less time going into engineering. They also establish a direct sales force more quickly, including more than the founder, carry out formal sales forecasts, and do analyses of potential markets, indicating in many ways their stronger focus on their customers.

Additional Financings

Not surprisingly, two thirds of the high-technology start-ups raise additional funds one or more times after their initial capitalization. Unfortunately, the data do not indicate how many more companies want but are not able to generate more funds, nor do they indicate how much more capital is desired by those that got some additional monies. Fund raising clearly takes much time, measured both in terms of overall effort as well as in elapsed duration, producing usually from several hundred thousand dollars up to several million dollars. The principal sources of later investments are typically different from the original investors, with personal funds and family and friends playing a very small role in follow-on funding. Private investors again have a large piece of the pie in both frequency and amounts of later investments but, more important, now they are joined by venture capital firms, the public stock market, and larger nonfinancial corporations, all of which are seldom the sources of initial financings.

Formal business plans are heavily used by the entrepreneurs as tools for raising capital, especially from financial institutions. Despite all the how-to books available on the subject, the research points out significant deficiencies in almost all plans investigated, with both marketing and financial plans especially weak and unsupported by data. The strength of the business plans as persuasion tools is evidenced significantly in the decisions to invest by venture capitalists. Follow-on studies, however, reveal little relationships between initial plans and later performance of these high-technology firms.

Wide variations in stated investment criteria exist among venture capital organizations (VCs) along many dimensions, including age of company, stage of its development, technical fields, and markets served. Almost all VCs do declare preferences to invest in teams rather than solo operations, and very few prefer seed round or even start-up investments. Remember from Chapter 1 that only Digital Equipment of the several cases described received venture capital investment as the first money in, and then at an extremely high price in terms of percentage of ownership. The research on actual investment decision making suggests that venture capitalists are seldom consistent with their stated preferences, for example, most investing at even later stages of new company development than they represent as their focus. The "funnel effect" is clearly evident at each venture capital company, with 1000 initial inquiries for funds being transformed typically into 100 to 200 that receive careful screening, producing ten to twenty enterprises in which the venture capitalist makes investments. But, fortunately for the persistent technological entrepreneur, one investor's turndown may well become another investor's financing, if sufficient numbers of potential investing individuals or funds exist in the geographic area to provide those options.

Going Public

The special financing decision of a public stock offering is taken by only a small fraction, perhaps 10 percent, of high-technology companies. These few firms apparently split 50-50 between going public at their foundings or early in their lives, often called "selling the sizzle", and going public after more substantial growth and performance, referred to in contrast as "selling the steak". Chapter 1 describes Tyco Laboratories' situation in going public early in its life, to secure its independence from its parent corporation, apparently selling its sizzle into a receptive stock market. Of examples mentioned specifically in this book, both EG&G and DEC went public at much later stages in their growth, with some substantial degree of success already established for both firms. The two clusters of public offerings are substantially different in company characteristics and underwriting process and costs, but overall important benefits accrue to both groups. The sizzle sales are generally by young small firms, which sometimes cannot secure an underwriter. When they do, they incur much higher direct costs paid out of underwriting proceeds while also suffering important dilution in their stockholdings from warrants issued to the underwriters. Occasionally their stock issues are withdrawn prior to sale, or terms modified, or stock only partially sold in nonguaranteed deals. In contrast, larger high-technology companies are better advised and prepared and clearly more in control of their decisions to go public. Most end up with high quality underwriters with national distribution capabilities.

But the after-issue differences between the small and large stock sellers are less strong. Longer term price appreciation of their stocks are more-or-less similar. Though larger high-technology companies perceive greater benefits of being public in regard to acquisitions of other companies and financial benefits to founders and employee stock option holders, the smaller companies often see the public capital as rescuing their corporate lives, a benefit they hold in high valuation.

Success and Failure

The last three chapters explore various possible explanations for success or failure of high-technology companies. Here is what I have learned from the research in terms of the key characteristics of the successful founders and of their firms.

Successful Founders

In some ways the most important finding about the technological entrepreneurs themselves is how few of their quantifiable personal characteristics relate to their later success. This should not surprise anyone. Indeed, what characteristics of any group of individuals predict later success? But meaningful data deny any statistical ties between birth order, family background, religious upbringing, parental occupation, even entrepreneurial heritage and the success or failure of the technology-based companies these individuals later create. Successful entrepreneurs are made, not born.

The "average" technical entrepreneur has achieved a master of science degree in some engineering field, and founds the "average performing" corporation. To be sure, this average performance is not bad, when compared with nontechnical firms. Chapter 9 shows that the actual failure rate of high-technology companies is only 15 to 30 percent over their first five years, a small fraction of the typical failures of nontechnology companies. Going beyond the master's degree to the Ph.D. leads to fewer, not more, successes, except in special cases. Ph.D. founders usually establish quite different and less successful firms than their less well-educated cohorts: Ph.D.-initiated firms are more often formed by sole founders, focusing on research and/or consulting activities, and with low initial capitalization. These dimensions all relate to and indeed influence less successful outcomes. The exceptions occur, in particular, in the biomedical field, now including biotechnology-oriented companies, where Ph.D. founders do outperform all others as a group.

The age of the entrepreneur at the time of company formation also does not relate to success, or to the raw amount of work experience, indicating that younger founders do as well statistically as older entrepreneurs (and vice versa). I indicated earlier that engineers who have risen into supervisory roles are more likely to become entrepreneurs. Now I further point out that those high-technology entrepreneurs who had served as engineering supervisors before setting up their firms do achieve significantly greater entrepreneurial success. This evidences either their greater talent, perhaps combined with managerial inclination, that led to their jobs as supervisors in the first place, or it reflects the useful human/market/technology insights gained from their time spent as supervisors, or both.

Although technological entrepreneurs are driven by many different sets of motives, the most successful have significantly higher psychological need for achievement. The high need for achievement founders create firms that grow on average two and one-half times more rapidly than their moderate need for achievement entrepreneurial cohorts. The very best high-tech performers combine high need to achieve with moderate need for personal power, no doubt providing leadership for their companies but with a participative style that enables others also to contribute to company success. High-technology entrepreneurs who have a high need for power typically become sole founders, are likely to run their firms in autocratic fashion, thereby preventing their subordinates from realizing their own ambitions in helping the company to grow. More targeted studies of the managerial styles of high-technology entrepreneurs and their resulting company outcomes are needed.

Finally and most important, successful entrepreneurs of high technology do not go it alone. They co-found their companies, the more the merrier. This team approach to building success works best when technological capabilities are balanced by sales and marketing experience in the founding group. Co-founder teams behave significantly differently from sole founders: at the outset (more capital, hardware product orientation) and throughout their growth (earlier and stronger marketing orientation from many perspectives, larger follow-on financing), all of their distinctions correlating significantly and independently with, and in my opinion contributing vitally to, company success.

Successful Firms

Beyond their founders' personal characteristics, successful high-technology enterprises have significant critical differences at their beginnings and as they mature.

At the Start. The unique *potential* competitive advantage of a high-technology new enterprise right from its start can be its technology. Advanced technology can be created by the technological entrepreneur, or often more easily can be transferred by that entrepreneur from his previous work environs. This permits a new firm to enter the market with a comparative advantage over its already established competitors, even those that are much larger. A new high-tech company is far less likely to be able to start with a competitive advantage in regard to marketing, finance, production capabilities, or other nontechnical resources. Therefore, it should come as no shock that a strong correlate of later company success is the extent to which the new firm enjoys a high degree of technology transfer from its prior source or incubator organization. When the source from which the entrepreneur emerges is an advanced technology laboratory, great opportunities exist for rapidly moving new ideas and skills that are at the leading edge of technology into the commercial marketplace. The MIT labs and academic departments from which many of the studied entrepreneurs had come did indeed provide much technology ripe for transfer at the very start of these companies. Similar rapid transfer occurs from government R&D labs as well as from numerous engineering departments of larger firms all around the Boston area. Delay in use of an advanced and attractive technology correlates significantly and negatively with company performance, the longer the delay the worse the eventual performance.

Another significant founding predictor of company success is a product orientation at time zero. Having an existing software or hardware product or one under development differentiates successful high-technology start-ups from their far less successful cohorts that begin with contract R&D or consulting efforts. My research base does not contain enough service-product companies to indicate their linkage to later success or failure, but my intuition and personal experience suggest that technology-based service firms also outperform companies that are primarily selling their personal technical capabilities.

More successful companies are founded with larger initial capital, but I am concerned here about the possible complications as to cause and effect. A group of prospective company founders and their initial ideas and plans constitute a basis for evaluation by potential investors, possibly before the entrepreneurs have put their own money into the company, as in the Chapter 1 examples of Digital and Meditech. A strongly positive evaluation is likely to lead to larger initial capital being raised. Do the attractive characteristics of the founders, which generate more capital, produce the eventual success or does their larger initial financing? Obviously both sets of influences are independently important, as the statistical research demonstrates.

The special situation of companies in heavily regulated fields like medical devices and pharmaceuticals needs mentioning. To the extent these firms engage in significant technological innovation at the start, the Food and Drug Administration regulations impose severe costs and delays before product income can be generated. Therefore, innovating biomedical firms are significantly more sensitive in their eventual results to having sufficiently large start-up capital to endure the long process of getting product approvals and bringing products to the market. This same finding may apply to other firms affected by strong regulatory controls.

A complex of start-up characteristics that turns out to be a critical success factor is the initial marketing orientation of the firm. This shows up in many ways, including the fact that high-technology companies with greater sales experience among their founders significantly outperform other enterprises. Entrepreneurial attitudes that reflect attentiveness toward their customers' inputs and awareness of their competitors' behavior, as measured by founder activities within their first few months of existence, relate significantly to both early and later growth and success. An initial or early marketing function in the firm also portends corporate success.

Finally, evident at the beginning of successful technological companies is a managerial orientation, reflected sometimes in the prior supervisory, managerial, or business experience among founding teams, shown also by an effort balance among technical, sales, manufacturing and administrative aspects of the new firm. Sensitivity to the company's cost structure and early recruitment of senior staff to fill primarily managerial needs also characterize the more successful performers.

Later Strategic Evolution. As the high-technology firm moves forward from its founding other more strategically oriented factors come into play as determinants of success. Critical is the maintenance of a "strategic focus" by the firm in regard to development of its later products and businesses. While undertaking a moderate degree of technological aggressiveness, successful companies do not engage in developing and applying widely diverse technologies and even more certainly avoid wide market variation. Significant success in product line development comes from steady advances in a company's core technology, exploited through market expansion based on regular introductions of new product functionality, movement toward related customer groups, and gradual addition of distribution channels.

"Super-successful" companies continue to strengthen their market orientation throughout their lives. Significantly they quickly add formal marketing techniques like sales forecasting and market analyses, carried out by formal marketing departments. More important, they aggressively forward integrate toward their ultimate customers, changing from selling primarily to manufacturers of original equipment toward selling and servicing their end-users. They shift their competitive advantages as they grow from initial technological uniqueness toward price/performance and customer service dimensions. The successful high-technology firm recognizes the importance of a continuing stream of new products and consequently invests substantial resources in research and development, but control of both the idea flow and the resultant development process indicates heavy marketing-orientation. The super-success has transformed itself from an initial technology-dominated strategy to an ongoing market-focused strategy that effectively utilizes technology as its underlying base.

In the process of super-successful growth and evolution two thirds of the founding CEOs of successful technological firms become victims of external events and pressures. Usually outside investors of these growth-seeking firms oust and replace the founder-CEO with an outsider with marketing background, who carries out the marketing transformation described earlier. Founding technical entrepreneurs survive under two very different sets of circumstances. Founders of less successful high-technology companies tend to survive, maintaining their personally dominant roles until they tire, retire, or sell out. Thus entrepreneur survival and company survival-but-not-success go hand in hand in many cases.

There is a second group of persisting technological founder-CEOs. They constitute one third of the super-successful high-technology growth companies—people like Bernie Gordon of Analogic, Bill Poduska of Apollo, Phil Villers of Computervision, Ken Olsen of DEC, all mentioned previously in this volume. Their companies, under their leadership from day 1, perform as well statistically as those super-successes that were founded by technologists but brought to fruition by replacement marketing oriented executives. The data do not permit further characterization of these unique founder-builders of high-technology enterprises. Perhaps they manifest an appropriate marketing orientation from the outset despite their technical trappings. Perhaps they grow and change personally as their companies evolve. Perhaps they find and follow a path toward success other than marketing oriented strategic transformation of their high-technology firms. The search for answers to these questions is left to other scholars and entrepreneurs.

The Future of Technological Entrepreneurship

What happens now? The U.S. economy is weakened and in decline. Japan, Singapore, Taiwan, and Korea are rapidly growing as centers of technology-based industry. Europe is experiencing increased political and economic consolidation, with prospects of heightened barriers to outside companies. Does this mean an end to high-technology entrepreneurship? The evidence suggests the opposite.

The Beat Goes On

The case histories of Chapter 1 are all well along in their life cycles, albeit with different degrees of success and size. In many ways their futures are certain, or at least circumscribed. They are based on various advanced technologies, ranging over optics, computers, materials, software, and electronics. Most high-tech firms are still young and many are just getting started. For them the future is totally uncertain. Yet I sense that for technological entrepreneurs what matters looks very much like what has already been observed earlier in this book. To illustrate let us examine the case of a new company, now just three years old, and in the newest of emerging technological fields, biotechnology. PerSeptive Biosystems is too young to "prove" anything, but its formative years again show the uncertainty and sometimes even turmoil that need resolution as a start-up finds its own direction and moves forward.

Noubar Afeyan and PerSeptive Biosystems, Inc. Beirut, Lebanon, is the starting point of the story. Noubar Afeyan was born there in 1962 in the Armenian section, the youngest of three sons. His father had had an architecture education in Bulgaria and worked at that profession in Lebanon during the day, but started an import/export business at night two years before Noubar was born. While Noubar was growing up, his father's business, now a full-time occupation, grew too and Noubar remembers well his childhood impressions of warehouses, the port, unloading ships, rounding up day workers by truck. His childhood was not exactly typical for a to-be biotechnology entrepreneur, with continuing outbreaks of violence eroding the quality of life. But in that polyglot community Noubar received a good education, becoming fluent in Arabic, Armenian, English, and French, while also developing a "street smart" sense. In 1975, the civil war in Lebanon had erupted again and Noubar's parents decided to move the family out of the country. They departed for Montreal, where cousins were already living. One indicator of the trauma they were experiencing is that the Beirut airport closed two days after the Afeyans departed, not to reopen for four years.

During their first year in Montreal, the boys entered Jesuit school while their father tried to start a manufacturing business. Months of frustration led to the elder Afeyan reestablishing his import/export enterprise, which five years later, in 1981, finally led into the manufacture of plastic-covered furniture and related products. Noubar, his brothers, and his mother actively worked at the business whenever needed.

Noubar thought about applying to MIT when he was finishing high school. But concern about leaving his then 86-year-old great-aunt, who had lived with the family throughout his childhood, led Noubar to enroll in McGill University, just one block down the street from his house. Choosing chemical rather than electrical engineering for his undergraduate major was influenced by his father's encouragement that chemical engineering might at least relate somewhat to his plastics activities. The first semi-serious discussions about starting his own company took place regularly with Noubar's undergraduate friend Dave Rich. Together they were going to create ARC—Afeyan-Rich Company, to undertake a variety of creative businesses but nothing real came of these many chats. Noubar's rather remarkable great-aunt, 97 years old in 1990, reminds him that he was always talking about running his own business even when he was a child.

Despite an attractive job offer from Dow Chemical Noubar decided, in 1983, to apply, and got accepted, to MIT's new Ph.D. program in biotechnology process engineering. His work primarily with Professors Daniel Wang and Charles Cooney exposed him to their growing networks of consulting relationships with large and small companies in the United States and abroad, as well as venture capital firms. Noubar generated lots of ideas, leading to impressive experimental findings and he published papers with several MIT faculty. But he had no interest at all in becoming a professor; he clearly wanted all along to be a key player in some company. One of the MIT faculty, Raymond Baddour, gave a seminar on all the alumni of the MIT Chemical Engineering department who had achieved outstanding success (several in association with Baddour's own multiple entrepreneurial efforts; see Appendix, Track 1). That seemed great to Noubar and he wanted to become part of that list.

In 1986, Afeyan enrolled in the New Enterprises course in the Sloan School of Management and had to prepare a business plan. He worked out a semiserious plan, without the financials, for one of his ideas on a protein purification system that he named CARE—Continuous Affinity Recycle Extraction—doing the project alone because he was afraid someone might try to take the idea and run with it. Noubar reports feeling that the Sloan School was entirely foreign to him—the jargon, costume, aspirations of business school students were almost alien to the engineer. Later that year he entered into negotiations with the giant Swedish firms Alfa-Laval and Pharmacia to turn that CARE idea into an actual company, but the deal evaporated when the two large companies had a falling out with each other. The negotiations were not without some lasting benefit; Noubar married one of the women Alfa-Laval had sent over to MIT during the discussions.

In spring 1987, as he was nearing completion of his Ph.D. dissertation, Noubar came to see me for permission to take my course on Corporate Strategies for New Business Development. I initially refused him, as he obviously lacked the formal prerequisite subjects. I told him that I doubted that he could write an acceptable term paper without lots of Sloan School background education. Afeyan's persistence overcame my stubbornness and in the end he prepared one of the best papers in the course, contrasting entry strategies into the biotechnology field of several major chemical companies.

At the same time Noubar had tired of trying to get his own ideas translated into the basis for a new firm. He reluctantly accepted the notion that he seemed to be hearing from everyone that he was too young and inexperienced and needed industry seasoning and was on the verge of accepting a job in industry. Fortuitously, Professor Danny Wang, his thesis chairman, introduced Afeyan to a much older experienced multicompany entrepreneur who was being pushed out by investors from his latest company creation. Noubar, then 24 years old, and the older entrepreneur, 60 years old and wiser from his several company start-ups, hit it off immediately, and by the end of one intensive day of discussions they had more-or-less agreed to start a new company in the area of biotechnology processing. On the very next day several people began warning Noubar that this relationship would not work out, that his partner was too inflexible and had a reputation of being tough to work with. Noubar had concluded that they were a good complementary match. Now he had the chance to translate his many ideas into a real company. Wistfully thinking back to that time, Noubar recalls, "I thought he knew his limitations and that therefore we'd be able to get along. I could create and promote and he could manage the technology development."

In August 1987, Noubar Afeyan became the first Ph.D. graduate from MIT's Center for Bioprocess Engineering, and he was already hard at work getting the new company underway with his partner. During October Afeyan brought me a copy of their first business plan, a proposal for launching Synosys Corporation as a rather generic developer and producer of biotechnology processing hardware, with one of Noubar's system ideas as product number 1. He invited me to become a director and an initial stockholder, but after reading the plan I declined. They hoped to enlist major corporations as their principal financiers and collaborators in a series of strategic alliances. Neither the plan nor its timing was great; the stock market's Black Monday occurred one week after I saw Noubar. Among other effects of the market crash, venture capital companies became more conservative and especially skeptical of new companies focused on capital equipment markets, such as Synosys. Despite the apparent problems Synosys was incorporated at the end of November, the two founders each owning half the company, with the bills being paid by loans to the firm from the older partner.

Months dragged by as they presented their plan to numerous companies and prospective   investors. In the meantime Professor Fred Regnier of Purdue University, one of the world's leading experts in separations technology, was recruited to become an advisor to Synosys and began working with Noubar on new approaches to porous materials for biotechnology separations processes. Noubar took a part-time job at MIT in January 1988, working as the technology transfer manager for the BioProcess Engineering Center under Danny Wang's directorship. Much more time was going to be needed to raise the capital to get going and they could use the income and the contacts the job would produce. They also rented a small office in American Twine Office Park, a converted old mill filled with MIT high-technology spin-offs, located just behind MIT in East Cambridge.

With all the critical feedback on their first business plan Noubar wrote a new one in March, aimed at venture capitalists not corporations as the potential funders and following to the letter the guidelines in Jeff Timmons' textbook, *New Venture Creation* (1985). This plan had a dual focus: the porous separations materials and a biotechnology processing hardware system, aimed at new product developers. Noubar started sending around the new plan and visiting venture capital firms. I received my copy in April 1988, read it and decided to visit the founders in their offices. As I entered I was offered a bench to sit on, just being hammered into completion by Danny Wang's son who was working there part time. The co-founders and I talked at length, especially about my feelings that the materials business was attractive by itself and that the hardware system was a confusing distraction. We also talked about the roles of the two co-founders and whether they would be willing to bring in a more experienced partner as CEO or Executive VP (reminiscent of AR&D's hesitation with DEC). Despite some reservations, upon return to my office I called the biotech specialist at First Stage Capital, a venture capital firm I co-founded and serve as a General Partner, and suggested he look into Synosys in depth. First Stage began working closely with Synosys, criticizing many aspects of their plans but also encouraging their overall efforts.

Danny Wang began to get more involved and technical progress was being made with the porous materials. But funding decisions dragged on. In June First Stage Capital turned down Synosys because of our unwillingness to fund the hardware part of the company. Noubar borrowed some money from his father and began to pay the bills; his partner was running out of funds and was looking for alternative employment in academia. In August, Wang and several friends invested $200,000 and suddenly things seemed a bit brighter. In September an agency of the Canadian government with which Noubar's father had good relationships indicated that it would be willing to invest $2 million in the company, provided that Synosys would move its hardware activities to Canada. Despite uncertainty as to whether Synosys should or would accept the Canadian funds, that change in prospects was enough to renew First Stage Capital's interest in financing the materials portion of the company.

By November, still without the major capital infusion needed to move ahead decisively, the Synosys team and its consultants developed "perfusion chromatography", an approach to protein purification that could produce a tenfold speed advantage over existing technologies. Finally, or so it seemed, in mid-January 1989 Synosys reached agreement with First Stage Capital and Noubar signed off on a detailed term sheet. Two days later his older partner suddenly announced he was through, claiming family pressures, thus beginning several months of fighting over stock, roles, compensation. Terry Loucks from Rothschild Ventures, who had met Noubar during the fund-raising period, agreed to come in as full-time chairman and CEO, with Noubar becoming president and chief technical officer. On April 1, hopefully not to be remembered later as April Fools' Day, the checks were signed for $1 million for one-third of the company, First Stage Capital being joined in the investment by Raytheon Ventures and 3i, a large British venture capital fund with an office in Boston. In search of a new identity the firm was reincorporated as PerSeptive Biosystems, trying to put the trauma of Synosys Corporation in the background. Two years had elapsed since the day Noubar and his now former older partner had agreed to start a company. The new team now featured Noubar Afeyan, Terry Loucks, and Fred Regnier from Purdue.

People began to be hired, several new Ph.D.s coming in from Purdue and MIT. Patents were filed on the concepts, the materials, and the designs for new processing equipment. The now named "Poros" materials began to generate amazing results in lab tests. In November the PerSeptive team stole the show at a technical symposium in Philadelphia, presenting several papers on the technology and its performance and generating instant sample orders from several large companies.

As of September 1990 PerSeptive Biosystems had 26 employees, a growing group of enthusiastic customers, and newly delivered checks from its second round of venture capital, $3.3 million from its original venture capital investors plus Venrock and Bessemer Securities, the funds needed for expansion of the company and for development of its second product line, an instrument system for automating use of the Poros materials in developmental applications. I asked Noubar what he wanted to accomplish now. "I want to create an analog of Hewlett Packard—tools for a new breed of engineers, bio-process engineers. If the industry grows the way it's expected to, we should be able to reach $40–50 million in sales in five to six years. By the way, becoming rich wouldn't hurt", he added. "There's nothing more motivating to succeed than having to beg for money to get started." Noubar Afeyan was 28 years old.

Compared with the cases presented throughout this book Afeyan shows strong continuity of the earlier patterns. Noubar's father was a professional and an entrepreneur, and Noubar gained much experience with the process of business development while he was growing up. After receiving his MIT degree he had worked only in an MIT lab before setting up his own company at a young age. In fact, Noubar really started PerSeptive on a part-time basis while he was still in graduate school, and his general interests in running his own firm stem from childhood. The technology was transferred directly from his MIT education and lab work as well as from Fred Regnier's work at Purdue. Initial funds came from his co-founder's savings, then from his family and friends, and only later in two large rounds of investments from venture capital firms. The company's initial focus is on its own internal product development. Beyond these characteristics of background and start-up that seem so similar to most previous high-technology entrepreneurs, Noubar's and PerSeptive's futures are a blank slate.

Grass Roots Growth

Let us also look at the trends in recent years, beginning at the grass roots. Students are showing far more interest in entrepreneurship courses and clubs. At the MIT Sloan School of Management students have worked with the MIT Technology Transfer Office to set up projects on possible commercialization of MIT technology. The Sloan School's graduate student New Venture Association has raised money in conjunction with the MIT Entrepreneurs Club, consisting mainly of engineering undergraduates, to provide awards for the best new business plans developed by student groups. At the Harvard Business School the several elective courses in entrepreneurial management, finance, and marketing muster over 25 percent of the total student body, breaking from that school's traditional concentration on the large corporation. This growth in student interest and enrollment in entrepreneurship subjects is a national phenomenon, manifested in parallel by the outbreak of national academic society meetings devoted to the same topic. Awards for papers or research on entrepreneurship are suddenly being provided by business and engineering schools across the country, further nourishing the interests and exposure.

Increasingly states and regions throughout the United States are championing the cause of high-technology entrepreneurship. While Massachusetts was long ago the first state to establish a venture capital organization to aid new firms, the successful and continuing Massachusetts Technology Development Corporation, many other states have joined its ranks, and with much greater political and economic commitment. Pennsylvania's

Ben Franklin Partnership has recruited four venture capital firms to begin efforts in different parts of the state, including Zero Stage Capital of Pennsylvania, located in State College, Pennsylvania, and working collaboratively with the main campus of Penn State University to transfer technology, build new companies, and stimulate the region's economy. Efforts underway nationwide include targeting state funds or state pension funds to invest in local oriented venture capital companies, formation of new company incubator organizations in various cities and on college campuses, developing tax legislation aimed at providing incentives to new or young firms and to their investors.

Indeed high-technology companies are being started at an increasing pace. The Bank of Boston, for example, finds that more new Massachusetts firms were organized by MIT alumni in the past decade than in any prior ten years (Bank of Boston, 1989). In biotechnology alone they created twenty new companies. A few years back writers like John Kenneth Galbraith came to the wrong conclusion that the age of entrepreneurship in the United States is dead (Galbraith, 1985). He argued that only giant corporations could survive in the present era. Today some writers are equally wrong in arguing that we have too much entrepreneurism in the United States, which they claim harmful to our competitiveness in world markets. These modern Luddites urge that government policies should be changed so as to discourage our high rate of new company formation, again claiming that only giant corporations can compete effectively (Ferguson, 1988). This naive argument totally ignores that entrepreneurs have been a unique source of U.S. innovation and economic growth for centuries (Gilder, 1988). As strongly evidenced in this book high-technology entrepreneurs have rapidly moved ideas from university and corporate research and development laboratories out to the market, where both they and society have benefited. In contrast many large corporations have excelled in generating new technologies but have failed to exploit them commercially. The policy issue that does need attention is not how to stifle independent entrepreneurs, but rather how to stimulate comparable corporate entrepreneurship.

One discouraging development is the apparent peaking in the late 1980s of new venture capital funds raised in the United States. According to Venture Economics, Inc. new monies added to U.S.-based venture capital funds declined from slightly more than $4 billion in 1987 to about $2.5 billion in 1989, and increasing fractions of these funds are targeted toward later stage companies, rather than seed-stage or early growth stage new enterprises. To put this into perspective, however, as recently as 1980 new commitments to venture capital funds were only $0.5 billion. The "overshoot" in funding during the six years from 1983 to 1988 in my judgment caused many inappropriate venture capital investments to be made, and sometimes at rather irresponsibly high evaluations. Money chased deals, and many companies got funded that would have been overlooked in other time periods. (In sharp contrast, remember that DEC was funded with $70,000 that purchased 78 percent of the new firm.) With the exception of these last few years the new venture capital funds available currently are quite comparable in magnitude with the past. Furthermore, the preference of most venture capitalists has been for later stage investments, in recent years financing even leveraged buyouts and buybacks of public stock. As a participant in seed funding, I do not notice a dramatically different investment climate at this stage from what has existed during most of the last twenty years, albeit the venture capital squeeze might get even worse. From another perspective the clear institutionalization of the U.S. venture capital industry probably means that it now will go through cycles that are comparable to those experienced in the IPO market and in the stock market in general. It is still too soon to tell how persistent or over what duration the venture capital cycle might be.

Three further aspects of financing deserve comment. (1) As indicated in Chapters 5 and 7 almost all new high-technology enterprises are initially financed by personal savings, family and friends, and informal investors. Later financing still prominently involves the informal investor, alone or in small groups. No evidence suggests that these sources are less available today than in the past. (2) Today, corporations are playing a more important role in high-technology financing than in prior years, especially foreign corporations. The increased activity of especially Japanese firms in providing early-round funds for U.S. high-technology companies has the favorable short-term impact of more "smart money" being available, with additional side-benefits to the entrepreneurs of more rapid growth into foreign markets, if not directly at least on a royalty basis. These vigorous direct foreign investments will in the longer run no doubt further strengthen the foreign companies' technological bases, through learning, licensing, alliances, and acquisitions. This will pose increased downstream competitive problems for larger U.S. corporations that are not actively linking to emerging high-technology enterprises. (3) In terms of other trends, on a worldwide basis, more venture capital funds are being developed, financed by the plentiful dollars available abroad, with intended investments partially in their own regions of the world and partially in the United States. And, although temporarily the market for new public stock issues of high-technology companies is merely lukewarm, I appraise the overall financing situation for start-up and growth-stage high-technology firms as quite reasonable.

The growth spurt in independent technological entrepreneurship is not limited to the United States. All around the world the symptoms of change are evident. Under the title "Britain's Ivory Tower Goes High Tech", *Science* magazine (Dickson, 1985) documents the recent development of "Silicon Fen" or the "Cambridge phenomenon". "Cambridge University in England ... has over the past few years catalyzed the rapid growth of a cluster of high-technology companies into one of Europe's most successful imitations of California's Silicon Valley or Boston's Route 128" (p. 1560), Four years later the same author sees this movement broadening to the Continent as well (Dickson, 1989). "Although their numbers are small compared to the United States, scientists in Europe are leaving academic research in growing numbers to start new high-tech companies... The obstacles in Europe have been formidable. A deep-rooted distaste for commerce in the university world has, for example, discouraged entrepreneurial thinking in the research community. And the secure, risk-free careers offered by large private corporations and public research organizations have had a powerful, negative influence on scientists' and engineers' willingness to set up in business on their own. Indeed, many blame the lack of individual entrepreneurism in Europe on a culture that discourages risk-taking in general.... But, ... whatever qualms may be expressed about their growing presence on the technology scene—there is little doubt that high-tech entrepreneurism is at last becoming an integral part of Europe's industrial landscape." And now the growth of venture capital funding in Europe has exceeded new venture capital funds in the United States (E.V.C.A./Peat Marwick McLintock, 1990). Similar evidences exist in Asia, where the still few new high-technology entrepreneurs are found to be similar in many respects to their American predecessors (Ray and Turpin, 1987).

Sticking my neck out a bit I see high-technology entrepreneurship, both in the United States and overseas, as entering a growth mode. From a couple of nodes in the United States, first Route 128 and then Silicon Valley, U.S. hubs of entrepreneurship have spread to Ann Arbor, Boulder, Minneapolis, Austin, Atlanta, Seattle, and myriad others. Each area has had its own initiating forces, not all dependent on a dominant technological university and its laboratories as forbearers. Each has had to go through its own period of start-up, getting to some successes, generating local visibility of role models for others, gradually building financial and industrial infrastructure, proliferating the positive feedback loops into more active new enterprise formation. This continues to take place throughout the United States, increasingly helped by the role of national media in making the experiences of one part of the country perceived and appreciated by other regions. But high-technology entrepreneurial growth is still primarily a local phenomenon. Only the very beginnings of this pattern are yet underway in Europe and Asia, with a long life ahead. It took over forty years for what has occurred in Greater Boston to reach this stage of documented publication. The next forty years should see far more technology-based entrepreneurship worldwide. Rest assured, while the future is never certain and storm clouds loom for some aspects of technological enterprise, high-technology entrepreneurship remains a continuing and ever more important part of the American dream and reality, increasingly shared by aspiring young technologists all over the world.

References

Bank of Boston. *MIT: Growing Businesses for the Future* (Boston: Economics Department, Bank of Boston, 1989).

D. Dickson. "Britain's Ivory Tower Goes High Tech", *Science*, 227, March 29, 1985, 1560–1562.

D. Dickson. "An Entrepreneurial Tree Sprouts in Europe", *Science*, 245, September 8, 1989, 1038–1040.

European Venture Capital Association / Peat Marwick McLintock. *Venture Capital in Europe, 1990 EVCA Handbook* (London: E.V.C.A./Peat Marwick McLintock, 1990).

C. H. Ferguson. "From the People Who Brought You Voodoo Economics", *Harvard Business Review*, May–June 1988, 55–62.

J. K. Galbraith. *The New Industrial State*, fourth edition (Boston: Houghton Mifflin, 1985).

G. Gilder. "The Revitalization of Everything: The Law of the Microcosm", *Harvard Business Review*, March–April 1988, 49–61.

D. M. Ray & D. V. Turpin. "Factors Influencing Entrepreneurial Events in Japanese High Technology Venture Business", in N. C. Churchill et al. (editors), *Frontiers of Entrepreneurship Research, 1987* (Wellesley, MA: Babson College, 1987), 557–572.

A Quarter Century of Research

A quarter century of my ongoing research on technological entrepreneurship forms the basis for this book. That research has focused upon the formation and growth of independent high-technology new enterprises. In parallel I have also been studying the related issues of internal entrepreneurship and other approaches to development of new product and new business ventures in already existing larger companies. The latter area of continuing research on corporate venturing and "intrapreneurship" is not treated in this book.

My research on new technical companies has investigated many aspects of their formation and growth, the background and characteristics of the founding entrepreneurs, the evolution of the companies from their early days, and the factors affecting their success and failure, including product strategy and overall strategic change. Several of these studies have focused on the entrepreneurs' search for venture capital and on the decision processes involved in funding new high-risk firms. Working with graduate research assistants and other students, over 40 research studies in all, clustered into five "Research Tracks", have been undertaken as part of a continuing comprehensive and integrated program that began in 1964 and is still underway in 1991.

Multitrack Research Program

The First Track: Spin-off Companies from MIT

The research began in 1964 as part of the MIT Research Program on the Management of Science and Technology, funded since 1962 for several years at the MIT Sloan School of Management by the U.S. National Aeronautics and Space Administration (NASA). My early concern was that NASA was misguided in the handling of its Technology Utilization Program, which was attempting to transfer technology developed in the space program to commercial users. NASA was employing massive reporting systems and large computer data bases to that end. Several of us at MIT believed that if and when technology transferred from its place of origin to another setting of use, it occurred primarily through people— both face-to-face contact and actual physical movement. My specific original hypothesis was that engineers and scientists who leave MIT labs and academic departments to set up new companies in fact transfer significant amounts of defense and space-related technology to the commercial market. My research studies, over many years, turn out strongly and consistently to support this basic hypothesis.

The initial studies were thus based on a conviction of the importance of people as "carriers" and "pushers" of technology, and were focused by strong personal interest in their activities related to new enterprise formation. Furthermore, I was committed to determining what makes a technical entrepreneur in the first place as well as what influences the success or failure of the new technical enterprises that entrepreneurs create.

The research started by examining the major MIT laboratories as possible sources of new companies. With Herbert A. Wainer, my first research assistant and then longer-term colleague in this research program, and other graduate students, we looked only at the post-World War II period and tried in this first phase to identify all those people who had been full-time employees of several major MIT laboratories but who had left the labs to participate as founders of wholly new companies, up to the time of the studies. The close collaborations with numerous graduate students over the years of this research lead me to adopt plural pronouns in describing the research design and implementation in this Appendix.

We were intentionally conservative in defining the "spin-off" firms that would be included in our samples. We excluded from our formal analyses all situations in which the ex-MIT lab employee, even though a key person in a new company, was not actually a founder of it. We excluded all situations in which the company, although new in terms of organizational form, was created as part of or as a subsidiary of another firm already in existence. We omitted a number of nonprofit organizations started by former MIT employees, such as The MITRE Corporation. Thus, we left out a number of activities that are rather similar to those for-profit enterprises where MIT staff members had, in fact, been founders.

We began our studies with the then two largest MIT labs, the defense oriented Lincoln Laboratory and the Instrumentation Lab (now the independent nonprofit Charles Stark Draper Laboratory, Inc.). We went back in the history of Lincoln Laboratory since its founding in the late 1940s and traced the people who left it. As shown in Table A–1, by this process we identified up to the time of our study the formation of no fewer than 50 new companies by people who had been full-time employees of the Lincoln Laboratory. In the case of the Instrumentation

Table A–1
Track 1: Research on Spin-off Companies Founded by MIT
Staff*

Sources of New Enterprises	New Companies Identified	Participated in Research Study
Major laboratories		
Electronic Systems Laboratory	13	11
Instrumentation Laboratory	30	23
Lincoln Laboratory	50	49
Research Laboratory for Electronics	14	13
Academic Departments (not including labs studied separately)		
Aeronautics and Astronautics	18	18
Chemical Engineering (Ph.D.s only)	18	18
Electrical Engineering	15	7
Materials Science	10	8
Mechanical Engineering	13	9
Totals	181	156

* In addition to my research associate Herbert A. Wainer, those research assistants and thesis students who contributed primarily to this research track include Erich K. Bender, Howard A. Cohen, Dean A. Forseth, John C. Ruth, and Paul V. Teplitz.

Laboratory, we identified 30 new companies formed by former staff members, from 1951 up to the time of our research study.

We then went beyond these organizations to study the MIT Research Laboratory for Electronics, itself a spin-off of the wartime MIT Radiation Laboratory, and identified another 14 companies. From former staff of the MIT Electronic Systems Laboratory (formerly named the Servomechanisms Laboratory; now the Laboratory for Information and Decision Systems) we found another 13 companies. Thus, from these four major laboratories alone we documented more than 100 companies that had been spun-off since World War II. We decided not to study a large number of other MIT laboratories as the findings from each of those we did examine seemed quite similar to the others.

As part of the Track 1-type research we later turned to study spin-off companies from the regular academic departments at MIT. We searched only in the engineering area where we expected to find the greatest amount of technology flow and entrepreneurship to have taken place. Starting with the Aeronautics and Astronautics Department we identified 18 companies that, by the time of our data collection, had been formed by present or former full-time research staff and faculty. From the Electrical Engineering Department we uncovered 19, four of which had already been identified in the lab studies; the Materials Science Department had already generated another ten firms; out of Mechanical Engineering thirteen companies were started; a special study of Chemical Engineering Department Ph.D. graduates turned up 18 new firms.

Despite these numerous studies we still only developed a limited view of MIT-based entrepreneurship. We had followed only the full-time staff members and faculty members (with the exception of the Chemical Engineering analysis where we traced Ph.D. graduates), looked only at the formation of totally new for-profit enterprises, excluded a number of the laboratories, completely excluded the natural and physical sciences, the social sciences, management and architecture, and even left out a number of the engineering departments. Despite these self-imposed constraints, by the completion of Track 1 of our research program we had already documented more than 180 companies formed from MIT.

What is most significant is that this first track included only our earliest studies of high-technology entrepreneurs. Had we extended the research to cover all MIT faculty and staff (excluding students and alumni) and brought the numbers to the present time, my best guess is that we would uncover about 500 companies founded by MIT faculty and staff since World War II. Indeed, a 1989 search by the MIT Resource Development Office, supported by analyses by the Economics Department of the Bank of Boston, uncovered over 600 firms that had been started by MIT alumni just in Massachusetts, with current revenues exceeding $39 billion and employment of over 300,000 (Bank of Boston, 1989). No doubt many of those were started by former faculty and staff. I am convinced that even that number is a gross underestimate of the total MIT spin-off firms and their impact.

Other Spin-Off Companies

In the next track of the research program, partially overlapping in time with the execution of the Track 1 research, we broadened our scope to permit comparison of the flow of new companies from MIT with the entrepreneurial flow that has taken place from other forms of institution, as listed in Table A–2. In doing this we took advantage of the fact that within the Greater Boston environs is The MITRE Corporation, a large nonprofit systems engineering organization working primarily with the U.S. Air Force. MITRE is itself a spin-off from the MIT Lincoln Laboratory, excluded by our tight definition from our earlier Track 1 studies by virtue of its not-for-profit status. Although we did not study MITRE itself, we did study as part of Track 2 the spin-off companies formed by people who have left the MITRE Corporation, thereby tracing another nine companies, of whom four had been identified earlier when we examined Lincoln Lab alumni. Next, we investigated a major government laboratory in the area, the U.S. Air Force Cambridge Research Laboratory (AFCRL). From AFCRL we identified, traced, and studied an additional 16 companies. Incidentally, many people had discouraged our study of AFCRL, asserting cynically that

Table A–2
Track 2: Research on Spin-off Companies Originating from
Other than MIT Sources[*]

Sources of New Enterprises	New Companies Identified	Participated in Research Study
Not-for-profit corporation		
MITRE Corporation	5	5
Government laboratory		
Air Force Cambridge Research Laboratory	16	15
Industrial firms		
Electronic systems company	45	39
Diversified technological company	58	23
Totals	124	82

[*] In addition to Herbert A. Wainer, those research assistants and thesis students who contributed primarily to this research track include Frederick L. Buddenhagen, Jerome Goldstein and Christopher L. Taylor.

we would find no instances of entrepreneurship arising from government employees.

The last part of this Track 2 series of comparative spin-off analyses is our study of the role of technical industry in providing incubating bases from which entrepreneurship has taken place. We identified all major corporations within the Greater Boston area that were somewhat similar in size and technologies to the major MIT laboratories studied in Track 1. We then selected two of those major firms and investigated the companies that were formed by individuals and groups who left those industrial settings to engage in this process of technological entrepreneurship. From one large Greater Boston area "Route 128" electronics firm we studied in-depth 39 new companies that spun off, of the total of 45 new enterprises we had been able to identify from that single firm. From another somewhat more diversified technological company we found 58 spin-off firms but collected data on only two dozen of them, limited in numbers primarily because of the national geographic spread of the spin-offs. Incidentally, in the Tracks 1 and 2 studies of spin-offs from 13 Greater Boston-area "sources", approximately 15 to 20 percent of the companies are located outside the area, some in neighboring areas but others scattered throughout the United States.

Independent Samples of New Enterprises
With over 200 so-called spin-off firms in the data base, my curiosity began to get the best of me. Departing from the approach of examining source organizations that had been incubators of new companies, I began

to encourage and direct research at special situations of interest. A series of industry-focused research samples was developed (in what I now label as Track 3) of several high-technology fields and, in marked contrast, of low-to-nontechnical consumer oriented manufacturing as well. An intensive study was carried out on the first three years of technical firms, and then several studies were launched to examine the later growth processes of high-technology firms. This cluster of continuing Track 3 research studies, shown in Table A–3, has thus far gathered comprehensive data on nearly 200 additional new enterprises and their founding entrepreneurs, with very little duplication of the preceding sets of firms, bringing the total company sample size to well over 400.

Special Studies of Characteristics of Entrepreneurs

Concurrent with all of these broad company-focused investigations have been two related research tracks. The first of these sought additional insights into the personal characteristics of the technical entrepreneurs. In particular, I wanted to contrast them with other engineers and scientists who seemed to have similar skills and opportunities but who had not become entrepreneurs. To this end, identified in Table A–4, two

Table A–3
Track 3: Research on Independent Samples of New
Enterprises[*]

Focus of Research	New Companies Identified	Participated in Research Study
High-technology industries		
Biomedical and pharmaceutical	106	29
Computer-related firms (3 studies)	105	53
Energy-related companies	40	17
Recently-formed high-technology firms	75	18
Consumer-oriented manufacturers	51	12
Young technical firms	20	18
Later growth issues		
Computer-related companies	10	10
Computer software companies	11	8
Computer terminals companies	10	10
High-growth-rate firms	34	21
Totals	462	196

[*]Those research assistants and thesis students who contributed primarily to this aspect of research include Jeffrey D. Bellin, Charles R. Bow, Oscar Hauptman, James A. Ishikawa, Michael W. Klahr, S. William Linko Jr., Steve Lipsey, Patricia S. McCarthy, Marc Meyer, Thomas E. Mullen, Ronald J. Pankiewicz, Barbara A. Plantholt, and Sheila M. Riordan.

Table A–4
Track 4: Supplemental Research on the Personal
Characteristics of Technological Entrepreneurs[*]

Focus of Research	Individuals who Participated in Research Study
Psychological analyses	
Thematic Apperception Tests (TATs) of high-technology entrepreneurs	51
Myers-Briggs tests of participants in MIT Enterprise Forum and 128 Venture Group	78
Total	129

	Individuals in Initial Sample	Participated in Research Study
Control Studies		
Staff of major MIT laboratories	391	299
MIT faculty (2 studies)	73	73
Totals	464	372

[*] Those research assistants and thesis students who contributed primarily to this research track include John W. Cuming, Andrew L. Gutman, and Donald H. Peters, as well as Herbert A. Wainer and Irwin M. Rubin, then a faculty colleague in the MIT Sloan School of Management.

sets of psychological studies were undertaken as well as several major control studies of MIT laboratory staff and faculty. The results of these studies are presented in Chapter 3.

Special Studies of the Financing of High-Technology Firms

The final concurrent research track has been the financing of new technical enterprises. Our numerous research studies in this area, noted in Table A–5, examine both the efforts and effectiveness of new companies in raising money and the decision processes and outcomes of financial institutions, mostly venture capital firms, in investing in these new enterprises. A total of 99 new enterprises, again largely nonduplicative of companies previously examined, were studied with specific focus on their fund-raising behaviors, as well as 106 varied financial institutions. Results of these studies are discussed in Chapters 5, 7, and 8.

During the same period as our U.S. research on technological entrepreneurship, I participated in MIT comparative research studies of 30 technology-based new companies in Israel and 60 new technical firms in Sweden, which built upon and lent support to our U.S. studies. The results of these

Table A–5
Track 5: Research on the Financing of New Enterprises[*]

Focus of Research	New Companies in Initial Sample	Participated in Research Study
Entrepreneurial search for funding		
Analyses of business plans submitted to venture capitalists (2 studies)	28	28
Search processes for raising venture capital (2 studies)	94	40
Went public (2 studies)	60	31
Totals	182	99

	Institutions that Participated in Research Study
Venture capital investment management	
In-depth analyses of venture capital investment decision processes (44 decisions)	2
Investment decision making by bank lending officers in three cities(48 people)	3
Underwriters of high-technology firms (2 studies)	21
Institutional investor attitudes and actions (75 in initial sample)	66
Performance of publicly-held venture capital firms (31 in initial sample)	14
Totals	106

[*] Those research assistants and thesis students who contributed primarily to this research track include Harold W. Bogle, Eugene F. Briskman, Richard H. Bullen Jr., Russell B. Faucett, Andrew L. Gutman, David R. Hall, Jonathan A. Marcus, Joel D. Mattox, Charles W. McLaughlin, Julian N. Nikolchev, John D. Proctor, Robert L. Sutherland, and Gary L. Wingo.

non-United States studies are mentioned occasionally as they apply to various chapters in this book.

Research Methods

The research methods employed in a multiphase twenty-five-year program of over 40 related studies not only evolve and change, but include a wide diversity of approaches. Psychological testing of entrepreneurs, for example, is carried out using methods different from those employed in the statistical analysis of venture capital portfolio performance. Yet Tracks 1 and 2 of the research, the studies of spin-off enterprises from MIT and other organizations, and a large portion of Track 3, the investigation of industry oriented groupings of new firms, were done with a high degree of similarity of methods and measurements. The research methods to be described in this section are used in those studies to develop the primary data and analyses of high-technology companies and their founding entrepreneurs. Techniques used in each of the other aspects of the research program are presented where their results are introduced in the book.

At the outset, the work on this research program was planned naively to take place in three stages: (1) finding the companies and deciding what information to collect, (2) interviewing the entrepreneurs to gather personal and company information, and (3) getting supplemental information from public records. Events that developed at the very beginning of the research forced a considerable mixing in time of these phases and the eventual magnification of the tasks. But this structure still represents a meaningful way to view the work.

Preparatory Work: Definition of Spin-off

The first problem to appear in the conduct of this research was to decide what constitutes a company spin-off from an MIT lab or department or from any other incubating source organization. Studies of product spin-offs from research and technology have been somewhat vague, and have invariably encountered serious problems of completeness of coverage. Finding all possible products and applications arising from a single source is an almost superhuman task. Researchers have tangled with fuzzy or grey areas—products whose true origins were not clear, or which were only partly derived from the source under study. To keep this study objective (and to keep it manageable in size), a firm line must be drawn beyond which all possibilities must be rejected, however interesting they may appear to be. Hopefully, also, the decision rule it fosters will be simple enough for rapid and repeated use.

The intentionally conservative working definition of spin-off used for the Track 1 and 2 research studies employed the following criteria:

1. One or more of the founders of the new company must have worked as a full-time employee of the source laboratory (university, government, industry, or nonprofit) or academic department being studied.
2. Those who worked only part time for the laboratory or as visitors from another organization or as consultants to the laboratory or academic department are excluded.
3. The individual must be considered as a founder of the company, by the company, to be included in the sample.
4. The individual need not have engaged in the founding of a new enterprise immediately after leaving the source laboratory or academic department (i.e., a time lag between termination with the source laboratory and company formation is permitted).

5. Nonprofit organizations that qualify under the previous four criteria as spin-offs are not included due to the basic differences between profit and nonprofit concerns.

Spin-off, as defined, refers to the person who has founded a company or to his or her company.

The above criteria deliberately exclude from study individuals who spun-off from any source organization into an already established firm. No matter how significant their impact was on the firms, these people were not considered in the research. This is entirely consistent with my desire in this aspect of the research to study only the entrepreneurs who actually founded a new company.

Identifying the Spin-offs

To gather data from each source organization's spin-offs, a list is needed of former employees who had formed companies. It is not surprising that no such report existed prior to our research. In lieu of this we were aided generously by the directors of the source organizations, their personnel departments, the MIT Public Relations Office, and the legal departments of the industrial organizations. From these people we obtained primary lists of people known to have left and believed to have founded a company.

Using these lists as a base in each case, we obtained additional names of possible spin-offs by two means: (1) through speaking with other staff members of the source organizations, and (2) through questioning of the known spin-offs. Each of these means greatly aided the discovery of other companies. (In the social sciences, this method is known as a "snowball" sampling technique, the sample building as the snowball rolls forward.)

Let me give an example of how this worked. In the case of the MIT Instrumentation Laboratory (now the Charles Stark Draper Laboratory), Mr. J. B. Feldman, executive officer of the lab, searched his memory and the laboratory records of the several hundred staff members who had left since 1946 and produced a list of about 20 possibilities, which formed the initial basis for the study. As the initial interviews with that group progressed, more possible entrepreneurial situations were uncovered. In all, 51 such leads were investigated. There is no guarantee that these 51 constituted an exhaustive list for Instrumentation Lab spin-offs until the time of the research, but it is likely that any further additions are minor ones. The most likely exclusions from the sample, if any exist, are very young companies started as part-time ventures by people who were still at the lab when the study took place. Some of these people feel a need to conceal their company ventures and may have escaped detection. Other sources of possible undetected enterprises are people who have formed companies several years after leaving the lab, particularly if they are in distant parts of the country. Our resulting lists of spin-off companies are, therefore, biased in the direction of understating total entrepreneurship from each source.

As another example, the original list of suggested names from MIT

Lincoln Laboratory identifies 58 people who Lincoln personnel staff believed had left to become involved in 31 new ventures. Further suggestions from the follow-up interviews, combined with the tight application of our definition of spin-off, caused 18 of these people to be excluded but added another 22 for a net gain of four. From a company viewpoint seven of the leads on the original Lincoln Lab list were excluded but 27 were added for a net gain of 20 companies. Incidentally, Lincoln Laboratory recently updated this list as part of its fortieth anniversary celebration, using slightly different criteria. Because my co-founder Jack Pugh used to work part time at Lincoln Lab, the new listing (but not my original sample) even includes Pugh-Roberts Associates, Inc., the management consulting firm we formed in 1963.

In every case senior managers from the source organization dramatically underestimated the extent of spin-off company formation. This was true even in the case of Stark Draper, head of the MIT Instrumentation Lab, who was enthusiastic about spin-off entrepreneurs from his lab and who energetically encouraged our research efforts. The worst underestimate occurred in the case of our first study of new enterprises from an industrial company. There, after two days of prior consideration and in-house discussion, the director of personnel came up with a list of only five possible spin-offs. Within six months we documented 45 companies as originating from that one large electronics firm. This "iceberg" phenomenon of entrepreneurial spin-off flow seems characteristic, with only the tip of the activity being visible to senior management. Organizations seldom do a good job of monitoring the extent and causes of regular job turnover, never mind the special situations of new company spin-off. Incidentally, during our research study the legal department of the electronics firm was readily able to identify ten spin-offs with whom they had had legal dealings, both friendly and otherwise.

In the Tracks 3, 4, and 5 research studies of focused clusters of new firms, samples were developed by means specific to each study. For example, for one of those analyses we decided to focus on manufacturers operating in the energy industry in eastern Massachusetts. Company names were collected from a number of sources including: Standard Industrial Codes (SIC) listings of manufacturers in Massachusetts; a publication of the Massachusetts Office of Energy Resources, Division of Solar and Conservation; a listing of energy control systems manufacturers from *Energy Users News*; and firms listed under ENERGY or SOLAR in the NYNEX Telephone Yellow Pages and the Massachusetts Department of Commerce's Listing of Corporations. The resulting list consisted of eighty-six firms located in eastern Massachusetts. These firms were contacted by phone to determine whether they had been properly identified for this research and to elicit the names of their founding entrepreneurs. Firms were deemed appropriate if they were involved in a manufacturing operation in eastern Massachusetts, which served the broadly defined energy industry and had been initiated between

1967 and 1977. The time frame selected for that particular study was intended to allow sufficient operational experience to measure performance without having to reach too far back to establish the firm's past. The screening of the list of eighty-six firms resulted in forty appropriate firms, of which seventeen were eventually included in the research.

Sample development efforts were quite intense and time consuming, in our efforts to be thorough. For example, the better part of a research assistant's year was devoted to compiling and verifying the most recent sample of young Massachusetts biomedical and pharmaceutical companies for just one of our studies in Track 3. Each of the resulting studies ends up with some clear biases and limitations due to the sample development procedures, and the interpretation of findings from the research has been carefully done to attempt to account for those limitations.

Obtaining the Information

The next stage of the research was obtaining much and detailed information about the entrepreneurs and their companies. In all of the Tracks 1, 2, and 3 company studies, as well as in many of the studies in Tracks 4 and 5, a highly structured interview, based on a detailed questionnaire, was used whenever possible.

The person collecting the data (the author, graduate research assistant or thesis student) would arrange for an interview with the spin-off entrepreneur and, in conducting the structured interview, would administer a detailed questionnaire. When it was impossible to meet with the entrepreneur personally, the researcher forwarded a questionnaire to him and administered the questions by telephone (less than 10 percent of the cases). A last method was to forward a questionnaire to the entrepreneur and have him answer the questions at his leisure. This method was used only when the previously mentioned approaches could not be (less than 1 percent of the cases).

The questionnaire, in all its variations from the original one used in the first studies in 1964 up to the present form, covers the following categories of information:

1. Entrepreneur's background, family, and education
2. Entrepreneur's laboratory employment and reasons for leaving the laboratory
3. Entrepreneur's work and business experience before time of company formation
4. Circumstances surrounding formation of company
5. The company's subsequent growth and problems
6. Company's policies relative to such areas as marketing, finance, and personnel
7. Company's utilization of laboratory technology

The interviews with the many company founders inevitably constitute both the most interesting and most time consuming phase of the data collection. They are interesting because of the chances to meet and talk with an extremely varied, striving and accomplished group of people, most of whom are very cooperative and anxious to help. However, the interviews are often quite time consuming, requiring typically from one to two and one-half hours to administer. Some interviews stretched to seven or eight hours over two or three sessions. This places a natural limit on the number of people with whom we could speak in a day. The entrepreneurs seem to enjoy reminiscing about their early days and discussing their business experiences. One man greeted the interviewer with, "You represent a serious threat to me," and ended up talking for more than two hours. Several persons mentioned that this was the first time they had stopped to look back on their business lives and size up their present position. One of them commented to the researcher, "You should look at yourself as almost performing a service to these people." In few cases did a high-technology company founder refuse to provide the desired information, and then often because of claimed proprietary reasons, rather than lack of sympathy for the research purpose.

Two minor problems arise in conducting these interviews. The first is the scheduling of an initial hour or more with a busy executive. A number of interviews were held at lunch or after hours to avoid this problem. The second difficulty is that of taking notes on everything said during the interview. Extensive writing disrupts the conversation and makes the subject ill at ease. It also drags out the interview. A tape recorder was used in many of the sessions, although it effectively doubles the amount of time required for interviewing and later transcribing. In addition it is difficult in some cases to track down or sometimes to gain the participation of the entrepreneur, particularly when the enterprise had already gone out of business. Indeed, some minor bias has no doubt crept into the various samples of companies studied in that it is likely that any companies not located are less successful than those traced. For example, in the research on spin-offs from the electronic systems firm (Table A–2), the six companies (entrepreneurs) that refused to cooperate (of 45 identified) included three defunct or sold companies whose founders had left the Northeast, two whose founders "did not have time", and one old established company whose founder was "away on business" for the duration of this study.

In general, the information collected with the questionnaires is complete and seemingly realistic. However, the wide span of companies with respect to age, type of work, and cooperation of the founder results in some spotty and incomplete information, especially financial. Additional information was gathered from public records. Many of the companies appear in standard sources, such as Standard and Poor's *Directory of Corporations* or in U.S. Small Business Administration Regulation A filings. Stock brokerage firms willingly supplied information on the relatively few public companies studied. All companies in Massachusetts are required to file annual reports, including balance sheets, that are on file in the State House Corporate

Records Department in Boston. These records, as well as the Articles of Incorporation and complete histories of stock issues, are open to the public and were collected on most companies included in our research. Dun and Bradstreet data and ratings were also obtained for many of the companies sampled.

A final source of information for the spin-off studies (Tracks 1 and 2 of the research program) is a "supervisor questionnaire" that was devised to be completed by an individual under whom the entrepreneur had worked during his association with the technology-source laboratory. It is designed to measure the supervisor's perception of the entrepreneur's technical ability, technical creativity, and managerial ability. The supervisor is also asked to classify the entrepreneur as regards the nature of his work while at the laboratory. Actually the laboratory interviews prove more valuable for the information they generated about the lab and its structure. A few supervisors were hesitant to talk about unpleasant circumstances of a former subordinate leaving the lab. Perhaps not surprisingly, the entrepreneurs show little hesitation in discussing these same events when they were later interviewed.

The background interviews in the source organizations also provide historical information on the growth and evolution of the parent organizations. The evolution of those sources is of interest for possible explanations of changing environment for innovation and for clues to reasons for entrepreneurs leaving, their backgrounds, and the number of spin-offs formed in any one year.

Measurement and Analysis Issues

Answers to the detailed questionnaires led easily to the quantification of information. Almost all of the answers to the questions were coded and appropriately arranged in computer data files. Incomplete information on some of the companies does not affect the analysis of data, as relevant codes were given to isolate missing information.

Two types of issues exist beyond the raw information: First is the problem of measuring certain key dimensions; second is the selection of analysis methodology. Measurement issues encountered in the central question of how to characterize company performance are discussed in Chapter 9, where success and failure are first evaluated. Similar questions of measuring the nature of the work that had been done previously by the entrepreneur and the extent of technology transferred by the entrepreneurs from their source laboratories are also discussed in the most relevant chapters (Chapters 3 and 4, respectively).

The data gathered in the various research tracks are categorized, cross classified, displayed, and assessed separately in each of the more than forty studies, with data aggregated across studies where appropriate. In general, analysis methods are chosen that do not overlook subtleties in the data, yet avoid complex multivariate techniques.

A profile of the average entrepreneur and his company, for example, might emerge from the means, variances, and frequency distributions of the quantitative and multiple choice answers on the questionnaires. A profile can prove valuable in identifying interesting patterns in the data but helps little in explaining the patterns.

Of greater interest may be the answers to questions about how various patterns came about or how they relate to company success. These questions require statistical testing of the relationships among multiple variables, in search of whether any apparent relations found might be caused by chance alone. To reject this chance hypothesis and conclude that the variables are related, the probability that the observed trends are caused by chance must be sufficiently low. This computed probability (of type 1 error) appears in parentheses throughout this book. As an example, "Those founders with more years of service at the parent had greater success with their spin-offs ($p = .01$)." This statement indicates that length of service apparently correlates with the degree of success of the spin-off companies, and that there is a 1 percent probability that the distribution of questionnaire answers leading to this correlation occurred by chance alone.

A test for statistical significance only indicates that two variables may be statistically dependent, and does not by itself give strength or form to the relationship. Furthermore, the statistical techniques required to test two variables depend on the levels of the variables tested. For example, "nominal" level variables classify objects only by category (e.g., yes–no, or type of organization). Variables are "ordinal" level when the categories have an order (e.g., low–moderate–high, or degree level). "Interval" level variables have distances between ranks (e.g., age, or years of education). Treating higher level variables (interval is higher than ordinal, which is higher than nominal) as lower level is permissible in statistical testing, but this is avoided whenever possible to use all of the information available from each variable.

The statistical tests are neither infallible nor do they actually prove explanatory importance. Underlying causes might explain the apparent relationships between two variables. However, reasonable assumptions and inference based on anecdotal information are used to build a strong case whenever possible.

Table A–6 describes the type of statistical test employed in dealing with the various classes of data. The Fisher-Exact Test (or Chi-Squared Test if sample is sufficiently large) is utilized whenever both variables are of the nominal variety (Siegel, 1956, pp. 96–104). Consider the following hypothetical contingency table for a small sample of companies:

		Used own funds?	
		Yes	No
Started	Yes	5	0
part-time?	No	3	6

By observation one notes a strong tendency for those who started their businesses part-time to have used their own funds. However, the same distribution might have occurred purely by chance. The Fisher-Exact test allows one to compute the probability that this distribution occurred by chance. By calculation, the distribution shown here is found to have a 2.8 percent probability of occurring by chance (the probability of type 1 error mentioned above). In statistical terminology, the conclusion drawn from the above distribution is: "At the .028 level of significance, those starting their businesses part-time are more likely to use their own funds." As indicated earlier, to eliminate the redundant use of the phrase "at the such-and-such level of significance", throughout the book I simply enclose in parentheses the level of statistical significance associated with the particular relation (e.g., "Those entrepreneurs starting their businesses on a part-time basis also tend to use their own personal funds to finance the enterprise (.028)"). Unless otherwise stated, the one-tail level of significance is given, simply implying that we had predicted the relation and its direction in advance of the testing.

As shown in Table A–6, the Mann-Whitney U test is used whenever one variable is nominal and the other ordinal or stronger. This test simply states whether or not there is a significant difference between the means of

Table A–6
Statistical Tests Utilized

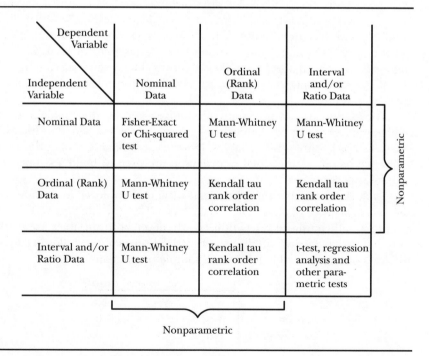

Independent Variable \ Dependent Variable	Nominal Data	Ordinal (Rank) Data	Interval and/or Ratio Data	
Nominal Data	Fisher-Exact or Chi-squared test	Mann-Whitney U test	Mann-Whitney U test	Nonparametric
Ordinal (Rank) Data	Mann-Whitney U test	Kendall tau rank order correlation	Kendall tau rank order correlation	
Interval and/or Ratio Data	Mann-Whitney U test	Kendall tau rank order correlation	t-test, regression analysis and other parametric tests	

Nonparametric

the rank (ordinal) variables in the two populations defined by the nominal variable (Siegel, pp. 116-127). Consider the statement: "Those companies starting part-time transfer a greater amount of technology (.036)". This means that all the companies in the sample are divided into two populations on the basis of the nominal variable, "start part time? yes or no". The average amount of technology transferred in the sample comprised of those firms that did start on a part-time basis is larger, at the .036 level of statistical significance.

The Kendall tau test, referred to in Table A–6, simply measures the degree to which two rank variables are correlated (Siegel, pp. 213-223). If two rank variables are significantly positively related, when one variable increases (decreases) the other variable is also expected to increase (decrease). (The converse is true when the variables are found to be negatively correlated.) Note that these variables are not said to be linearly correlated, or even that a specific change in one variable predicts the magnitude of the change in the other. It simply says that they "move" together. (In calculating the levels of significance, the "tau not corrected for ties" is used consistently. As such, the level of significance stated is always conservative.)

The reader is cautioned against associating relationship with causality. Frequent mention is made that a strong relationship or correlation exists between two variables. This does not, however, imply that one is caused by the other. Occasionally this inference is drawn based on reasonable assumptions, but it is usually impossible to prove causality mathematically.

Finally, I wish to point out a fact of life in the science of statistics. All other things being equal, the strength of the statement that can be made concerning a relationship increases approximately as the square root of the sample size. While this is so for very legitimate mathematical reasons, it does not detract from the fact that due to relatively small samples in individual studies throughout the research program, the strength of the conclusions that can be stated is much less than would be the case were the samples larger.

References

Bank of Boston. *MIT: Growing Businesses for the Future* (Boston: Economics Department, Bank of Boston, 1989).

S. Siegel. *Non-parametric Statistics for the Behavioral Sciences* (New York: McGraw-Hill Book Company, 1956).

INDEX OF FOUNDERS AND FIRMS

SUBJECT INDEX